Hugo Hens

Building Physics
Heat, Air and Moisture

Building Physics and Applied Building Physics

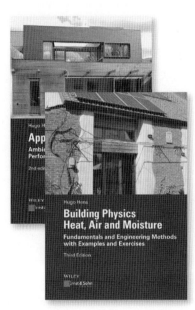

As with all engineering sciences, Building Physics is oriented towards application, which is why, a first book treats the fundamentals and a second volume examines performance rationale and performance requirements as well as energy efficient building design and retrofitting.

- content well structured combining theory with typical building engineering practice
- solved examples and applications
- author with international appreciation (Scandinavia, Benelux, USA, Canada)

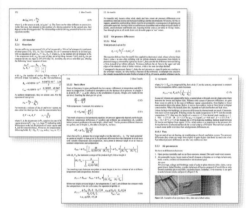

Hugo S. L. C. Hens
Package: Building Physics and Applied Building Physics
2017. approx. 600 pages.
approx. € 99,–*
ISBN 978-3-433-03209-1
Also available as **ebook**

BUNDLE ebook + Print!
Building Physics 3rd edition
approx. € 79,–*
ISBN 978-3-433-03199-5

Order online:
www.ernst-und-sohn.de

Recommendations:

- Performance Based Building Design I
- Performance Based Building Design II
- Journal Bauphysik

Ernst & Sohn
Verlag für Architektur und technische
Wissenschaften GmbH & Co. KG

Customer Service: Wiley-VCH
Boschstraße 12
D-69469 Weinheim

Tel. +49 (0)6201 606-400
Fax +49 (0)6201 606-184
service@wiley-vch.de

* € Prices are valid in Germany, exclusively, and subject to alterations. Prices incl. VAT, excl. shipping. 1056136_dp

Hugo Hens

Building Physics
Heat, Air and Moisture

Fundamentals and Engineering Methods
with Examples and Exercises

Third Edition

Professor Hugo S. L. C. Hens
University of Leuven (KULeuven)
Department of Civil Engineering
Building Physics
Kasteelpark Arenberg 40
3001 Leuven
Belgium

Cover: Extremely low energy house in Belgium, newly built as an extension of an old farm.

Photo: Hugo Hens

This third edition is a revision of the second, published in 2012. The text has been upgraded and extended where appropriate.

All books published by **Ernst & Sohn** are carefully produced. Nevertheless, authors, editors, and publisher do not warrant the information contained in these books, including this book, to be free of errors. Readers are advised to keep in mind that statements, data, illustrations, procedural details or other items may inadvertently be inaccurate.

Library of Congress Card No.: applied for

British Library Cataloguing-in-Publication Data
A catalogue record for this book is available from the British Library.

Bibliographic information published by the Deutsche Nationalbibliothek
The Deutsche Nationalbibliothek lists this publication in the Deutsche Nationalbibliografie; detailed bibliographic data are available on the Internet at <http://dnb.d-nb.de>.

© 2017 Wilhelm Ernst & Sohn, Verlag für Architektur und technische Wissenschaften GmbH & Co. KG, Rotherstraße 21, 10245 Berlin, Germany

All rights reserved (including those of translation into other languages). No part of this book may be reproduced in any form – by photoprinting, microfilm, or any other means – nor transmitted or translated into a machine language without written permission from the publishers. Registered names, trademarks, etc. used in this book, even when not specifically marked as such, are not to be considered unprotected by law.

Print ISBN: 978-3-433-03197-1
ePDF ISBN: 978-3-433-60857-9
ePub ISBN: 978-3-433-60856-2
Mobi ISBN: 978-3-433-60855-5
oBook ISBN: 978-3-433-60854-8

Coverdesign: Sophie Bleifuß, Berlin, Germany
Typesetting: Thomson Digital, Noida, India
Printing and Binding: Strauss GmbH, Mörlenbach, Germany

Printed in the Federal Republic of Germany

Printed on acid-free paper

3. revised edition

To my wife, children and grandchildren

*In remembrance of Professor A. De Grave
who introduced Building Physics as a new discipline
at the University of Leuven (KU Leuven), Belgium in 1952*

Table of Contents

Preface		XIII
Acknowledgements		XV
0	**Introduction**	1
0.1	Subject of the book	1
0.2	Building physics	1
0.2.1	Definition	1
0.2.2	Constraints	2
0.2.2.1	Comfort	2
0.2.2.2	Health and wellbeing	2
0.2.2.3	Architecture and materials	3
0.2.2.4	Economy	3
0.2.2.5	Sustainability	3
0.3	Importance	3
0.4	History	5
0.4.1	Applied physics	5
0.4.1.1	Heat, air and moisture	5
0.4.1.2	Acoustics	5
0.4.1.3	Lighting	6
0.4.2	Thermal comfort and indoor air quality	6
0.4.3	Building services	7
0.4.4	Building design and construction	8
0.4.5	The situation at the University of Leuven (KULeuven) and elsewhere	8
0.5	Units	9
0.6	Symbols	10
	Further reading	12
1	**Heat transfer**	15
1.1	Overview	15
1.1.1	Heat	15
1.1.2	Temperature	15
1.1.3	Sensible and latent heat	15
1.1.4	Why are heat and temperature so compelling?	16
1.1.5	Some definitions	17
1.2	Conduction	17
1.2.1	Conservation of energy	17
1.2.2	The conduction laws	18
1.2.2.1	First law	18
1.2.2.2	Second law	20
1.2.3	Steady state	20
1.2.3.1	One-dimensional flat assemblies	20
1.2.3.2	Two dimensions, cylinder symmetric	28
1.2.3.3	Two and three dimensions: thermal bridges	29

1.2.4	Transient	33
1.2.4.1	Periodic boundary conditions: flat assemblies	34
1.2.4.2	Any boundary conditions: flat assemblies	44
1.2.4.3	Two and three dimensions: thermal bridges	47
1.3	Heat exchange at surfaces	48
1.4	Convection	49
1.4.1	In general	49
1.4.2	Typology	51
1.4.2.1	Driving forces	51
1.4.2.2	Flow types	51
1.4.3	Quantifying the convective surface film coefficient	52
1.4.3.1	Analytically	52
1.4.3.2	Numerically	52
1.4.3.3	Dimensionally	53
1.4.4	Values for the convective surface film coefficient	55
1.4.4.1	Flat surfaces	55
1.4.4.2	Cavities	58
1.4.4.3	Pipes	59
1.5	Radiation	60
1.5.1	In general	60
1.5.2	Definitions	61
1.5.3	Reflection, absorption and transmission	61
1.5.4	Radiant bodies	64
1.5.4.1	Black	64
1.5.4.2	Grey	71
1.5.4.3	Coloured	73
1.5.5	Simple formulae	74
1.6	Building-related applications	75
1.6.1	Surface film coefficients and reference temperatures	75
1.6.1.1	Indoors	76
1.6.1.2	Outdoors	78
1.6.2	Steady state: flat assemblies	80
1.6.2.1	Thermal transmittance of envelope parts and partitions	80
1.6.2.2	Average thermal transmittance of parts in parallel	82
1.6.2.3	Electrical analogy	84
1.6.2.4	Thermal resistance of an unvented cavity	84
1.6.2.5	Interface temperatures	86
1.6.2.6	Solar transmittance	87
1.6.3	Local inside surface film coefficients	90
1.6.4	Steady state: two and three dimensions	92
1.6.4.1	Pipes	92
1.6.4.2	Floors on grade	93
1.6.4.3	Thermal bridges	94
1.6.4.4	Windows	98
1.6.4.5	Building envelopes	99
1.6.5	Heat balances	101

1.6.6	Transient	101
1.6.6.1	Periodic: flat assemblies	101
1.6.6.2	Periodic: spaces	102
1.6.6.3	Thermal bridges	106
1.7	Problems and solutions	106
	Further reading	121
2	**Mass transfer**	**125**
2.1	Generalities	125
2.1.1	Quantities and definitions	125
2.1.2	Saturation degrees	127
2.1.3	Air and moisture transfer	127
2.1.4	Moisture sources	129
2.1.5	Air and moisture in relation to durability	130
2.1.6	Links to energy transfer	132
2.1.7	Conservation of mass	132
2.2	Air	133
2.2.1	Overview	133
2.2.2	Air pressure differentials	134
2.2.2.1	Wind	134
2.2.2.2	Stack	135
2.2.2.3	Fans	136
2.2.3	Air permeances	136
2.2.4	Airflow in open-porous materials	140
2.2.4.1	The conservation law adapted	140
2.2.4.2	One dimension: flat assemblies	141
2.2.4.3	Two and three dimensions	143
2.2.5	Airflow across assemblies with air-open layers, leaky joints, leaks and cavities	144
2.2.6	Air transfer at the building level	145
2.2.6.1	Definitions	145
2.2.6.2	Thermal stack	145
2.2.6.3	Large openings	146
2.2.6.4	The conservation law applied	147
2.2.6.5	Applications	149
2.2.7	Combined heat and air flow in open-porous materials	151
2.2.7.1	Heat balance equation	151
2.2.7.2	Steady state: flat assemblies	152
2.2.7.3	Steady state: two and three dimensions	156
2.2.7.4	Transient: flat assemblies	156
2.2.7.5	Transient: two and three dimensions	157
2.2.7.6	Air permeable layers, joints, leaks and cavities	157
2.2.7.7	Vented cavity	157
2.3	Water vapour	160
2.3.1	Water vapour in the air	160
2.3.1.1	Overview	160

2.3.1.2	Quantities	161
2.3.1.3	Vapour saturation pressure	161
2.3.1.4	Relative humidity	166
2.3.1.5	Changes of state in humid air	167
2.3.1.6	Enthalpy of humid air	167
2.3.1.7	Measuring air humidity	168
2.3.1.8	Vapour balance indoors	168
2.3.1.9	Relative humidity at a surface	171
2.3.2	Vapour in open-porous materials	172
2.3.2.1	Different compared with air?	172
2.3.2.2	Sorption/desorption isotherm and specific moisture ratio	173
2.3.3	Vapour transfer in the air	177
2.3.4	Vapour flow by diffusion in open-porous materials and assemblies	179
2.3.4.1	Flow equation	179
2.3.4.2	Mass conservation	181
2.3.4.3	Applicability of the <equivalent> diffusion concept	182
2.3.4.4	Steady state: flat assemblies	182
2.3.4.5	Steady state: two and three dimensions	192
2.3.4.6	Transient regime	193
2.3.5	Vapour flow by diffusion and convection in open-porous materials and assemblies	195
2.3.6	Surface film coefficients for diffusion	201
2.3.7	The surface film coefficient for diffusion applied	204
2.3.7.1	Diffusion resistance of an unvented cavity	204
2.3.7.2	Do vented cavities enhance drying?	204
2.3.7.3	Surface condensation and the vapour balance indoors	206
2.4	Moisture	207
2.4.1	Overview	207
2.4.2	Water flow in a pore	208
2.4.2.1	Capillarity	208
2.4.2.2	Poiseuille's law	210
2.4.2.3	Isothermal water flow in a pore contacting water	212
2.4.2.4	Isothermal water flow in a pore after water contact	218
2.4.2.5	Non-isothermal water flow in a pore after water contact	219
2.4.2.6	Remark	219
2.4.3	Vapour flow in a pore that contains water isles	219
2.4.3.1	Isothermal	220
2.4.3.2	Non-isothermal	220
2.4.4	Moisture flow in a pore that contains water isles	221
2.4.5	Moisture flow in materials and assemblies	221
2.4.5.1	Transport equations	221
2.4.5.2	Moisture permeability	223
2.4.5.3	Mass conservation	224
2.4.5.4	Starting, boundary and contact conditions	224
2.4.5.5	Remarks	225
2.4.6	Simplified moisture flow model	225

2.4.6.1	How it looks	225
2.4.6.2	Applying the simplified model	227
2.5	Problems and solutions	240
	Further reading	263

3	**Combined heat, air and moisture flow**	**267**
3.1	Introduction	267
3.2	Material and assembly level	267
3.2.1	Assumptions	267
3.2.2	Solution	267
3.2.3	Conservation of mass	268
3.2.4	Conservation of energy	269
3.2.5	Flux equations	272
3.2.5.1	Heat	272
3.2.5.2	Mass, air	272
3.2.5.3	Mass, moisture	273
3.2.6	Equations of state	273
3.2.6.1	Enthalpy and vapour saturation pressure versus temperature	273
3.2.6.2	Relative humidity versus moisture content	273
3.2.6.3	Suction versus moisture content	274
3.2.7	Starting, boundary and contact conditions	274
3.2.8	Two examples of simplified models	274
3.2.8.1	Non-hygroscopic, non-capillary materials	274
3.2.8.2	Hygroscopic materials at low moisture content	276
3.3	Whole building level	277
3.3.1	Balance equations	277
3.3.1.1	Vapour	277
3.3.1.2	Air	279
3.3.1.3	Heat	279
3.3.1.4	Closing the loop	281
3.3.2	Sorption-active surfaces and hygric inertia	282
3.3.2.1	Generalities	282
3.3.2.2	Sorption-active thickness	283
3.3.2.3	Zone with one sorption-active surface	285
3.3.2.4	Zone with several sorption-active surfaces	287
3.3.2.5	Harmonic analysis	287
3.3.3	Consequences	288
3.4	Problems and solutions	291
	Further reading	305

Postscript	**309**

Preface

Until the first energy crisis of 1973, building physics was a dormant beauty within building engineering, with seemingly limited applicability in practice. While soil mechanics, structural mechanics, building materials, building construction and HVAC were perceived as essential, designers only demanded advice on room acoustics, moisture tolerance, summer comfort or lighting when really needed or in case problems arose. Energy was of no concern, while thermal comfort and indoor environmental quality were presumably guaranteed thanks to infiltration, window operation and the HVAC system. The energy crises of the 1970s, persisting moisture problems, complaints about sick buildings, thermal, visual and olfactory discomfort, and the move towards greater sustainability changed all this. Societal pressure to diminish energy consumption without degrading building usability activated the notion of performance-based design and construction. As a result, building physics and its potential to quantify performance moved to the front line of building innovation.

As for all engineering sciences, building physics is oriented towards application. This demands a sound knowledge of the basics in each of its branches: heat and mass transfer, acoustics, lighting, energy and indoor environmental quality. Integrating the basics on heat and mass transfer is the main objective of this book, with mass limited to air, (water) vapour and moisture. It is the result of 38 years of teaching architectural, building and civil engineers, coupled with some 50 years of experience in research and consultancy. Where needed, information and literature from international sources has been used, which is why each chapter concludes with an extended reading list.

In an introductory chapter, building physics is presented as a discipline. The first chapter then concentrates on heat transport, with conduction, convection and radiation as main topics, followed by concepts and applications typical for building physics. The second chapter treats mass transport, with air, vapour and moisture as main components. Again, much attention is devoted to the concepts and applications related to buildings. The last chapter discusses combined heat, air and moisture transport. All three chapters are followed by exercises.

The book uses SI units. It should be suitable for those undertaking undergraduate and graduate studies in architectural and building engineering, although mechanical engineers, studying HVAC, and practising building engineers who want to refresh their knowledge, may also benefit. It is presumed that the reader has a sound knowledge of calculus and differential equations, along with a background in physics, thermodynamics, hydraulics, building materials and building construction.

Acknowledgements

A book reflects the work of many, not only the author, who writes by standing on the shoulders of those who have gone before. Therefore, I would like to thank the thousands of students I have encountered during 38 years of teaching. They gave me the opportunity to test the content. Although I started my career as a structural engineer, my predecessor Professor Antoine de Grave planted the seeds that fed my interest in building physics. The late Bob Vos of TNO, the Netherlands, and Helmut Künzel of the Fraunhofer Institut für Bauphysik, Germany, showed the importance of experimental work and field testing to understand building performance, while Lars Erik Nevander of Lund University, Sweden, taught that solving problems in building physics does not always need complex modelling, mainly because reality in building construction is much more complex than any model could simulate.

Several researchers and PhD students have been involved during the four decades at the Laboratory of Building Physics. I am very grateful to Gerrit Vermeir (now Emeritus Professor), Staf Roels, Dirk Saelens and Hans Janssen, who became colleagues at the Department of Civil Engineering, Faculty of Engineering Sciences at KULeuven; Jan Carmeliet, now professor at ETH-Zürich; Piet Standaert, principal at Physibel Engineering; Jan Lecompte at Bekaert NV; Filip Descamps, a principal at Daidalos Engineering and part-time professor at the Free University Brussels (VUB); Arnold Janssens, professor at Ghent University (UG); Rongjin Zheng, associate professor at Zhejiang University, China; Bert Blocken, full professor at the Technical University Eindhoven (TU/e); Griet Verbeeck, now professor at Hasselt University, and Wout Parys, now a principal at BauPhi Engineering and part-time professor at the Faculty of Engineering Technology of KULeuven, who all contributed by their work. The experiences gained as a structural engineer and building site supervisor at the start of my career, as building assessor over the years, as researcher and operating agent of four Annexes of the IEA, Executive Committee on Energy in Buildings and Communities, forced me to rethink the engineering-based performance approach each time. The many ideas I exchanged and received in Canada and the USA from Kumar Kumaran, the late Paul Fazio, Bill Brown, William B. Rose, Joe Lstiburek and Anton Ten Wolde were also of great help.

Finally, I thank my family, my wife Lieve, who has managed living together with a busy engineering professor, my three children who had to live with that busy father, and my grandchildren.

Hugo S.L.C. Hens
Leuven, July 2017

0 Introduction

0.1 Subject of the book

This is the first volume in a series of four books:

- **Building Physics: Heat, Air and Moisture, Fundamentals and Engineering Methods with Examples and Exercises**
- Applied Building Physics: Ambient Conditions, Building Performance and Material Properties
- Performance Based Building Design 1: From Below Grade Construction to Outside Walls with Transparent Insulation
- Performance Based Building Design 2: From Low-Slope Roofs to Finishes and Risk.

This volume discusses the physics behind the heat, air and moisture, also called hygrothermal response of materials, building assemblies and whole buildings. The second volume on applied building physics deals with the ambient conditions indoors and outdoors, the performance rationale, and the heat, air and moisture metrics at the levels of the whole building and the building assembly. In addition, extended tables with material properties are added. The third and fourth volumes on performance-based building design use these metrics and the requirements related to structural mechanics, acoustics, lighting, fire safety, economics and sustainability, to design and construct whole buildings and their composite parts.

Note that the term 'building physics' is hardly used in the English-speaking world, where 'building science' is more common. Yet building science differs as, on the one hand it does not encompass acoustics and lighting, while on the other hand it includes more practice-related topics ranging from HVAC issues to organizational concerns.

0.2 Building physics

0.2.1 Definition

As an applied science, building physics studies the hygrothermal, acoustic and visual performance of materials, building assemblies, spaces, whole buildings and, be it under the name urban physics, the built environment. The constraints faced are the user demands related to overall comfort, health and safety, together with architectural facts and figures, durability issues, economic restrictions and sustainability-related requirements.

The term 'applied' indicates the field is a tool directed towards problem solving. Topics tackled in the heat, air and moisture subfield are air-tightness, thermal insulation, transient thermal response, moisture tolerance, thermal bridging, salt transport, temperature and humidity-related stress and strain, net energy demand, gross energy demand, end energy use, primary energy consumption, ventilation, thermal comfort and indoor air quality. In the building acoustics subfield, the topics discussed include the air- and structure-borne noise transmission by outer walls, floors, partition walls, party walls, glazing and roofs, room acoustics and the abatement of installation and ambient noise. The lighting subfield includes daylighting, artificial lighting, and the impact that both have on human wellbeing and primary energy consumption. Urban physics finally looks to the thermal, acoustic, visual and wind comfort outdoors, wind and rain patterns in cities, the spread of air pollution in cities, the heat island effect, and all aspects related to energy management at the city level.

0.2.2 Constraints

0.2.2.1 Comfort

Comfort is typically defined as a condition of mind that expresses satisfaction with the surroundings. Attainment of comfortable conditions depends on what humans need to feel thermally, acoustically and visually at ease: not too cold, not too warm, not too noisy, no large contrasts in luminance, and so on.

Thermal comfort engages physiology and psychology. As exothermal creatures with a constant core temperature of about 37 °C (310 K), humans must be able to lose heat to the environment under any circumstance, whether by conduction, convection, radiation, perspiration, transpiration or breathing. Air temperature, its gradient, the radiant temperature, radiant asymmetry, contact temperatures, relative air velocity, air turbulence and relative humidity in the direct environment will fix the heat exchanged. For a given activity and clothing, humans will quote certain combinations of the named ambient parameters as comfortable, and others not, although adaptation influences satisfaction.

Acoustic comfort strongly relates to mental awareness. Physically, young adults can hear sounds with frequencies between 20 and 16 000 Hz. In terms of sound intensity, however, humans scale logarithmically with better hearing for higher frequencies. Acoustics therefore uses the logarithmic unit the decibel (dB), with 0 dB for audibility and 140 dB for the pain threshold. Undesired noises such as neighbours, traffic, industry and aircraft will disturb people, generate complaints and often create long-lasting disputes.

Visual comfort combines mental with physical aspects. Physically, the eye sees electromagnetic waves with wavelengths between 0.38 and 0.78 µm. The maximum sensitivity lies near 0.58 µm, the yellow-green range. But overall sensitivity adapts to the average luminance. When dark it increases 10 000 times compared with daytime, but the eyes perceive that change logarithmically. Great differences in brightness will disturb, while well-adapted lighting creates a feeling of cosiness.

0.2.2.2 Health and wellbeing

Wellbeing is determined not only by the absence of illness, but also an absence of neuro-vegetative complaints, psychological stress or physical unease. Dust, fibres, (S)VOCs, radon, CO, viruses and bacteria, moulds and mites, too much noise, thermal discomfort and great luminance contrasts are all disturbing to the users of buildings.

0.2.2.3 Architecture and materials

Building physics faces architectural and material restrictions. Façade and roof form, aesthetics pursued and the materials chosen will all shape the building, while their design engages a multitude of metrics and requirements. Conflicting structural and physical issues often complicate solutions. Necessary thermal cuts interfere with the strength and stiffness of the connections required. The creation of waterproof and vapour-permeable structures are not always compatible. The necessary acoustic absorption could interfere with vapour tightness, for example.

0.2.2.4 Economy

Not only must the building costs respect budget limits, but the total expenditure over the timespan a building is owned or used should be the lowest achievable. The initial investment, energy used, maintenance, future necessary upgrades and replacements play a decisive role. A building designed and constructed according to the metrics of building physics and all other advanced fields of study will generally incur lower costs than if done without consideration of fitness for purpose.

0.2.2.5 Sustainability

The environmental impact of human activity has increased substantially over recent decades with worrying consequences. Locally, building use produces solid, liquid and gaseous waste. Countrywide, construction and occupancy accounts for 35–40% of the end energy used. Fossil fuels still deliver the major part, meaning that the CO_2 produced by their burning overwhelms all other greenhouse gas releases.

The increasing impact of life-cycle inventory and analysis (LCIA) and the use of certification tools reflect the pursuit of sustainability. In LCIA, buildings are evaluated in terms of environmental impact from 'cradle to cradle', that is, from material production through construction and occupancy to demolition and re-use. For each stage, all material, energy and water inflows and polluting outflows are quantified, and the impact on human wellbeing and the environment is assessed. Certification programmes in turn focus on fitness for purpose that new buildings, retrofitted buildings and urban environments should offer.

0.3 Importance

The necessity of creating a comfortable indoor environment protected from the weather gave birth to the field now called 'building physics'. As various ambient loads such as sun, rain, wind and noise, but also temperature, vapour and air pressure differentials, burden the building enclosure, an appropriate design should annihilate their impact when needed and use it when aiding comfort and wellbeing, while lowering source energy use as far as possible.

In earlier days, experience was the guide. Former generations disposed of a limited range of materials – wood, straw, loam, brick, natural stone, lead, copper, cast iron, blown glass – for which uses increased over the centuries. Standard solutions for roofs, roof edges and outer walls existed. From the size and orientation of the windows to the overall layout, everything was conceived to limit heating in the winter and overheating in summer. Because noise sources outside urban centres were scarce, acoustics was not a consideration, while a lifestyle adapted to the seasons saved energy.

That era ended with the industrial revolution. New materials flooded the market, such as steel, reinforced and pre-stressed concrete, nonferrous metals, synthetics, bitumen and insulation materials. More advanced technologies turned existing materials into innovative products: cast and float glass, rolled metal products and pressed bricks, for example. Advances in structural mechanics allowed designs of any form and span. Due to the widespread exploitation of fossil fuels such as coal, petroleum and natural gas, energy became cheap. Construction exploded and turned into a demand/supply market. The result was mass building of too often minimal quality.

The early twentieth century saw a 'modern school' of architects surfacing that experimented with alternative structural solutions, simple details and new materials. The buildings they designed were neither energy-efficient nor exemplary cases of good quality. Typical was the profuse use of steel, concrete and glass, all difficult materials from a hygrothermal point of view, and a reduction in overhangs and façade relief. The results were obvious failures that necessitated premature restoration, which a better knowledge of building physics could have prevented. Figure 0.1 shows a non-insulated Le Corbusier house built in 1926 that lacked moisture tolerance, so had to be rebuilt with insulation in the late 1980s. Previously, heating all rooms to a comfortable temperature required 20 000 litres of fuel a year. This

Fig. 0.1 House designed by Le Corbusier after restoration

reduced to 4000 litres a year after renovation, with the inhabitants heating the rooms only when being used.

To realise high-performance buildings fit for purpose, a good knowledge and correct application of all metrics related to building physics is essential. This replaces the time-consuming learning by trial and error of the past, for which building technologies and architectural fashions are evolving too rapidly.

0.4 History

Building physics surged at the crossroads of several disciplines: applied physics, comfort and health, building services, building design and construction.

0.4.1 Applied physics

0.4.1.1 Heat, air and moisture

Up until the early twentieth century, the subject of heat transmission was the main study area. Vapour diffusion gained interest in the late 1930s, when Teesdale of the US Forest Products Laboratory published a study on 'Condensation in Walls and Attics'. In 1952, a paper by J.S. Cammerer entitled 'Die Berechnung der Wasserdampfdiffusion in den Wänden' ['Calculation of water vapour diffusion in walls'] appeared in *Der Gesundheitsingenieur*. By the end of the 1950s the same journal published H. Glaser's upgraded calculation method for interstitial condensation by vapour diffusion in cold storage walls. Others, among them K. Seiffert, applied that method to the evaluation of building assemblies. His book *Wasserdampfdiffusion im Bauwesen* [*Water Vapour Diffusion in Buildings*] led to the use of vapour barriers becoming more or less inevitable. The main cause of interstitial condensation, air flow in and across assemblies, was largely overlooked. The impact was first noticed in Canada, a country with a timber-frame tradition. In 1961, A.G. Wilson of the National Research Council wrote:

> One of the most important aspects of air leakage in relation to the performance of Canadian buildings is the extent to which it is responsible for serious condensation problems. Unfortunately this is largely unrecognized in the design and construction of many buildings, and even when failures develop, the source of moisture is often incorrectly identified.

From the 1960s on, many researchers studied combined heat, air and moisture transport, among them O. Krischer, J.S. Cammerer and H. Künzel in Germany, A. De Vries, B.H. Vos and E. Tammes in the Netherlands, L.E. Nevander in Sweden and A. Tveit in Norway.

0.4.1.2 Acoustics

In the early twentieth century, physicists started showing an interest in noise control in buildings. In 1912, Berger submitted a PhD thesis at the Technische Hochschule München entitled *Über die Schalldurchlässigkeit* [*About Sound Transmission*]. In

1920, Sabine published his reverberation time formula. In the years thereafter, room acoustics became a favourite, with studies on speech intelligibility, optimal reverberation times, and reverberation times in anechoic rooms. A decade later, L. Cremer initiated a breakthrough in airborne sound transmission. In his paper *Theorie der Schalldämmung dünner Wände bei schrägem Einfall* [*Theory of sound insulation of thin walls at oblique incidence*], he recognized that coincidence between the sound waves in the air and the bending waves on a wall played a major role in degrading sound insulation. Later, his studies on structure-borne noise in floors advanced the idea of floating screeds as a solution. K. Gösele and M. Heckl fully established the link between building acoustics and construction through easy-to-use rules on how to build floors, walls and roofs with excellent air- and structure-borne sound attenuation. In the US, Beranek published his book *Noise and Vibration Control* in 1970, a remake of *Noise Reduction* of 1960, which became a standard reference for engineers who had to solve noise problems.

0.4.1.3 Lighting

Lighting came later. In 1931, a study was completed at the Universität Stuttgart, dealing with *Der Einfluss der Besonnung auf Lage und Breite von Wohnstraßen* [*The influence of solar irradiation on the location and width of residential streets*]. Later, physicists used radiation theory to calculate the illumination of surfaces and the luminance contrasts in the surroundings. By the end of the 1960s, the daylight factor was introduced as a quantity to evaluate the illumination indoors by natural light. More recently, since the energy crises of the 1970s, the relationship between artificial lighting and primary energy use has surged as a topic.

0.4.2 Thermal comfort and indoor air quality

By the nineteenth century, engineers were working on housing and urban hygiene. Max von Pettenkofer (1818–1901) (Figure 0.2) was the first to evaluate the impact of ventilation on the indoor CO_2 concentration. The 1500 ppm acceptability limit is attributed to him, as is the notion of a 'breathing material', the result of a link made later between finding more health complaints in stone buildings and fewer in brick dwellings, due to brick not completely blocking the passage of air. The true reason, of

Fig. 0.2 Max von Pettenkofer

0.4 History

Fig. 0.3 P.O. Fanger

course, was the poorer thermal performance of stone constructions and the related increase in mould and humidity.

Thermal comfort moved to the forefront in the twentieth century. Research by Yaglou in the 1930s, sponsored by the American Society of Heating and Ventilation Engineers (ASHVE), a predecessor of ASHRAE, led to the notion of 'operative temperature'. Originally, his definition overlooked radiation, which changed after A. Missenard, a French engineer, critically reviewed the data and saw the impact of the radiant temperature. The late P.O. Fanger (Figure 0.3) firmly founded the relation between perceived thermal comfort and all parameters intervening in his book *Thermal Comfort*, published in 1970. Based on physiology, heat exchange between the clothed body and the ambient surroundings and the differences in comfort perception between individuals, he developed a steady-state thermal model for the active, clothed human. Since then, his Predicted Mean Vote (PMV) versus Predicted Percentage of Dissatisfied (PPD) curve forms the kernel of all comfort standards worldwide. After 1985, the adaptive model gained support, being a refinement of Fanger's work.

Concerns about indoor air quality led to the cataloguing of a multitude of pollutants with their impact in terms of health risks. Over the years, 'sick building syndrome' (SBS) accompanied the move to fully air-conditioned builds, which reinforced the need for an even better understanding. Nonetheless, 'better' did not always result from a sound interpretation of the facts. Too often, discontent with the job was overlooked. In this area also, P.O. Fanger had an impact with his work on perceived indoor air quality based on bad smells and air enthalpy.

0.4.3 Building services

In the nineteenth century, building services technicians were searching for methods to calculate the heating and cooling load. The knowledge that had developed in physics, which provided concepts such as the thermal transmittance of a flat assembly, helped a lot. Quite early, organizations such as ASHVE and the Verein Deutscher Ingenieure (VDI) had technical committees dealing with the topic. An active member of ASHVE was W.H. Carrier (1876–1950), recognized in the US as the father of air conditioning. He was the first to publish a usable psychometric chart.

For H. Rietschel, professor at the Technische Universität Berlin and author of a comprehensive book on *Heizung und Lüftungstechnik* [*Heating and Ventilation Techniques*], heat loss and heat gain through ventilation was of concern. Along with others, he noted that well-designed ventilation systems malfunctioned when the envelope lacked air-tightness. This sharpened the interest in air transport. Vapour in the air became a worry once air-conditioning (HVAC) gained popularity, while previously humidity had been of concern because of possible health effects. Sound attenuation troubled the HVAC community because of system noisiness, while lighting became a factor because HVAC engineers got contracts for lighting design. Since 1973, energy efficiency has dominated.

0.4.4 Building design and construction

The many complaints about noise and moisture, with which modernists had to wrestle, slowly changed the field of building physics into one that helped to avoid the construction failures of some 'state-of-the art' solutions. Concern about moisture tolerance began in the early 1930s when peeling and blistering of paints on insulated timber-frame facades became an issue. Insulation materials were new at that time. It motivated Teesdale to study interstitial condensation. Some years later, ventilated attics with insulation at ceiling level were tested by F. Rowley, Professor in Mechanical Engineering at the University of Minnesota. The results were instructions about vapour retarders and attic ventilation. In Germany, the Freiland Versuchsstelle Holzkirchen, founded in 1951, used building physics as a driver for upgraded construction quality. When, after 1973, energy efficiency became a hot topic and insulation a necessity, the knowledge gathered proved extremely useful for the construction of high-quality, well-insulated buildings and the development of glazing with better insulating properties, lower solar and better visual transmittance. In the 1990s, the need for better quality resulted in the performance rationale of IEA EBC Annex 32, 'Integral Building Envelope Performance Assessment'.

0.4.5 The situation at the University of Leuven (KULeuven) and elsewhere

At KULeuven, lecturing in building physics started in 1952, when the field became compulsory for architectural engineering and optional for civil engineering. Antoine de Grave, a civil engineer who was the head of the building department at the Ministry of Public Works, was nominated as professor. He taught the course until his passing away in 1975. He published two books: *Bouwfysica* [*Building Physics*] and *Oliestook in de woning* [*Oil heating in Houses*]. By 1975, building physics had also become compulsory for civil engineering. Shortly after, together with Gerrit Vermeir, then a researcher and after 1992 professor in building acoustics, we started a working group with the aim to push research and consultancy in building physics. In 1978 that group gave birth to the Laboratory of Building Physics, to become the Unit of Building Physics within the Department of Civil Engineering in 1990. Over the years, the topics tackled included the physical properties of building and insulating materials, the performance of well-insulated building assemblies, end energy use in buildings, indoor environmental quality, air- and structure-borne noise attenuation and room acoustics. In 1997, urban physics was added. A basic motivation behind the research and consultancy done was upgrading the quality of the built environment in cooperation with the building sector. Since my retirement in 2008, Staf Roels is teaching performance-based building design now, Dirk Saelens building services, energy use and indoor environmental quality, and Hans Janssen the basics. In 2011, first Arne Dijckmans and then Edwin Reynders succeeded Gerrit Vermeir in the field of building acoustics. In 2013, Bert Blocken, a former researcher at the Unit and now professor at the TU/e (Eindhoven), joined as part-time, specializing in urban physics.

Ghent University (UGent) waited until 1999 to nominate Arnold Janssens, then a postdoctoral researcher at the Unit, as full-time professor in building physics. He did a tremendous job there in organizing and heading research in that field. At the Free University Brussels (VUB) a part-time professorship was offered to F. Descamps, also a former researcher at the Unit and manager of an engineering office in building physics.

Before the Second World War in the Netherlands, Professor Zwikker gave lectures at the Technical University Delft (TU-Delft) analysing buildings from a physical point of view, but a course named 'Building Physics' only started in 1955 with Professor Kosten of the Applied Physics Faculty as chair-holder. In 1963 Professor Verhoeven took over. Shortly after, in Eindhoven, a technical university (TU/e) was founded in 1969 with Professor P. De Lange as first holder of the Chair of Building Physics at the Faculty of the Built Environment.

0.5 Units

The book uses the SI system (internationally mandatory since 1977), with base units of the metre (m), the kilogram (kg), the second (s), the kelvin (K), the ampere (A) and the candela. Derived units of importance when studying building physics are:

Force:	newton (N);	$1\,N = 1\,kg.m.s^{-2}$	
Pressure:	pascal (Pa);	$1\,Pa = 1\,N/m^2$	$= 1\,kg.m^{-1}.s^{-2}$
Energy:	joule (J);	$1\,J = 1\,N.m$	$= 1\,kg.m^2.s^{-2}$
Power:	watt (W);	$1\,W = 1\,J.s^{-1}$	$= 1\,kg.m^2.s^{-3}$

The ISO standards (International Standardization Organization) for symbols are applied. If a quantity is not included, the CIB-W40 recommendations (International Council for Building Research, Studies, and Documentation, Working Group 40 on Heat and Moisture Transfer in Buildings) and the Annex 24 list edited by the IEA EBC (International Energy Agency, Executive Committee on Energy in Buildings and Communities) are followed.

0.6 Symbols

Table 0.1 List of symbols and quantities

Symbol	Meaning	SI units
a	Acceleration	m/s^2
a	Thermal diffusivity	m^2/s
b	Thermal effusivity	$W/(m^2.K.s^{0.5})$
c	Specific heat capacity	$J/(kg.K)$
c	Concentration	kg/m^3, g/m^3
e	Emissivity	–
f	Specific free energy	J/kg
g	Specific free enthalpy	J/kg
g	Acceleration due to gravity	m/s^2
g	Mass flux	$kg/(m^2.s)$
h	Height	m
h	Specific enthalpy	J/kg
h	Surface film coefficient for heat transfer	$W/(m^2.K)$
k	Mass-related permeability (mass may be moisture, air, salt)	s
l	Length	m
l	Specific enthalpy of evaporation or melting	J/kg
m	Mass	kg
n	Ventilation rate	s^{-1}, h^{-1}
p	Partial pressure	Pa
q	Heat flux	W/m^2
r	Radius	m
s	Specific entropy	$J/(kg.K)$
t	Time	s
u	Specific latent energy	J/kg
v	Velocity	m/s
w	Moisture content	kg/m^3
x,y,z	Cartesian coordinates	m
A	Water sorption coefficient	$kg/(m^2.s^{0.5})$
A	Area	m^2
B	Water penetration coefficient	$m/s^{0.5}$
D	Diffusion coefficient	m^2/s
D	Moisture diffusivity	m^2/s
E	Irradiation	W/m^2
F	Free energy	J
G	Free enthalpy	J
G	Mass flow (mass = vapour, water, air, salt)	kg/s
H	Enthalpy	J
I	Radiation intensity	J/rad
K	Thermal moisture diffusion coefficient	$kg/(m.s.K)$

0.6 Symbols

Table 0.1 (Continued)

Symbol	Meaning	SI units
K	Mass permeance	s/m
K	Force	N
L	Luminosity	W/m^2
M	Emittance	W/m^2
P	Power	W
p	Thermal conductance	W/(m^2.K)
P	Total pressure	Pa
Q	Heat	J
R	Thermal resistance	m^2.K/W
R	Gas constant	J/(kg.K)
S	Entropy	J/K
S	Saturation degree	–
T	Absolute temperature	K
T	Period (of a vibration or a wave)	s, days
U	Latent energy	J
U	Thermal transmittance	W/(m^2.K)
V	Volume	m^3
W	Air resistance	m/s
W	Work	J
X	Moisture ratio	kg/kg
Z	Diffusion resistance	m/s
α	Thermal expansion coefficient	K^{-1}
α	Absorptivity	–
β	Surface film coefficient for diffusion	s/m
β	Volumetric thermal expansion coefficient	K^{-1}
η	Dynamic viscosity	N.s/m^2
θ	Temperature	°C
λ	Thermal conductivity	W/(m.K)
λ	Wavelength	m
μ	Vapour resistance factor	–
ν	Kinematic viscosity	m^2/s
ρ	Density	kg/m^3
ρ	Reflectivity	–
σ	Surface tension	N/m
$\rfloor$	Thermal pulsation	J/(m^2.K)
τ	Transmissivity	–
ϕ	Relative humidity	–
α, ϕ, Θ	Angle	rad
ξ	Specific moisture capacity	kg/kg per unit of moisture potential
Ψ	Porosity	–
Ψ	Volumetric moisture ratio	–
Φ	Heat flow	m^3/m^{3}W

Table 0.2 List of currently used suffixes

Symbol	Meaning	Symbol	Meaning
Indices			
A	Air	m	Moisture, maximal
c	Capillary, convection	r	Radiant, radiation
e	External, outdoors	sat	Saturation
h	Hygroscopic	s	Surface, area, suction
i	Inside, indoors	rs	Resulting
cr	Critical	v	Water vapour
CO_2, SO_2	Chemical symbols for gases	w	Water, wind
		φ	Relative humidity

Table 0.3 List of notations used

Notation	Meaning
[ ], bold	Matrix, array, value of a complex number
dash (e.g. $\bar{a}$)	Vector

Further reading

Beranek, L. (ed.) (1971) *Noise and Vibration Control*, vol. **369**, McGraw-Hill Book Company, New York.

CIB-W40 (1975) Quantities, Symbols and Units for the Description of Heat and Moisture Transfer in Buildings: Conversion Factors. IBBC-TNP, Report no. BI-75-59/03.8.12, Rijswijk.

De Freitas, V.P. and Barreira, E. (2012) *Heat, Air and Moisture Transfer Terminology, Parameters and Concepts*, CIB publication.

Donaldson, B. and Nagengast, B. (1994) *Heat and Cold: Mastering the Great Indoors*, ASHRAE Publications, Atlanta.

Hendriks, L. and Hens, H. (2000) *Building Envelopes in a Holistic Perspective, IEA-ECBCS Annex 32*, ACCO, Leuven.

Hens, H. (2008) Building Physics: from a dormant beauty to a key field in building engineering. Proceedings of the Building Physics Symposium, Leuven.

ISO-1000 (1981) SI units and recommendations for the use of their multiples and of certain other units. ISO.

Kumaran, K. (1996) *Task 3: Material Properties, Final Report IEA EXCO ECBCS Annex 24*, ACCO, Leuven.

Künzel, H. (2001) *Bauphysik. Geschichte und Geschichten*, Fraunhofer IRB-Verlag, Stuttgart.

Northwood, T. (ed.) (1977) *Architectural Acoustics*, Dowden, Hutchinson & Ross, Inc., Stroudsburg, PA.

Rose, W. (2003) The rise of the diffusion paradigm in the US, in *Research in Building Physics* (eds J. Carmeliet, H. Hens and G. Vermeir), A.A. Balkema, pp. 327–334.

USGBC (2008) LEED 2009 for New Construction and Major Renovations. No. 200371. Washington, DC.

Winkler Prins Technische Encyclopedie (1976) Article on Building Physics [in Dutch]. Elsevier, Amsterdam, pp. 157–159.

1 Heat transfer

1.1 Overview

1.1.1 Heat

A first description of what heat is comes from thermodynamics. That discipline describes how systems and their environment interchange energy. Anything can be a system: a material, a building assembly, a building, part of a HVAC system, even a whole city. Energy transmitted as 'work' is purposeful and organized, whereas as 'heat' it is diffuse and chaotic. A second description resides in particle physics, where the statistically distributed kinetic energy of atoms and free electrons stands for heat. In any case, heat is the least noble, most diffuse form of energy to which each nobler form degrades; consider the second law of thermodynamics.

1.1.2 Temperature

The temperature reflects the quality of the heat. Higher values reflect the increased kinetic energy of atoms and free electrons, resulting in higher exergy and the potential to convert more heat via a cyclic process into work. Lower temperatures and therefore less kinetic energy of atoms and free electrons result in less exergy. Higher temperatures require warming up, lower temperatures cooling down of a system. Like any potential, temperature is a scalar, which, as heat, cannot be measured directly. It is sensed and because many material properties depend on it, indirectly quantifiable. A mercury thermometer uses the volumetric expansion of mercury when heated and the contraction when cooled. In a Pt100 thermometer, the electrical resistance of platinum wire changes with temperature. Temperature logging with thermocouples uses the varying contact potential between metals.

The SI system advances two temperature scales, one empiric in degrees Celsius, °C, with the symbol θ, and one thermodynamic in degrees kelvin, K, with the symbol T. Zero °C coincides with the triple point of water, and 100 °C with the boiling point of water at 1 atmosphere pressure. Zero K instead stands for the point of absolute zero, and 273.15 K coincides with the triple point of water. Temperature differences are given in K, temperatures in °C or K, with the following relationship between the two:

$$T = \theta + 273.15$$

Instead of degrees Celsius, the US generally uses degrees Fahrenheit (°F):

$$°F = 32 + 9/5\,°C$$

1.1.3 Sensible and latent heat

Sensible heat transfer, whether by conduction, convection or radiation, requires temperature differences. Conduction refers to the heat exchanged when vibrating

Building Physics: Heat, Air and Moisture: Fundamentals and Engineering Methods with Examples and Exercises. Third Edition. Hugo Hens.
© 2017 Ernst & Sohn GmbH & Co. KG. Published 2017 by Ernst & Sohn GmbH & Co. KG.

atoms collide and free electrons move. Transmission between solids at different temperatures in ideal contact and among points at different temperature in the same solid is conduction-based. Conduction also intervenes in gases and liquids, whether or not they are in contact with solids. According to the second law of thermodynamics, conduction always goes in the direction of lower temperatures. A medium is required, but conduction induces no movement.

Convection instead is the result of macroscopic motion in liquids and gases wherein temperatures differences exist, included contact with colder or warmer solids. Whether it is an external force, a difference in density or both that fosters movement, will define the type of convection: forced, natural or mixed. Also convection needs a medium.

Radiation, finally, concerns the heat transferred due to the emission and absorption of electromagnetic waves by surfaces. At temperatures above 0 K, every surface emits. If two or more are at different temperatures, the result is heat exchange. Radiation does not need a medium, while the governing laws differ strongly from those describing conduction and convection.

Latent heat directly links to changes of state. Its release or absorption requires no temperature differences but affects the whole thermal picture. To give an example, water evaporating absorbs a quantity of sensible heat equal to the heat of evaporation, so acts as a heat sink. When the water vapour formed moves to a colder spot where it condenses again, that heat of evaporation is re-emitted, so forms a heat source. These sources and sinks not only impact the temperatures in materials and assemblies, but also the latent heat transferred.

1.1.4   Why are heat and temperature so compelling?

Why heat? The main motivation is that energy use in buildings matters. In fact, thermal comfort indoors requires an operative temperature at the desired level, which in cold and temperate climates means heating. Related fossil fuel burning still represents a substantial share of the overall end energy consumed and CO_2 emitted, such that energy efficiency in buildings became mandatory, among others by minimizing the heat traversing the enclosure.

Why temperature? There are many reasons. Inside surface temperatures close to the air temperature indoors improve thermal comfort, whereas low values increase mould and surface condensation risks. Excessive summer temperatures indoors affect usability, while higher temperature differences across envelope assemblies foster air and moisture movement, thermal stress and crack risk. Too many swings from above to below freezing point may damage frost-sensitive porous materials that stay wet. High temperature gradients favour combined moisture and dissolved salt displacement, while high temperatures accelerate the chemical breakdown of synthetics. Whether these effects remain controllable depends on how building assemblies are designed and built.

1.1.5 Some definitions

Amount of heat, symbol Q, units J	Quantifies the energy exchanged as heat. As energy is a scalar, so is the amount of heat.
Heat flow, symbol Φ, units J/s = W	Heat migrating per unit of time. Heat flow is a measure for 'power', thus a scalar.
Heat flux, symbol q, units W/m²	Heat migrating per unit of time across a unit surface normal to the flow. Flux is a vector with the same direction as the surface vector. The components in Cartesian coordinates are q_x, q_y, q_z, in polar coordinates q_R, q_ϕ, q_Θ.

Solving a heat transfer problem means determining a scalar temperature (T) and vector heat flux (q) field. The calculation thus requires a scalar and a vector equation.

1.2 Conduction

1.2.1 Conservation of energy

A first relationship between heat flux (q) and temperature (T) follows from the conservation of energy axiom. If the system is an infinitely small material volume dV and everything that is around it the environment, then, knowing that conduction does not displace mass, the energy balance becomes:

$$d\Phi + d\Psi = dU + dW \tag{1.1}$$

with $d\Phi$ the resulting heat flow between system and environment, $d\Psi$ the heat dissipated uniformly in the system, dU the change in the system's internal energy and dW the work exchanged with the environment, all three per unit of time. Dissipation can include heat produced by an exothermic chemical reaction, heat absorbed thanks to an endothermic chemical reaction, the Joule effect due to an electric current passing through, latent heat released or absorbed, and so on. The work exchanged equals:

$$dW = Pd(dV) = Pd^2V$$

with P being the pressure in Pa. The balance states that the heat exchanged ($= d\Phi$), released or absorbed modifies the internal energy of the material volume while generating an exchange of work with the environment. If isobaric, the balance can be rewritten as:

$$d(U + PdV) = dQ + dE$$

with $U + PdV$ being the enthalpy H. Writing the resulting heat flow, the enthalpy change and the heat dissipated as:

$$d\Phi = -\text{div}(q)dV \qquad dH = \left|\frac{\partial(\rho c_p T)}{\partial t}\right|dV \qquad d\Psi = \Phi' dV$$

where c_p is the specific heat capacity at constant pressure of the material (J/(kg.K)), ρ is its density (kg/m^3) and Φ' is the dissipated heat per unit of time and volume, positive for a source, negative for a sink, which makes the conservation law:

$$\left(\operatorname{div}(q) + \Phi' + \frac{\partial(\rho c_p T)}{\partial t}\right) dV = 0 \tag{1.2}$$

For solids and liquids the specific heat capacity depends little on the change of state. One value, symbol c, can be used, with the product ρc equal to the volumetric specific heat capacity. For gases, the value varies according to the change of state, giving as the relationship between the specific heat capacity at constant pressure (c_p) and constant volume (c_v):

$$c_p = c_v + R$$

with R the specific gas constant (in Pa.m^3/(kg.K)). Because conservation of energy holds for any infinitely small material volume, the relationship between heat flux (q) and temperature (T) finally becomes:

$$\operatorname{div}(q) = -\frac{\partial(\rho c T)}{\partial t} - \Phi' \tag{1.3}$$

1.2.2 The conduction laws

1.2.2.1 First law

The first law is the empirical conduction equation introduced by the French physicist Fourier in 1822 (Figure 1.1):

$$q = -\lambda \operatorname{grad} T = -\lambda \operatorname{grad} \theta \tag{1.4}$$

Fig. 1.1 The French physicist Fourier

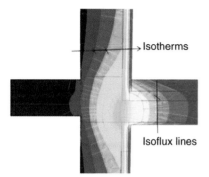

Fig. 1.2 Lines of equal temperature (isotherms) and equal heat flow rates (isoflux lines)

This is a vector relation, which states that the conductive heat flux anywhere in a solid, liquid or gas varies proportionally to the temperature gradient there. The multiplier λ is the thermal conductivity, units W/(m.K). The minus sign indicates that flux and temperature gradient, positive from colder to warmer, oppose each other. Thermodynamics in fact teaches that, if not forced externally, heat always moves in the direction of lower temperatures. Otherwise entropy would decrease without energy input from the environment, which is impossible.

The following observation supports the law. With the surfaces of equal temperature, called isotherms, in a material traced and the heat fluxes visualized, it can be observed that these develop perpendicular to the isotherms and increase where these are near to one another (Figure 1.2).

At the same time the fluxes keep proportionality with the thermal conductivity, which for reasons of simplicity is assumed to be scalar and constant, even though for building and insulating materials such assumptions do simplify reality. The value in fact depends on temperature, moisture content, thickness and sometimes age. In anisotropic materials, it is even a tensor. Fortunately, for first-order calculations, 'scalar and constant' often suffices.

In right-angled Cartesian coordinates [x,y,z], the heat fluxes along the axes become:

$$q_x = \boldsymbol{q_x u_x} = -\lambda \frac{\partial T}{\partial x} \qquad q_y = \boldsymbol{q_y u_y} = -\lambda \frac{\partial T}{\partial y} \qquad q_z = \boldsymbol{q_z u_z} = -\lambda \frac{\partial T}{\partial z}$$

Often the temperature in °C replaces K. The heat flow across a surface area dA with direction n then equals:

$$d\Phi_n = \boldsymbol{q d A_n} = -\lambda \frac{\partial \theta}{\partial n} d A_n \boldsymbol{u_n^2} = -\lambda \frac{\partial \theta}{\partial n} d A_n$$

Along the three axes, one gets:

$$d\Phi_x = -\lambda \frac{\partial \theta}{\partial x} d A_x \qquad d\Phi_y = -\lambda \frac{\partial \theta}{\partial y} d A_y \qquad d\Phi_z = -\lambda \frac{\partial \theta}{\partial z} d A_z$$

1.2.2.2 Second law

To arrive at a second law, the conduction equation is implemented in the conservation of energy expression:

$$\text{div}(\lambda \text{ grad } T) = \frac{\partial(\rho c T)}{\partial t} - \Phi' \tag{1.5}$$

The result is a scalar equation that allows the calculation of temperature fields. In the case that the thermal conductivity and the volumetric specific heat capacity are constant, the expression simplifies to what is known as Fourier's second law:

$$\nabla^2 T = \left(\frac{\rho c}{\lambda}\right) \frac{\partial T}{\partial t} - \frac{\Phi'}{\lambda} \tag{1.6}$$

with ∇^2 being the Laplace operator, in Cartesian coordinates equal to:

$$\nabla^2 = \frac{\partial^2}{\partial x^2} + \frac{\partial^2}{\partial y^2} + \frac{\partial^2}{\partial z^2} \tag{1.7}$$

Further discussions focus on solving both laws for a series of building-related cases.

1.2.3 Steady state

Steady state indicates that the temperatures and heat fluxes remain time-independent. For that, constant boundary conditions, constant material properties and constant energy dissipation are needed. When all are invariable, then:

$$\partial T / \partial t = 0$$

With the temperature in °C, Fourier's second law so simplifies to:

$$\nabla^2 \theta = -\Phi'/\lambda \tag{1.8}$$

1.2.3.1 One-dimensional flat assemblies

In one dimension with temperature changes normal to the surface, the equation further reduces to:

$$\frac{d^2\theta}{dx^2} = -\frac{\Phi'}{\lambda} \tag{1.9}$$

Without dissipation, it becomes:

$$\frac{d^2\theta}{dx^2} = 0$$

with, as a solution:

$$\theta = C_1 x + C_2 \tag{1.10}$$

where C_1 and C_2 are the integration constants, fixed by the boundary conditions. That simple equation governs conduction across flat assemblies with the end faces at constant but different temperatures. Buildings include numerous flat assemblies such as low-slope roofs, sloped roof pitches, outer walls, floors, partition walls, glass surfaces, and so on. In cross-section, these can be single-layer or composite.

Single-layer

If single-layer, the thermal conductivity of the material and the layer thickness (d) has to be known. The boundary conditions are: $x = 0$: $\theta = \theta_{s1}$; $x = d$: $\theta = \theta_{s2}$ with θ_{s1} and θ_{s2} being the different temperatures at the end faces, whereby θ_{s1} is assumed to be colder than θ_{s2}. The integration constants then become

$$C_2 = \theta_{s1} \qquad C_1 = (\theta_{s2} - \theta_{s1})/d$$

which changes the temperatures in the layer into:

$$\theta = \frac{\theta_{s2} - \theta_{s1}}{d} x + \theta_{s1} \tag{1.11}$$

Or, at steady state, the temperatures in a flat, single-layer assembly with neither heat source nor sink will vary linearly between the values at both end faces (Figure 1.3).

Related heat flux equals:

$$q = -\lambda \operatorname{grad} \theta = -\lambda \frac{d\theta}{dx} = -\lambda \left(\frac{\theta_{s1} - \theta_{s2}}{d} \right) \tag{1.12}$$

In absolute terms, this equation becomes:

$$q = \lambda \left(\frac{\theta_{s2} - \theta_{s1}}{d} \right) \tag{1.13}$$

showing that the flux is proportional to the thermal conductivity of the material and the temperature difference between both end faces and inversely proportional to the layer thickness. For given thickness and temperature difference, a lower thermal

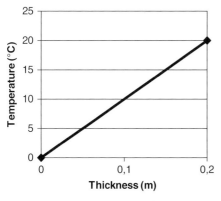

Fig. 1.3 Temperatures in a single-layer assembly

conductivity reduces the flux, meaning less heat being lost or gained. Materials with very low thermal conductivity are therefore called insulation materials. Rearranging gives:

$$q = \frac{\Delta \theta}{d/\lambda} \qquad (1.14)$$

with d/λ the thermal resistance of the flat single-layer assembly, symbol R, units m².K/W. The higher that number, the lower the heat flux for a given temperature difference between both end faces, or, the better the assembly insulates. Higher values require larger thicknesses or, for a given thickness, better insulating materials. The inverse of the thermal resistance is the thermal conductance, symbol P, units W/(m².K), a quantity telling how much heat per unit of time and surface passes across the assembly for a 1 °C or 1 K temperature difference between both end faces.

Composite

Composite assemblies consist of two or more plane-parallel layers. In buildings, most flat parts are composite, an example being the filled cavity wall of Figure 1.4, showing the inside leaf, cavity fill and veneer wall.

In steady state, without heat sources or sinks, the flux must be the same in each layer. Otherwise, thermal storage or discharge will make the regime transient. Suppose the temperature of end face 1 is θ_{s1}, and for end face 2, θ_{s2} ($\theta_{s1} < \theta_{s2}$). If the thermal conductivities and thicknesses of all layers are known and possible contact resistances between layers can be neglected, which fits for non-metallic materials, the heat flux q being constant means:

Layer 1 $\quad q = \lambda_1 \dfrac{\theta_1 - \theta_{s1}}{d_1}$

Layer 2 $\quad q = \lambda_2 \dfrac{\theta_2 - \theta_1}{d_2} \ldots$

Fig. 1.4 Filled cavity wall

1.2 Conduction

Layer $n-1$ $\quad q = \lambda_{n-1} \dfrac{\theta_{n-1} - \theta_{n-2}}{d_{n-1}}$

Layer n $\quad q = \lambda_n \dfrac{\theta_{s2} - \theta_{n-1}}{d_n}$

with $\theta_1, \theta_2, \ldots, \theta_{n-1}$ the unknown interface temperatures. Rearrangement and summing gives:

$q \dfrac{d_1}{\lambda_1} = \theta_1 - \theta_{s1}$

$+q \dfrac{d_2}{\lambda_2} = \theta_2 - \theta_1$

$+ \ldots$

$+q \dfrac{d_n}{\lambda_n} = \theta_{s2} - \theta_{n-1}$

$q \sum\limits_{i=1}^{n} \left(\dfrac{d_i}{\lambda_i} \right) = \theta_{s2} - \theta_{s1}$

or:

$$q = \dfrac{\theta_{s2} - \theta_{s1}}{\sum\limits_{i=1}^{n} (d_i / \lambda_i)} \tag{1.15}$$

The sum $\sum (d_i / \lambda_i)$, symbol R_T, units m² K/W, is called the total thermal resistance of the assembly, and the ratio d_i / λ_i, symbol R_i, same units as R_T, the thermal resistance of layer i. The higher the total thermal resistance, the lower the steady-state heat flux and the better the assembly insulates. A high thermal resistance requires a sufficiently thick insulation layer. Since the energy crises of the 1970s, thermal insulation of high performance became a prime measure to lower the net heating demand and CO_2 release where heating systems burn fossil fuels.

How the assembly is designed fixes its total thermal resistance. As a whole, the commutation property applies and the layer sequence does not matter. Inside insulation should perform as well as outside insulation. From a whole building physics perspective, that conclusion is somewhat deceptive: the same thermal resistance, yes, but both diverge in overall performance, a fact not highlighted in the steady state. The analogy with the current (I) in an electrical circuit subjected to a voltage difference ΔV when the electrical resistances (R_{ei}) stay in series is instructive:

$I = \Delta V / \sum R_{ei}$

Clearly, temperature replaces voltage, heat flux the current, and thermal resistance the electrical resistance. This allows the conversion of heat conduction problems into an electrical analogy.

With both end face temperatures known, all interface temperatures follow from rearranging the heat flux equation per layer:

$$\theta_1 = \theta_{s1} + q\frac{d_1}{\lambda_1} = \theta_{s1} + (\theta_{s2} - \theta_{s1})\frac{R_1}{R_T}$$

$$\theta_2 = \theta_1 + q\frac{d_2}{\lambda_2} = \theta_1 + q R_2 = \theta_{s1} + (\theta_{s2} - \theta_{s1})\frac{(R_1 + R_2)}{R_T} \ldots$$

$$\theta_{n-1} = \theta_{n-2} + q\frac{d_{n-1}}{\lambda_{n-1}} = \theta_{s1} + (\theta_{s2} - \theta_{s1})\frac{\sum_{i=1}^{n-1} R_i}{R_T}$$

Writing $\theta_i = \theta_{i-1} + qd_i/\lambda_i$ as $(\theta_i - \theta_{i-1})/d_i = q/\lambda_i$ underlines that the temperature gradient in a layer is inversely proportional to its thermal conductivity. Hence, gradients are large in insulation and small in conductive layers. Each layer equation can yet be rewritten as:

$$\theta_x = \theta_{s1} + q R_{s1}^x \tag{1.16}$$

where R_{s1}^x is the thermal resistance between end face s1 and interface x in the assembly. When the calculation starts at end face s2, the equation becomes:

$$\theta_x = \theta_{s2} - q R_{s2}^x \tag{1.17}$$

In a [R,θ] plane with the thermal resistance R as abscissa and the temperature θ as ordinate, both represent a straight line linking both end face temperatures $(0, \theta_{s1})$ and (R, θ_{s2}) with the heat flux as slope. Effectively a composite assembly responds as if it is single-layered. To construct the temperature line in the thickness graph, first draw the assembly in the [R,θ] plane with the layers as thick as their thermal resistance. Mark the temperature θ_{s1} on end face s1, temperature θ_{s2} on end face s2, and trace the straight line in between. The thickness-related course then follows from transposing all intersections with the successive interfaces into the thickness graph and then linking them per layer as shown in Figure 1.5. Of course, the correct layer sequence must be kept.

For single-layer as well as composite assemblies, the product of the heat flux with the surface area (A) gives the heat flow:

$$\Phi = q A \tag{1.18}$$

Special cases
Three special cases demand consideration. The first concerns a single-layer assembly where the thermal conductivity changes with temperature or moisture content (w), so along the ordinate x (Figure 1.6).

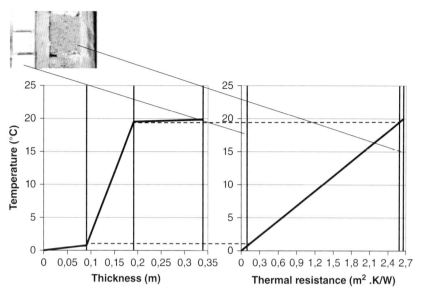

Fig. 1.5 Temperatures in a composite assembly, graphic construction

Should the thermal conductivity vary linearly with temperature ($\lambda = \lambda_o + a\theta$), then the heat flux will follow from ($x = 0$, $\theta = \theta_{s1}$; $x = d$, $\theta = \theta_{s2}$):

$$\int_{\theta_{s1}}^{\theta_{s2}} (\lambda_0 + a\theta)\,d\theta = \int_0^d q\,dx$$

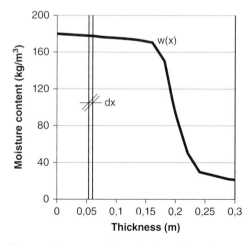

Fig. 1.6 Moisture profile with thermal conductivity as a function of x

giving as a solution:

$$\lambda_0(\theta_{s2} - \theta_{s1}) + \frac{a(\theta_{s2}^2 - \theta_{s1}^2)}{2} = q\,d \quad \text{or} \quad q = \lambda(\theta_m)\frac{\theta_{s2} - \theta_{s1}}{d} \quad (1.19)$$

with $\lambda(\theta_m)$ the thermal conductivity at the average temperature in the layer. The thermal resistance is $d/\lambda(\theta_m)$, while the temperature curve changes into a parabola:

$$a\theta^2/2 + \lambda_0\theta = q\,x + C$$

For an assembly loaded with a varying moisture content $w(x)$, so that $\lambda = F(w(x)) = f(x)$, the heat flux becomes ($x=0$: $\theta=\theta_{s1}$; $x=d$: $\theta=\theta_{s2}$):

$$q\int_0^d \frac{dx}{\lambda(x)} = \int_{\theta_{s1}}^{\theta_{s2}} d\theta$$

In case the moisture distribution is such that the thermal conductivity increases proportionally to x ($\lambda = \lambda_0 + ax$), the integrals give:

$$q = \frac{\theta_{s2} - \theta_{s1}}{\frac{1}{a}\ln\left(\frac{\lambda_0 + a\,d}{\lambda_0}\right)} \quad (1.20)$$

Again, the denominator stands for the thermal resistance. The temperature course now becomes:

$$\theta_x = \theta_{s1} + q\left[\frac{1}{a}\ln\left(\frac{\lambda_0 + a\,x}{\lambda_0}\right)\right] \quad (1.21)$$

The second special case concerns a single-layer assembly that dissipates or absorbs Φ joules of heat per unit of time and volume. If this happens uniformly over the thickness, the steady-state balance equation changes to:

$$\frac{d^2\theta}{dx^2} = -\frac{\Phi'}{\lambda}$$

with, as boundary conditions: $x=0$, $\theta=\theta_{s1}$; $x=d$, $\theta=\theta_{s2}$ ($\theta_{s1} < \theta_{s2}$). The solution is:

$$\theta = -\frac{\Phi'}{2\lambda}x^2 + C_1 x + C_2 \quad (1.22)$$

a parabolic temperature curve: convex for a heat source, and concave for a heat sink (Figure 1.7).

The heat flux changes to:

$$q = -\lambda\frac{d\theta}{dx} = \Phi'\,x - C_1\lambda \quad (1.23)$$

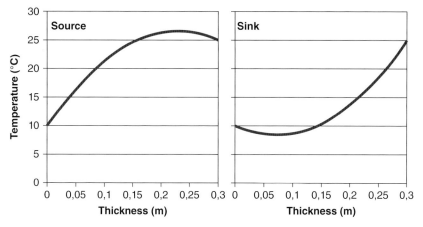

Fig. 1.7 Temperature in a uniformly spread heat source or sink in a single-layer assembly

Instead of no heat dissipated, the heat flux now changes from point to point across the assembly. The boundary conditions give as integration constants:

$$C_1 = \frac{\theta_{s2} - \theta_{s1}}{d} + \frac{\Phi d'}{2\lambda} \qquad C_2 = \theta_{s1}$$

The third special case concerns a composite assembly with a local heat source or sink, for example, due to a deposit or drying of condensate in an interface between two layers. The heat flux dissipated this way is q', while the temperature θ_{s1} at one end face passes θ_{s2} at the other end face. Then, a steady-state heat balance at the interface (x) with the source or sink gives the temperatures in the assembly. Presuming that the heat flows from both end faces to x:

End face s1: $\quad q_{s1}^x = \dfrac{\theta_{s1} - \theta_x}{R_{s1}^x}$

End face s2: $\quad q_{s2}^x = \dfrac{\theta_{s2} - \theta_x}{R_{s2}^x}$

In both equations, θ_x is the unknown temperature in x. Setting the sum of the two heat fluxes and the dissipated heat zero, with the dissipated heat negative for drying and positive for condensation, gives:

$$\theta_x = \frac{R_{s2}^x \theta_{s1} + R_{s1}^x \theta_{s2} + q' R_{s1}^x R_{s2}^x}{R_{s1}^x + R_{s2}^x} \tag{1.24}$$

Introducing that temperature in the flux equations from both end faces results in:

$$q_{s1}^x = \left(\frac{\theta_{s1} - \theta_{s2}}{R_T}\right) - \frac{q' R_{s2}^x}{R_T} \qquad q_{s2}^x = -\left[\left(\frac{\theta_{s1} - \theta_{s2}}{R_T}\right) + \frac{q' R_{s1}^x}{R_T}\right] \tag{1.25}$$

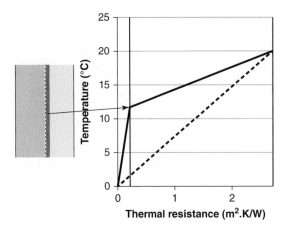

Fig. 1.8 Composite assembly, with heat source in the interface between two layers

with R_T the total thermal resistance of the assembly, equal to:

$$R_T = R_{s1}^x + R_{s2}^x$$

For a heat source the incoming flux drops and the outgoing flux increases compared with no source (Figure 1.8). For a sink, it is the inverse.

1.2.3.2 Two dimensions, cylinder symmetric

In cylinder coordinates, cylinder symmetric cases behave as one-dimensional, such as hung heating pipes. Of interest are the heat loss per metre run, the pipe temperature and the insulation efficiency. Consider a pipe with an inside radius r_1 and outside radius r_2. The temperature at the inside face is θ_{s1}, and at the outside face θ_{s2} (Figure 1.9). In steady state with no dissipation, the same heat flow must pass each cylinder concentric to the pipe's centre. With that centre as origin, the flow per metre run is:

$$\Phi = -\lambda(2\pi r)d\theta/dr = C^t$$

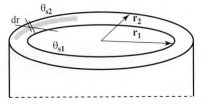

Fig. 1.9 The pipe problem

1.2 Conduction

Integration gives:

$$\Phi \int_{r_1}^{r_2} \frac{dr}{r} = -2\pi\lambda \int_{\theta_{s1}}^{\theta_{s2}} d\theta \quad \text{or} \quad \Phi = \frac{\theta_{s1} - \theta_{s2}}{\left(\frac{\ln(r_2/r_1)}{2\pi\lambda}\right)} \quad (1.26)$$

The denominator is the equivalent of the thermal resistance of a flat assembly. This is the thermal resistance per metre pipe, with units m.K/W.

For a composite pipe, the same reasoning as for a composite flat assembly gives the heat flow per metre run as:

$$\Phi = \frac{\theta_{s1} - \theta_{s2}}{\sum_{i=1}^{n} \left[\frac{\ln(r_{i+1}/r_i)}{2\pi\lambda_i}\right]} \quad (1.27)$$

The temperatures then follow from:

$$\theta_{i+1} = \theta_{s1} + \Phi \sum_{i=1}^{i} \left[\frac{\ln(r_{i+1}/r_i)}{2\pi\lambda_i}\right] \quad (1.28)$$

1.2.3.3 Two and three dimensions: thermal bridges

When looking in detail at outside walls, roofs, floors and partition walls, the assumption of 'flat' does not apply everywhere. What about lintels above windows? What about window reveals? What about junctions between two outside walls? What about the corners between two outer walls and a low-slope roof? Also, the end faces of flat assemblies are not necessarily isothermal (Figure 1.10). Studying steady-state

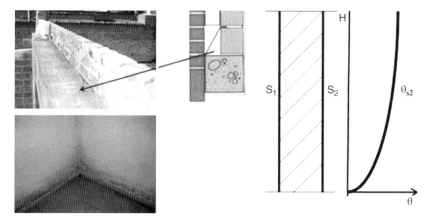

Fig. 1.10 Lintel above a window; corner between two outside walls and the floor; flat assembly with non-isothermal inside face

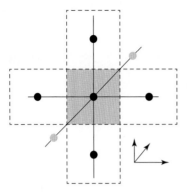

Fig. 1.11 Central and neighbouring control volumes

heat transfer now requires solving:

$$\frac{\partial^2 \theta}{\partial x^2} + \frac{\partial^2 \theta}{\partial y^2} + \frac{\partial^2 \theta}{\partial z^2} = \pm \frac{\Phi'}{\lambda}$$

or, without dissipation:

$$\frac{\partial^2 \theta}{\partial x^2} + \frac{\partial^2 \theta}{\partial y^2} + \frac{\partial^2 \theta}{\partial z^2} = 0$$

For some very elementary cases, with one material, easy geometry and simple boundary conditions, this partial differential equation can be solved analytically. In the majority of building-related cases, only a numeric approach, for example using the control volume method (CVM), offers a solution. Many building details yet consist of rectangular material volumes. In CVM, such detail is meshed in cubic or beam-like control volumes and the sum of the steady-state heat flows coming from the six adjacent volumes is set to zero (Figure 1.11).

If the meshing coincides with the interfaces between layers, then all control volumes remain material-homogenous but the calculations do not give the interface temperatures. The preference therefore goes to control volumes whose centres lie on the interfaces. They are not material-homogenous but the calculations give the interface temperatures. Along the x, y and z axes, the sides of the control volumes around each centre-point equal the sum of half the distance in the negative and positive directions to the adjacent centre points. If all seven centre points are situated in the same material and the mesh width is equal along all three axes, the heat flow from the centre point in the nearest by volume $(l-1,m,n)$ to the centre point in the central volume (l,m,n) equals:

$$\Phi_{l-1,m,n}^{l,m,n} = \lambda \frac{(\theta_{l-1,m,n} - \theta_{l,m,n})a^2}{a} = a\lambda(\theta_{l-1,m,n} - \theta_{l,m,n})$$

1.2 Conduction

with a being the mesh width. The other five give:

$$\Phi_{l+1,m,n}^{l,m,n} = \lambda \frac{(\theta_{l+1,m,n} - \theta_{l,m,n})a^2}{a} = a\lambda(\theta_{l+1,m,n} - \theta_{l,m,n})$$

$$\Phi_{l,m-1,n}^{l,m,n} = \lambda \frac{(\theta_{l,m-1,n} - \theta_{l,m,n})a^2}{a} = a\lambda(\theta_{l,m-1,n} - \theta_{l,m,n})$$

$$\Phi_{l,m+1,n}^{l,m,n} = \lambda \frac{(\theta_{l,m+1,n} - \theta_{l,m,n})a^2}{a} = a\lambda(\theta_{l,m+1,n} - \theta_{l,m,n})$$

$$\Phi_{l,m,n-1}^{l,m,n} = \lambda \frac{(\theta_{l,m,n-1} - \theta_{l,m,n})a^2}{a} = a\lambda(\theta_{l,m,n-1} - \theta_{l,m,n})$$

$$\Phi_{l,m,n+1}^{l,m,n} = \lambda \frac{(\theta_{l,m,n+1} - \theta_{l,m,n})a^2}{a} = a\lambda(\theta_{l,m,n+1} - \theta_{l,m,n})$$

Adding and setting to zero results in:

$$\theta_{l-1,m,n} + \theta_{l+1,m,n} + \theta_{l,m-1,n} + \theta_{l,m+1,n} + \theta_{l,m,n-1} + \theta_{l,m,n+1} - 6\theta_{l,m,n} = 0$$

Unknowns in this linear equation are the temperatures in the central and six adjacent control volumes.

In two dimensions, each control volume has only four adjacent volumes, so the seven heat balances reduce to five:

$$\theta_{l-1,m} + \theta_{l+1,m} + \theta_{l,m-1} + \theta_{l,m+1} - 4\theta_{l,m} = 0$$

If the central control volume bridges an interface between two materials, if that interface is parallel to the $[x, y]$ plane and if the thermal conductivities of the materials at both sides are λ_1 and λ_2, then for a mesh width a along the three axes, the heat flow from volume $(l-1,m,n)$ to the central volume (l,m,n) equals:

$$\Phi_{l-1,m,n}^{l,m,n} = \lambda_1 \frac{(\theta_{l-1,m,n} - \theta_{l,m,n})a^2}{2a} + \lambda_2 \frac{(\theta_{l-1,m,n} - \theta_{l,m,n})a^2}{2a}$$

or:

$$\Phi_{l-1,m,n}^{l,m,n} = \frac{a(\lambda_1 + \lambda_2)(\theta_{l-1,m,n} - \theta_{l,m,n})}{2}$$

The other five give:

$$\Phi_{l+1,m,n}^{l,m,n} = \frac{a(\lambda_1 + \lambda_2)(\theta_{l+1,m,n} - \theta_{l,m,n})}{2}$$

$$\Phi_{l,m-1,n}^{l,m,n} = \frac{a(\lambda_1 + \lambda_2)(\theta_{l,m-1,n} - \theta_{l,m,n})}{2}$$

$$\Phi_{l,m+1,n}^{l,m,n} = \frac{a(\lambda_1 + \lambda_2)(\theta_{l,m+1,n} - \theta_{l,m,n})}{2}$$

$$\Phi_{l,m,n-1}^{l,m,n} = a\lambda_1(\theta_{l,m,n-1} - \theta_{l,m,n})$$

$$\Phi_{l,m,n+1}^{l,m,n} = a\lambda_2(\theta_{l,m,n+1} - \theta_{l,m,n})$$

Adding and setting to zero results in:

$$\frac{(\lambda_1 + \lambda_2)(\theta_{l-1,m,n} + \theta_{l+1,m,n} + \theta_{l,m-1,n} + \theta_{l,m+1,n})}{2} + \lambda_2 \theta_{l,m,n-1} + \lambda_1 \theta_{l,m,n+1}$$

$$- 3(\lambda_1 + \lambda_2)\theta_{l,m,n} = 0$$

which is a linear equation with the temperatures in the central and six adjacent control volumes as the seven unknowns. Two dimensions give:

$$\frac{(\lambda_1 + \lambda_2)(\theta_{l-1,m} - \theta_{l+1,m})}{2} + \lambda_2 \theta_{l,m-1} + \lambda_1 \theta_{l,m+1} - 2(\lambda_1 + \lambda_2)\theta_{l,m,n} = 0$$

If the central control volume lies on the intersection between three materials with the interfaces parallel to the $[x, y]$ and $[y, z]$ planes and the thermal conductivities of the three materials are λ_1, λ_2 and λ_3, then the sum equals:

$$(\lambda_2 + \lambda_3)\frac{\theta_{l-1,m,n}}{2} + (\lambda_2 + \lambda_3)\frac{\theta_{l+1,m,n}}{2} + \lambda_3 \theta_{l,m-1,n} + (\lambda_1 + \lambda_2)\frac{\theta_{l,m+1,n}}{2}$$

$$+ (\lambda_1 + \lambda_2 + \lambda_3)\frac{\theta_{l,m,n-1} - \theta_{l,m,n+1}}{4} - (3\lambda_1 + 3\lambda_2 + 6\lambda_3)\frac{\theta_{l,m,n}}{2} = 0$$

which is again a linear equation with seven unknowns. Two dimensions give:

$$\lambda_3 \theta_{l-1,m} + (\lambda_1 + \lambda_2)\frac{\theta_{l+1,m}}{2} + (\lambda_2 + \lambda_3)\frac{\theta_{l,m-1}}{2} + (\lambda_1 + \lambda_3)\frac{\theta_{l,m+1}}{2} - (\lambda_1 + \lambda_2 + \lambda_3)\theta_{l,m} = 0$$

All other cases are solved the same way. In three dimensions, a control volume may contain eight materials, while in two dimensions, four materials. All generate a linear equation with seven or five unknowns. For p control volumes, the result is a system of p equations with p unknown temperatures, except those figuring as boundary conditions. Solving the system gives the temperature distribution. Then, the above equations allow calculation of the heat flux components along the axes.

To generalize the algorithm, suppose P_s is the surface-linked thermal conductance between two adjacent control volumes (units W/K). If in the same material, then:

$$P_s = (\lambda/d)A$$

In different materials, the named conductance consists of a serial and/or a parallel circuit of separate conductances (Figure 1.12).

$$\Phi_{l-1,m,n}^{l,m,n} = P_{sl-1,m,n}^{l,m,n}(\theta_{l-1,m,n} - \theta_{l,m,n}) \tag{1.29}$$

For the sum of the heat flows per control volume, the following equation, which demands the calculation of all surface-linked conductances P_s, applies:

$$\sum_{\substack{i=l,m,n \\ j=\pm 1}} [P_{s,i+j}\theta_{i+j}] - \theta_{l,m,n} \sum_{\substack{i=l,m,n \\ j=\pm 1}} P_{s,i+j} = 0 \tag{1.30}$$

1.2 Conduction

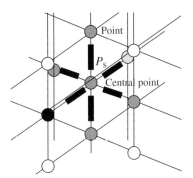

Fig. 1.12 Thermal conductances around the centre point in a meshing volume

Transferring the control volumes with known temperature to the right converts any two- or three-dimensional building detail into a system of equations:

$$[P_s]_{p,p}[\theta]_p = [P_{s,i,j,k}\theta_{i,j,k}]_p$$

where $[P_s]_{p,p}$ is a p rows, p columns conductance matrix, $[\theta]_p$ a column matrix of the p unknown temperatures, and $[P_{s,i,j,k}\theta_{i,j,k}]_p$ a column matrix of the p known temperatures. The accuracy of a CVM calculation depends on the meshing applied. The finer the mesh, the closer the solution will approach the exact one. However, the consequence is more equations. Infinitely fine meshes produce the exact solution, but the price paid is an infinite number of equations. Therefore, a compromise between accuracy and processing time must been sought. Fine meshes where large temperature gradients are expected, and less fine where small temperature gradients are expected, will minimize the divergence.

Today, powerful software packages for two- and three-dimensional heat transfer exist (Trysco^R, Heat^R, etc.).

1.2.4 Transient

Transient means that the temperatures and heat fluxes change with time. Varying material properties, time-dependent heat dissipation and time-dependent boundary conditions are responsible for that. If only the last intervenes, Fournier's second law can be written as:

$$\nabla^2 \theta = \frac{\rho c}{\lambda} \frac{\partial \theta}{\partial t}$$

The end face temperatures and heat fluxes often vary either periodically or suddenly. Outdoors, the air temperature fluctuates over periods of a year, n days, one day. Heating up in turn causes a jump in temperature and heat flux (Figure 1.13).

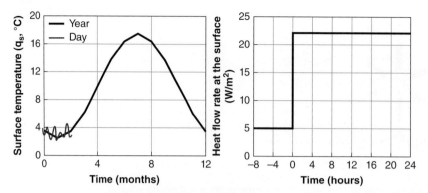

Fig. 1.13 Left: periodic change; right: non-periodic change

1.2.4.1 Periodic boundary conditions: flat assemblies

For flat layers, the second law simplifies to:

$$\frac{\partial^2 \theta}{\partial x^2} = \frac{\rho c}{\lambda} \frac{\partial \theta}{\partial t}$$

or, with $a = \lambda/\rho c$:

$$a \frac{\partial^2 \theta}{\partial x^2} = \frac{\partial \theta}{\partial t}$$

The ratio a, units m²/s, stands for the thermal diffusivity of the material. Its value indicates how easily a local temperature change spreads over any material. The higher it is, the faster this happens. A high thermal diffusivity requires a lightweight material with high thermal conductivity, or a heavyweight one with low volumetric heat capacity. None exists. Lightweights have a low thermal conductivity while the volumetric heat capacity of heavyweights is high. Many materials, except the metals, therefore have similar thermal diffusivities.

Substituting the thickness x in the equation by the thermal resistance R ($= x/\lambda$), which means multiplying both terms with λ^2, changes the formula in:

$$\frac{\partial^2 \theta}{\partial R^2} = \rho c \lambda \frac{\partial \theta}{\partial t} \tag{1.31}$$

with as heat flux:

$$q = -\frac{\partial \theta}{\partial R}$$

The temperatures at both end faces now fluctuate periodically with time. Transforming such periodic signals with base period T into a Fourier series gives for these temperatures:

$$\theta_s = \frac{B_{s0}}{2} + \sum_{n=1}^{\infty} \left[A_{sn} \sin\left(\frac{2n\pi t}{T}\right) + B_{sn} \cos\left(\frac{2n\pi t}{T}\right) \right]$$

1.2 Conduction

with:

$$A_{sn} = \frac{2}{T}\int_0^T \theta_s(t) \sin\left(\frac{2n\pi t}{T}\right) dt \qquad B_{sn} = \frac{2}{T}\int_0^T \theta_s(t) \cos\left(\frac{2n\pi t}{T}\right) dt$$

The constant $B_{so}/2$ represents the average value over the base period T, $A_{s1}, A_{s2} \ldots$, $A_{sn}, B_{s1}, B_{s2} \ldots, B_{sn}$, the harmonics of the 1st, 2nd, $\ldots$, nth order. Rewriting in a complex form, using Euler's formulas:

$$\sin(x) = \frac{\exp(ix) - \exp(-ix)}{2i} \qquad \cos(x) = \frac{\exp(ix) + \exp(-ix)}{2}$$

where $x = 2n\pi t/T$, the series becomes:

$$\theta_s(t) = \frac{1}{2}\sum_{n=-\infty}^{\infty}\left[\alpha_{sn}\exp\left(\frac{2in\pi t}{T}\right)\right] = \frac{1}{2}\sum_{n=-\infty}^{\infty}\left[(A_{sn} + iB_{sn})\exp\left(\frac{2in\pi t}{T}\right)\right] \quad (1.32)$$

where the nth complex temperature α_{sn} has:

Amplitude: $\sqrt{B_{sn}^2 + A_{sn}^2}$
Phase shift: $\operatorname{atan}(-A_{sn}/B_{sn})$

The amplitude indicates how large half the temperature variation is, while the phase shift fixes the time delay compared with a cosine function with period $T/(2n\pi)$ (radians). Consider the monthly mean outdoor air temperatures of Figure 1.14 The thick line proxy has an equation:

$$\theta_e = 9.45 + 7.18 \cos\left(\frac{2\pi t}{365.25} - (-2.828)\right)$$

with t time in days, being 365.25 in a year included the leap years, 9.45 the mean annual temperature (θ_{em}), 7.18 the annual amplitude (θ_{e1}) and -2.828 the related phase shift (ϕ_{e1} in radians). The time-dependent term resembles a rotating vector with value 7.18 starting -2.828 radians away from the real axis. As Fourier series, this gives:

$$\theta_e = \frac{B_o}{2} + \theta_{e1}\left[A_1 \sin\left(\frac{2\pi t}{365.25}\right) + B_1 \cos\left(\frac{2\pi t}{365.25}\right)\right]$$

or, with $B_o = 18.9$, $A_1 = \theta_{e1} \sin(\phi_{e1}) = -2.21$ and $B_1 = \theta_{e1} \cos(\phi_{e1}) = -6.83$:

$$\theta_e = 9.45 - 2.21 \sin\left(\frac{2\pi t}{365.25}\right) - 6.83 \cos\left(\frac{2\pi t}{365.25}\right)$$

The related complex value is $\alpha_{e1} = -6.83 - i\,2.21$ with amplitude $7.18\,°C$ and phase shift -2.828 radians.

Assume now that from time zero one or both end faces endure a periodic temperature and heat flux. The response will be twofold: a transient that dies slowly

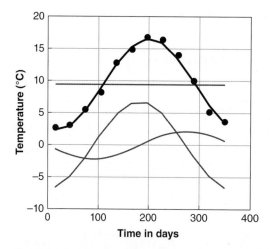

Fig. 1.14 Monthly mean temperatures. The black circles are the measured averages between 1901 and 1930. The thick line gives the result of a Fourier analysis with one harmonic. The thin lines are the annual mean, the sine term and cosine term

and a lasting periodic. As assemblies can neither compress nor extend thermal signals, the periodic must contain the same harmonics as the surface signal, but related temperature and heat flux amplitudes will dampen and gradually run behind, so building up a phase shift during the traverse. So the two will behave as complex, or the solution will be a Fourier series with the thermal resistance as an independent variable:

$$\theta(R,t) = \frac{1}{2} \sum_{n=-\infty}^{\infty} \left[\alpha_n(R) \exp\left(\frac{2in\pi t}{T}\right) \right]$$

$$q(t) = -\frac{d\theta(R,t)}{dR} = \frac{1}{2} \sum_{n=-\infty}^{\infty} \left[\alpha'_{sn}(R) \exp\left(\frac{2in\pi t}{T}\right) \right]$$

The prime for α'_{sn} reminds us that, mathematically, the complex heat flux is the first derivative of the complex temperature with respect to the thermal resistance:

Amplitude: $\sqrt{B'^2_{sn} + A'^2_{sn}}$
Phase shift: $\operatorname{atan}(-A'_{sn}/B'_{sn})$

Single layer
Inserting the complex temperature into Fourier's second law gives:

$$\sum_{n=-\infty}^{\infty} \left\{ \left[\frac{d^2 \alpha_n(R)}{dR^2} - \frac{2\rho c \lambda i n \pi}{T} \alpha_n(R) \right] \exp\left(\frac{2in\pi t}{T}\right) \right\} = 0 \qquad (1.33)$$

1.2 Conduction

Setting to zero presumes that all coefficients of the time exponentials must be zero, or:

$$\frac{d^2 a_n(R)}{dR^2} - \frac{2\rho c \lambda i n \pi}{T} a_n(R) = 0 \qquad (1.34)$$

The second-order partial differential equation so breaks into $2\infty + 1$ second-order differential equations, with the complex temperature as the dependent variable and the integer n moving from $-\infty$ past 0 to $+\infty$. Because the solutions for n positive and negative mirror each other, only infinity plus one ($\infty + 1$) equations have to be solved. As n figures only as a parameter and all are otherwise identical, one solution suffices:

$$a_n(R) = C_1 \exp(\omega_n R) + C_2 \exp(-\omega_n R) \qquad (0 \le n \le \infty) \qquad (1.35)$$

where:

$$\omega_n^2 = \frac{2\rho c \lambda i n \pi}{T}$$

The quantity ω is called the thermal pulsation. Using Euler's formulas then gives as complex temperatures and heat fluxes in a single-layer flat assembly:

$$a_n(R) = (C_1 - C_2)\sinh(\omega_n R) + (C_1 + C_2)\cosh(\omega_n R)$$
$$a'_n(R) = \frac{da}{dR} = \omega_n[(C_1 - C_2)\cosh(\omega_n R) + (C_1 + C_2)\sinh(\omega_n R)]$$

The integration constants $(C_1 - C_2)$ and $(C_1 + C_2)$ follow from the boundary condition that at the face $R=0$, the complex temperature and heat flux are $a_{sn}(0)$ and $a'_{sn}(0)$, or:

$$a_{sn}(0) = (C_1 - C_2)0 + (C_1 + C_2) = C_1 + C_2$$
$$a'_{sn}(0) = \omega_n[(C_1 - C_2) + (C_1 + C_2)0] = \omega_n(C_1 - C_2)$$

This converts the complex temperature and heat flux equations into:

$$a_n(R) = a_{sn}(0)\cosh(\omega_n R) + a'_{sn}(0)\frac{\sinh(\omega_n R)}{\omega_n} \qquad (1.36)$$

$$a'_n(R) = a_{sn}(0)\omega_n \sinh(\omega_n R) + a'_{sn}(0)\cosh(\omega_n R) \qquad (1.37)$$

The equations means two unknowns can be solved, whose inclusion in the Fourier series then gives the time functions. Of interest is the relationship between the complex quantities at both end faces. Suppose end face $R=0$ is the inside. For envelopes R_T then is the outside end face, for partitions the end face at the other side. The system thus becomes ($R=R_T$):

$$[A_{sn}(R_T)] = [W_n][A_{sn}(0)] \qquad (1.38)$$

with $[A_{sn}(0)]$ the column matrix of the unknown complex quantities at one end face, and $[A_{sn}(R_T)]$ the column matrix of the known complex quantities at the other end face. The system matrix $[W_n]$ for the nth harmonic solely depends on the thickness, the material properties, the base period and which harmonic is considered. As an alternative for the thermal resistance, it contains much more information.

For $n=0$, $\alpha_{so}/2$ and $\alpha'_{so}/2$ are the average temperature and heat flux on the inside end face over the period considered (T), with as thermal pulsation:

$$\omega_o^2(n=0) = 2i\rho c\lambda 0\pi/T = 0$$

This turns the equations for the complex temperatures and heat fluxes into:

$$\alpha_{so}(R_T) = \alpha_{so}(0) + \alpha'_{so}(0)\frac{0}{0} \qquad \alpha'_{so}(R_T) = \alpha_{so}(0)0 + \alpha'_{so}(0) = \alpha'_{so}(0)$$

The incoming and outgoing average fluxes look the same. As this holds for any thermal resistance between 0 and R_T, thus at any spot in the single layer, for $n=0$ the heat flux is a constant as in steady state. The ratio 0/0 in the temperature equation is solved using de l'Hopitâl's rule:

$$\lim_{\omega_0 \to 0} \left[\sinh(\omega_o R_T)/\omega_o\right] = \lim_{\omega_0 \to 0} [R_T \cosh(\omega_o R_T)] = R_T$$

which allows us to write:

$$\alpha_{so}(R_T) = \alpha_{so}(0) + \alpha'_{so}(0)R_T.$$

Anywhere in the single layer the same result comes out. Or, in the $[R,\theta]$ plane, the average temperatures lie on a straight line with the slope equal to the heat flux, again as in steady-state conditions. Both results extend the steady state from invariable to average over a sufficiently long time span.

Considering the harmonics, suppose that at the inside surface temperature remains constant. Then there no complex temperatures intervene, which gives at the other end face:

$$\alpha_{sn}(R_T) = \alpha'_{sn}(0)\frac{\sinh(\omega_n R_T)}{\omega_n} \quad \text{or} \quad \frac{\alpha_{sn}(R_T)}{\alpha'_{sn}(0)} = \frac{\sinh(\omega_n R_T)}{\omega_n} \qquad (1.39)$$

The function on the right reflects the ratio between the complex temperature at the outside face or the end face at the other side, and the complex heat flux at the inside face. In steady state, this ratio defines the thermal resistance. The name here is dynamic thermal resistance for the nth harmonic, symbol D_q^n, units m^2.K/W, with as amplitude the size and as argument (ϕ_q^n) the time shift between the complex temperature at the one and the complex heat flux at the other end face:

$$\phi_q^n = \arg\left[\sinh(\omega_n R)/\omega_n\right]$$

The assumption made looks purely theoretical. However, in reality the case is applicable to buildings with constant indoor temperature. For an infinitely long period the amplitude and phase shift become:

$$\left[D_q^n\right] = \lim_{n \to 0}\left[\frac{\sinh(0)}{0}\right] = \lim_{n \to 0}[R_T \cosh(0)] = R_T \qquad \phi_q^n = \lim_{n \to 0}\left[\arg\frac{\sinh(0)}{0}\right] = \infty$$

1.2 Conduction

Or, the dynamic thermal resistance then equals the steady-state value. For a really fast pulsation, period nil, the amplitude and phase shift change to:

$$[D_q^n] = \lim_{n\to\infty}\left[\frac{\sinh(\infty)}{\infty}\right] = \lim_{n\to\infty}[R_T\cosh(\infty)] = \infty \qquad \phi_q^n = \lim_{n\to\infty}\left[\arg\frac{\sinh(\infty)}{\infty}\right] = 0$$

Ever-faster fluctuations so push the dynamic thermal resistance to infinity, meaning the assembly dampens the signal completely. The dynamic value thus is always higher than the steady state one. Or, imposing a large thermal resistance suffices to get a high dynamic one.

Next, assume the heat flux at the inside face does not change. So with no complex heat fluxes there, the complex temperature at the other end face becomes:

$$\alpha_{sn}(R_T) = \alpha_{sn}(0)\cosh(\omega_n R_T) \quad \text{or} \quad \frac{\alpha_{sn}(R_T)}{\alpha_{sn}(0)} = \cosh(\omega_n R_T) \qquad (1.40)$$

The function on the right gives the ratio between the complex temperature at the outside face or the face at the other side and the complex temperature at the inside face. This is termed the temperature damping for the nth harmonic, symbol D_θ^n, no units. The amplitude again gives the size, and the argument (ϕ_θ^n) the time shift:

$$\Phi_\theta^n = \arg[\cosh(\omega_n R)]$$

The case anyhow looks fictitious. The property has no equivalent in steady state. Yet, in practice, it reflects the ability of an assembly to moderate the indoor impact of temperature changes outdoors. With few heat gains indoors and restricted ventilation, an enclosure with high temperature damping will keep the indoor temperature stable, a fact turning the number into a performance metric. For an infinitely long period the amplitude and phase shift become:

$$[D_\theta^n] = \lim_{n\to 0}[\cosh(0)] = 1 \qquad \phi_\theta^n = \lim_{n\to 0}\left[\arg[\cosh(0)]\right] = \infty$$

Then temperature damping nears 1. In fact, in steady state, without heat gains or losses, the temperatures on both end faces must be equal. For a zero period the amplitude and phase shift become:

$$[D_\theta^n] = \lim_{n\to\infty}[\cosh(\infty)] = \infty \qquad \phi_\theta^n = \lim_{n\to\infty}\left[\arg[\cosh\infty]\right] = 0$$

For ever faster fluctuations, temperature damping thus approaches infinity.

Finally, the temperature on the outside or the other face being constant converts the complex temperature on the inside face into:

$$0 = D_\theta^n \alpha_{sn}(0) + D_q^n \alpha'_{sn}(0) \quad \text{or} \quad \frac{\alpha'_{sn}(0)}{\alpha_{sn}(0)} = -\frac{D_\theta^n}{D_q^n} = -\omega_n \coth(\omega_n R) \qquad (1.41)$$

with D_θ^n the temperature damping and D_q^n the dynamic thermal resistance. Their ratio, equal to the complex heat flux on the inside face divided by its complex temperature, is named the 'admittance' for the nth harmonic, symbol Ad^n, units W/(m².K), with as amplitude the value of that ratio, and as argument (ϕ^n_{Ad}) the time shift between the two:

$$\phi^n_{Ad} = \phi^n_\theta - \phi^n_q$$

The admittance also looks like an impossible quantity. However, it relates to how easily single-layer assemblies pick up heat from within when the inside surface temperature fluctuates. The higher the amplitude, the more effective the heat storage. The statement 'capacitive' indicates façade walls or inside partitions with high admittance. As the thermal pulsation ω_n also writes:

$$\omega_n = \sqrt{i}\sqrt{\rho c \lambda}\sqrt{2n\pi/T}$$

a high admittance requires a large square root of the product of volumetric heat capacity and thermal conductivity, a value called the contact coefficient or effusivity of a material, symbol b, units J/(m².s$^{-1/2}$.K). The larger the contact coefficient, the more active that material as a heat storage medium. Heavy materials without an insulating finish inside have high effusivities, and thus large admittances. The amplitude and time shift for a period nearing infinity are:

$$[Ad^n] = \lim_{n \to 0}\left(\frac{D^n_\theta}{D^n_q}\right) = \frac{1}{R_T} \qquad \phi^n_{As} = \lim_{n \to 0}\left[\arg\left(-\frac{D^n_\theta}{D^n_q}\right)\right] = \infty$$

For really slow fluctuations, the admittance thus approaches the thermal conductance (P). In steady state the heat flux entering or leaving an assembly equals $P\Delta\theta_s$. For a very fast pulsation, the amplitude and phase shift equal:

$$[Ad^n] = \lim_{n \to \infty}\left(\frac{D^n_\theta}{D^n_q}\right) = \infty \qquad \phi^n_{As} = \lim_{n \to \infty}\left[\arg\left(\frac{D^n_\theta}{D^n_q}\right)\right] = 0$$

Or, ever-faster fluctuations move the admittance to infinity.

With the dynamic thermal resistance and temperature damping known, the complex system matrix of a single-layer assembly becomes:

$$[W_n] = \begin{bmatrix} D^n_\theta & D^n_q \\ \omega^2_n D^n_q & D^n_\theta \end{bmatrix}$$

Moving to real numbers requires first that the thermal pulsation is reformulated:

$$\omega_n = \sqrt{i}\,b\sqrt{\frac{2n\pi}{T}} = \frac{(1+i)\,b\sqrt{\frac{2n\pi}{T}}}{\sqrt{2}} = (1+i)\,b\sqrt{\frac{n\pi}{T}}$$

1.2 Conduction

whereby the conversion of $\sqrt{i}$ is based on $(1+i)^2 = 2i$ or $\sqrt{i} = (1+i)/\sqrt{2}$. With the thermal pulsation redressed, the product $\omega_n R$ becomes:

$$\omega_n R = \frac{(1+i)bx\sqrt{\frac{n\pi}{T}}}{\lambda} = (1+i)x\sqrt{\frac{n\pi}{aT}}$$

with x the ordinate from zero to the full thickness of the single-layer and $a = \lambda^2/b^2$ its thermal diffusivity. For $X_n = x\sqrt{n\pi/aT}$, the thermal pulsation equals $(1+i)X_n/R$. Thus temperature damping D_θ^n is given by:

$$D_\theta^n = \cosh(\omega_n R) = \cosh[(1+i)X_n] = \cosh(X_n)\cosh(iX_n) + \sinh(X_n)\sinh(iX_n)$$

or, with $\cosh(iX_n) = \cos(X_n)$ and $\sinh(iX_n) = i\sin(X_n)$:

$$\cosh(\omega_n R) = \cosh(X_n)\cos(X_n) + i\sinh(X_n)\sin(X_n)$$

Analogously, the dynamic thermal resistance D_q^n and the admittance Ad^n become:

$$D_q^n = \frac{\sinh(\omega_n R)}{\omega_n} = \frac{R}{2X_n}\left\{\begin{array}{l}[\sinh(X_n)\cos(X_n) + \cosh(X_n)\sin(X_n)] \\ + i[\cosh(X_n)\sin(X_n) - \sinh(X_n)\cos(X_n)]\end{array}\right\}$$

$$Ad^n = \omega_n \sinh(\omega_n R) = \frac{X_n}{R}\left\{\begin{array}{l}[\sinh(X_n)\cos(X_n) - \cosh(X_n)\sin(X_n)] \\ + i[\cosh(X_n)\sin(X_n) + \sinh(X_n)\cos(X_n)]\end{array}\right\}$$

As six functions G_{n1} to G_{n6} are defined:

$G_{n1} = \cosh(X_n)\cos(X_n)$ $\qquad\qquad\qquad G_{n2} = \sinh(X_n)\sin(X_n)$
$G_{n3} = [\sinh(X_n)\cos(X_n) + \cosh(X_n)\sin(X_n)]/(2X_n)$ $\quad G_{n4} = [\cosh(X_n)\sin(X_n) - \sinh(X_n)\cos(X_n)]/(2X_n)$
$G_{n5} = X_n[\sinh(X_n)\cos(X_n) - \cosh(X_n)\sin(X_n)]$ $\quad G_{n6} = X_n[\cosh(X_n)\sin(X_n) + \sinh(X_n)\cos(X_n)]$

They allow the expression of the temperature damping, the dynamic thermal resistance and the other term in the system matrix as:

$$\cosh(\omega_n R) = G_{n1} + iG_{n2} \qquad \frac{\sinh(\omega_n R)}{\omega_n} = R(G_{n3} + iG_{n4}) \qquad \omega_n \sinh(\omega_n R) = \frac{G_{n5} + iG_{n6}}{R}$$

The amplitude and time shift for the three properties then become:

Amplitude	Phase shift
$[D_q^n] = R_T\sqrt{G_{n3}^2 + G_{n4}^2}$	$\phi_q^n = \text{bgtg}(G_{n4}/G_{n3})$
$[D_\theta^n] = \sqrt{G_{n1}^2 + G_{n2}^2}$	$\phi_\theta^n = \text{bgtg}(G_{n2}/G_{n1})$
$[Ad^n] = [D_\theta^n]/[D_q^n]$	$\phi_{Ad}^n = \phi_\theta^n - \phi_q^n$

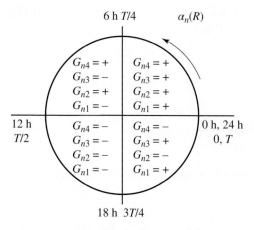

Fig. 1.15 Phase shift: sign of the G-functions

Per harmonic n, the complex 2×2 system matrix for a single-layer assembly herewith converts into a real 4×4 matrix:

$$W_n = \begin{bmatrix} G_{n1} & G_{n2} & R_T G_{n3} & R_T G_{n4} \\ -G_{n2} & G_{n1} & -R_T G_{n4} & R_T G_{n3} \\ G_{n5}/R_T & G_{n6}/R_T & G_{n1} & G_{n2} \\ -G_{n6}/R_T & G_{n5}/R_T & -G_{n2} & G_{n1} \end{bmatrix}$$

For the phase shift, Figure 1.15 shows which quarter-circle corresponds with which sign of the G-functions.

Composite

Composite assemblies figure as series-connected single-layers, each with a system matrix $[W_{n,i}]$, now called the layer matrix. The whole has as system matrix $[W_{nT}]$ (Figure 1.16):

$$[A_{sn}(R_T)] = W_{nT}[A_{sn}(0)] \tag{1.42}$$

with $[A_{sn}(R_T)]$ and $[A_{sn}(0)]$ the column matrices of the nth complex temperature and heat flux at each end face.

The relationship between the complex temperatures and heat fluxes at interface j and $j+1$ is:

$$[A_{n,j+1}] = W_{n,j}[A_{n,j}],$$

1.2 Conduction

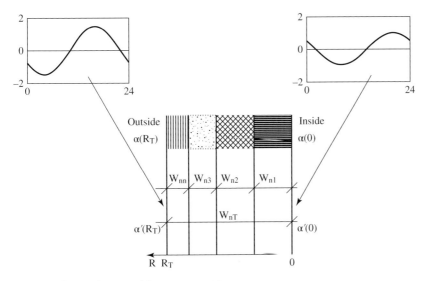

Fig. 1.16 Composite assembly, system matrix

an equation that holds per layer. Starting at the end face $R=0$ which coincides with inside, the series becomes:

$$[A_{n,1}] = W_{n,1}[A_{s,n}(0)]$$
$$[A_{n,2}] = W_{n,2}[A_{n,1}]$$
$$\ldots$$
$$[A_{n,m-1}] = W_{n,m-1}[A_{n,m-2}]$$
$$[A_{s,n}(R_T)] = W_{n,m}[A_{n,m-1}]$$

Transposing each preceding into the next equation gives:

$$[A_{s,n}(R_T)] = W_{n,m}W_{n,m-1}\ldots W_{n,2}W_{n,1}[A_{s,n}(0)]$$

or, the system matrix of a composite assembly equals:

$$[W_{nT}] = \prod_{j=1}^{m}[W_{n,j}] \qquad (1.43)$$

Multiplication starts on the inside, to end at the layer outside or at the layer on the other side. Each next-layer matrix is multiplied by the product of the preceding ones. Because the commutation property does not apply for a product of matrixes, that rule must be respected. So, in contrast to the thermal resistance, the transient response of a composite assembly changes with the layer sequence, which is very important in practice. Of course, each layer matrix keeps the same formulation and meaning as for a single-layer assembly. When two of the four complex temperatures and heat fluxes at the end faces are known, the others follow from:

$$[A_{sn}(R_T)] = [W_{nT}][A_{sn}(0)]$$

The complex temperatures and heat fluxes in the interfaces are then found by ascending or descending the layer equations in the correct sequence.

To get the complex temperatures and heat fluxes in a single-layer assembly, it is parcelled in m sub-layers with thickness Δx, layer matrix $[W_{\Delta x}]$ and system matrix $[W_n] = (W_{\Delta x})^m$. Per interface between each Δx, the same approach applies as for composite assemblies.

1.2.4.2 Any boundary conditions: flat assemblies

When the temperature or heat flux at an end face stays zero except for an infinitesimally short period dt, when it jumps to 1 (Figure 1.17), this is called a Dirac impulse.

Return to Fourier's second law now, and assume that the temperature θ_{s1} on end face s1 pulses this way. As a result, the heat flux there ($q_{s1}(t)$) and the temperature and heat flux at the other end face s2 ($\theta_{s2}(t)$, $q_{s2}(t)$) will vary. With constant material properties, the functions $q_{s1}(t)$, $\theta_{s2}(t)$ and $q_{s2}(t)$ are called the response factors and are written as:

$$I_{\theta_{s1} q_{s1}} \quad I_{\theta_{s1} \theta_{s2}} \quad I_{\theta_{s1} q_{s2}}$$

Analogously:

$$I_{q_{s1} \theta_{s1}} \quad I_{q_{s1} \theta_{s2}} \quad I_{q_{s1} q_{s2}}$$
$$I_{\theta_{s2} q_{s1}} \quad I_{\theta_{s2} \theta_{s1}} \quad I_{\theta_{s1} q_{s2}}$$
$$I_{q_{s2} q_{s1}} \quad I_{q_{s2} \theta_{s1}} \quad I_{q_{s2} \theta_{s2}}$$

Depending on the layer sequence, their value changes with the flow direction, from end face s1 to s2 or vice versa. Consider a temperature pulse with value θ_{so} at end face

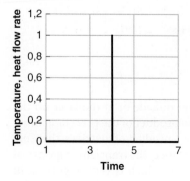

Fig. 1.17 Dirac impulse

1.2 Conduction

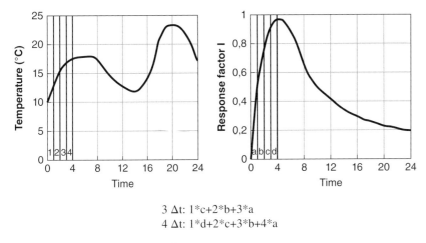

3 Δt: 1*c+2*b+3*a
4 Δt: 1*d+2*c+3*b+4*a

Fig. 1.18 Convolution, the principle

s1. With the response factors known, the temperature and heat flux at end face s2 and the heat flux at end face s1 become:

Impulse θ_o at face s1	Impulse q_o at face s1
$q_{s1} = \theta_o I_{\theta_{s1} q_{s1}}$	$\theta_{s1} = q_o I_{q_{s1} \theta_{s1}}$
$\theta_{s2} = \theta_o I_{\theta_{s1} \theta_{s2}}$	$\theta_{s2} = q_o I_{q_{s1} \theta_{s2}}$
$q_{s2} = \theta_o I_{\theta_{s1} q_{s2}}$	$q_{s2} = q_o I_{q_{s1} q_{s2}}$

A pulse at end face s2 results in analogue relationships. Any random signal $\theta_{s1}(t)$ can now be split into a continuous series of pulses $\theta_{s1}(t)\Delta t$. For the temperature response (θ_{s2}) at end face s2 on a signal θ_{s1} at end face s1, the following applies (Figure 1.18):

$t = 0 \quad \theta_{s2}(0) = 0$

$t = \Delta t \quad \theta_{s2}(\Delta t) = \theta_{s1}(t = 0)I_{\theta_{s1}\theta_{s2}}(t = \Delta t)$

$t = 2\Delta t \quad \theta_{s2}(2\Delta t) = \theta_{s1}(t = 0)I_{\theta_{s1}\theta_{s2}}(t = 2\Delta t) + \theta_{s1}(t = \Delta t)I_{\theta_{s1}\theta_{s2}}(t = \Delta t)$

$t = 3\Delta t \quad \begin{aligned}\theta_{s2}(3\Delta t) &= \theta_{s1}(t = 0)I_{\theta_{s1}\theta_{s2}}(t = 3\Delta t) + \theta_{s1}(t = \Delta t)I_{\theta_{s1}\theta_{s2}}(t = 2\Delta t) \\ &\quad + \theta_{s1}(t = 2\Delta t)I_{\theta_{s1}\theta_{s2}}(t = \Delta t)\end{aligned}$

...

$t = n\Delta t \quad \begin{aligned}\theta_{s2}(n\Delta t) &= \theta_{s1}(t = 0)I_{\theta_{s1}\theta_{s2}}(t = n\Delta t) + \theta_{s1}(t = \Delta t)I_{\theta_{s1}\theta_{s2}}(t = (n-1)\Delta t) \\ &\quad + \theta_{s1}(t = 2\Delta t)I_{\theta_{s1}\theta_{s2}}(t = (n-2)\Delta t) + \ldots\end{aligned}$

or:

$$\theta_{s2}(n\Delta t) = \sum_{j=0}^{n-1} \theta_{s1}(j\Delta t) I_{\theta_{s1}\theta_{s2}}((n-j)\Delta t)$$

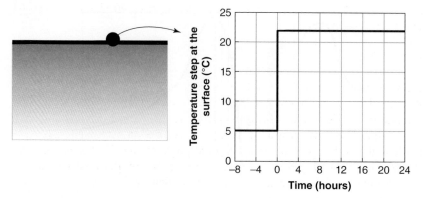

Fig. 1.19 Temperature step at the surface of a semi-infinite medium

As an integral:

$$\theta_{s2}(t) = \int_0^t \theta_{s1}(\tau) I_{\theta_{s1}\theta_{s2}}(t-\tau)\mathrm{d}\tau$$

This integral of the product of the signal scanned clockwise and the response factor scanned counter-clockwise is called the *convolution* integral of temperature θ_{s2} at end face s2 for the signal $\theta_{s1}(t)$ at end face s1. The same approach, be it with the correct response factors, holds for the two heat fluxes and for all situations where changing boundary conditions intervene. Response factors and convolution integrals have to be calculated numerically.

Now consider a temperature step at the surface of a semi-infinite medium. At time zero the medium is still uniformly warm, temperature θ_{s0}, but then the surface temperature suddenly jumps to a value $\theta_{s0} + \Delta\theta_{s0}$ (Figure 1.19).

A separation of variables allows the solution of Fourier's second law with $t=0$, $0 \le x \le \infty$, $\theta = \theta_{s0}$ as initial conditions, and $t \ge 0$, $x=0$, $\theta_s = \theta_{s0} + \Delta\theta_{s0}$ as boundary conditions. The outcome is:

$$\theta(x,t) = \theta_{s0} + \Delta\theta_{s0} \left(\frac{2}{\sqrt{\pi}} \int_{q=\frac{x}{2\sqrt{at}}}^{\infty} \exp(-q^2)\mathrm{d}q \right) \tag{1.44}$$

The term between brackets is the inverse error function, with a the thermal diffusivity of the medium. The heat flux at the surface ($x=0$) equals:

$$q = -\lambda \frac{\mathrm{d}\theta}{\mathrm{d}x} = -\lambda \left[\frac{2\Delta\theta_{s0}}{2\sqrt{\pi at}} \exp\left(-\frac{x^2}{4at}\right) \right]_{x=0} = -\frac{\Delta\theta_{s0} b}{\sqrt{\pi t}} \tag{1.45}$$

1.2 Conduction

with b the contact coefficient of the medium. Applying the definition of response factor to this equation gives:

$$I_{\theta_{s1} q_{s1}} = -b/\sqrt{\pi t}$$

Semi-infinite mediums do not exist, but the soil or very thick material layers are close. In any case, the heat flux equation illustrates what the contact coefficient does. A high value means a rapid uptake of heat when the surface temperature rises and a rapid release when it drops, whereas a low value stands for a slow uptake and release. Materials with high contact coefficients act as storage media. If, in passive solar buildings, walls and floors could not store the solar gains temporarily, the indoor conditions would become intolerable. Partitions should be thick enough and consist of heavy materials with a high contact coefficient.

The heat per m² flowing in or out of a semi-infinite medium beyond time zero equals:

$$Q = \int_0^t q \, dt = \frac{2b \Delta \theta_{s0}}{\sqrt{\pi}} \sqrt{t} = A_q \sqrt{t} \qquad (1.46)$$

with the quantity A_q called the heat absorption coefficient, units $J/(m^2.K.s^{½})$.

When two materials make contact and the one is at temperature θ_1, the other at temperature θ_2, the heat flux in the contact will equal ($\theta_1 > \theta_2$):

Material 1: $q_{s1} = (\theta_1 - \theta_c) b_1 / \sqrt{\pi t}$
Material 2: $q_{s2} = (\theta_c - \theta_2) b_2 / \sqrt{\pi t}$

Since both fluxes must be identical in absolute value, the contact temperature θ_c becomes:

$$\theta_c = \frac{b_1 \theta_1 + b_2 \theta_2}{b_1 + b_2} \qquad (1.47)$$

This instant value depends on the temperature and the contact coefficient of the two materials in contact, a reality affecting the impression when humans touch a material. A capacitative one with a high contact coefficient will feel cold or hot, while a material with low contact coefficient will feel comfortably warm. Indeed, in the first case the skin temperature will suddenly move from 32–33 °C to the material temperature, while in the second case the contact will adapt to the skin. So, touching concrete and aluminium is unpleasant, while touching wood is pleasant. To indicate that feeling, the term cold or warm material is used. The contact coefficient is an important metric when choosing floorings or chair finishes.

1.2.4.3 Two and three dimensions: thermal bridges

Where heat is transmitted two- or three-dimensionally, Fourier's second law for transient conduction applies. Analytical solutions for building details do not exist, which is why CVM is used. The transient heat capacity $\rho c \Delta V$ of each control volume intervenes. Without dissipation, the resulting heat flow now equals the change in heat content of each central control volume, or:

$$\left(\sum \Phi_m\right) \Delta t = \rho c \, \Delta V \, \Delta \theta$$

with Φ_m the average heat flow from each nearest volume during the time step Δt. If set equal to a weighted average of the flow at time t ($= 1$) and time $t + \Delta t$ ($= 2$), then:

$$\Phi_m = p \, \Phi_{t+\Delta t} + (1-p) \Phi_t \qquad (1 > p > 0)$$

The balance now writes as ($i = l,m,n$ and $j = \pm 1$):

$$p\left[\sum (P_{s,i+j}\theta_{i+j}) - \theta_{l,m,n} \sum P_{s,i+j}\right]^{(2)} + (1-p)\left[\sum (P_{s,i+j}\theta_{i+j}) - \theta_{l,m,n} \sum P_{s,i+j}\right]^{(1)}$$

$$= \rho c \Delta V \frac{\theta^{(2)}_{l,m,n} - \theta^{(1)}_{l,m,n}}{\Delta t}$$

For $p = 0$, the system only contains forward difference equations, and for $p = 1$, only backward difference equations. For $p = 0.5$, which means using the arithmetic average, a choice called the Cranck-Nicholson scheme, rearrangement gives:

$$\left[\sum (P_{s,i+j}\theta_{i+j}) - \theta_{l,m,n}\left(\sum P_{s,i+j} + \frac{2\rho c \Delta V}{\Delta t}\right)\right]^{(2)}$$

$$= \left[-\sum (P_{s,i+j}\theta_{i+j}) + \theta_{l,m,n}\left(\sum P_{s,i+j} - \frac{2\rho c \Delta V}{\Delta t}\right)\right]^{(1)}$$

with the actual temperatures $^{(2)}$ unknown and the preceding ones $^{(1)}$ known. This way, each time step gives a system of as many equations as there are control volumes at unknown temperature. For given initial and boundary conditions, solving the system per time step is possible. Questions to decide upon beforehand are how to mesh (finer in materials with high thermal diffusivity) and the time step to use: in accordance with mesh density, in accordance with the step in boundary condition values, or in accordance with the information density required. Poor choices can induce instability in the solution generated.

1.3 Heat exchange at surfaces

Up to now, end face temperatures were assumed to be known. However, in most cases, it is the air temperatures and sometimes the heat flux at an end face that are the quantities figuring as known boundary conditions. Every weather station, in fact, registers the air temperature outdoors, while measuring the value indoors is much simpler than logging surface temperatures. Therefore the focus now shifts from heat transfer face-to-face to heat transfer between the ambient at both sides. But, how does heat reach a surface? Two mechanisms are responsible for that: convection between the air and the surface, and radiant exchanges with all surfaces seen. The

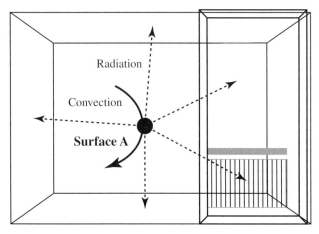

Fig. 1.20 Heat exchange at surface A by convection to and from the room air and radiation to and from all other surfaces

inside surface A in Figure 1.20, for example, radiates to and receives radiation from all the partitions, the radiator and the furniture, while exchanging convective heat with the air in the room.

1.4 Convection

1.4.1 In general

The term 'convective' aims at describing the heat transfer between fluids and surfaces. The heat flux and flow exchanged are then given as:

$$q_c = h_c(\theta_{fl} - \theta_s) \qquad \Phi_c = h_c(\theta_{fl} - \theta_s)A \qquad (1.48)$$

where θ_{fl} is the temperature in the undisturbed fluid, θ_s the surface temperature and h_c the convective surface film coefficient, units W/(m².K). The equations are known under the name of Newton's law. Convection thus seems to be linearly related to the driving temperature difference $(\theta_{fl} - \theta_s)$. The equations, however, rather serve as a definition of the surface film coefficient, which accommodates the whole complexity of the heat and mass flow in a fluid touching a surface. Together, the scalar law of mass conservation and the vector law of momentum conservation fix the mass flow at the surface:

$$\text{Mass (scalar) } div\,(\rho \mathbf{v}) = 0$$
$$\text{Momentum (Navier-Stokes, vector) } \frac{d(\rho \mathbf{v})}{dt} = \rho \mathbf{g} - \text{grad}\,P + \mu \nabla^2 \mathbf{v}$$

In these equations, ρ is the density and μ the dynamic viscosity of the fluid, P the total pressure and $\rho \mathbf{g}$ the gravity gradient. Unknown are the velocity components v_x, v_y, v_z and the total pressure. When the mass flow turns turbulent, the turbulence equations must be added. Often used is the (k, ε) model, where k is the turbulent kinetic energy and ε its dissipation. Both give an extra scalar equation. Otherwise said, turbulent flow requires three scalar equations, while the velocity components in the x, y, and z directions turn the vector momentum equation into another three scalar equations.

For constant properties, negligible kinetic energy and negligible friction, conservation of energy in a moving fluid simplifies to:

$$a\nabla^2\theta = \frac{d\theta}{dt} \tag{1.49}$$

with a being the fluid's thermal diffusivity. In the Navier-Stokes equation as well as in the energy equation, d/dt is a total derivative:

$$\frac{d}{dt} = \frac{\partial}{\partial x}\frac{\partial x}{\partial t} + \frac{\partial}{\partial y}\frac{\partial y}{\partial t} + \frac{\partial}{\partial z}\frac{\partial z}{\partial t} + \frac{\partial}{\partial t} = \underbrace{\frac{\partial}{\partial x}v_x}_{(1)} + \underbrace{\frac{\partial}{\partial y}v_y}_{(2)} + \underbrace{\frac{\partial}{\partial z}v_z}_{(3)} + \frac{\partial}{\partial t}$$

where in the energy equation the terms (1), (2) and (3) multiplied with $\rho c\theta$ quantify the enthalpy that a fluid at temperature θ moving at velocity **v** with components v_x, v_y, v_z carries with it. Energy conservation so introduces temperature θ as a seventh unknown.

All information on convection follows from solving this system of seven scalar partial differential equations, which computational fluid dynamics (CFD) allows us to do numerically. Once the temperature field is known, the heat fluxes at a surface follow from Fourier's first law, because right at the contact, convection becomes conduction across a laminar boundary layer.

The following relationship thus holds for the surface film coefficient:

$$h_c = -\lambda_{fl}\frac{(\text{grad }\theta)_s}{\theta_{fl} - \theta_s} \tag{1.50}$$

where $(\text{grad }\theta)_s$ is the temperature gradient in the boundary layer and θ_{fl} is the temperature at a particular location in the undisturbed fluid.

Solving the system analytically is possible for natural convection along a semi-infinite vertical surface and for forced convection along a semi-infinite horizontal surface, in both cases as long as the flow is laminar. With air as the fluid, velocity and temperature follow from similar equations:

$$\frac{\partial v_x}{\partial x} + \frac{\partial v_y}{\partial y} = 0 \qquad v_x\frac{\partial v_x}{\partial x} + v_y\frac{\partial v_y}{\partial y} = v_{fl}\frac{\partial^2 v_x}{\partial y^2} \qquad v_x\frac{\partial \theta}{\partial x} + v_y\frac{\partial \theta}{\partial y} = a_{fl}\frac{\partial^2 \theta}{\partial y^2}$$

Moreover, if the thermal diffusivity equals the kinematic viscosity, the equations become identical. The heat transfer and mass flow patterns then coincide, so solving gives as the surface film coefficient:

$$h_{c,x} = \frac{0.664\lambda_{fl}}{x}\sqrt{\frac{v_\infty x}{v_{fl}}}$$

where x is the distance to the surface's free edge and v_∞ the velocity in the undisturbed fluid. Thus the surface film coefficient changes along the surface.

However, any trial to calculate the convective heat transfer along wall faces in a room, along the outside face of a building or of any other construction analytically, is doomed to fail. Even in the most simple situations, convection depends upon all parameters and properties determining the mass and heat transfer:

Fluid properties	Flow parameters
Thermal conductivity (λ_{fl})	Geometry
Density (ρ_{fl})	Surface roughness
Specific heat capacity (c_{fl})	Temperature difference ($\theta_{fl} - \theta_s$)
Kinematic viscosity (μ_{fl}/ρ_{fl})	Nature and direction of the flow
Volumetric expansion coefficient	Velocity components v_x, v_y, v_z

1.4.2 Typology

1.4.2.1 Driving forces

If differences in fluid density due to temperature and concentration gradients are the sole driving force, then natural convection with flow patterns reflecting the related temperature and concentration fields will develop, as is most often the case indoors. If, instead, the driving force is an imposed pressure difference, for example wind, then forced convection will be the result with a flow pattern independent of temperature and concentration. But the two types are not clearly separated. Forced convection at low velocities adds a natural component, as is the case with wind around buildings, inducing pressure differences that are generally too weak to eliminate buoyancy. Pure natural convection also rarely occurs. Opening and closing doors in a building, for example, causes some mixed convection.

1.4.2.2 Flow types

Flow can be laminar, turbulent or transitory. Diverging and converging streamlines that never cross, and wherein particle and overall flow velocity coincide, characterize laminar flow. In turbulent flow, chaotic momentum creates whirling eddies with particle velocities different from the flow velocity. Turbulent kinetic energy (k) builds up in the eddies, while turbulent dissipation (ε) continuously extinguishes eddy motion. An exact description of turbulent flow is practically impossible, even applying a fractional eddy approach. Transient flow fills the gap between laminar and turbulent. In fact, small disturbances along the flow path suffice to switch from laminar to turbulent, after which the eddies die and the flow turns laminar again. Intense turbulent mixing favours heat transfer more than laminar flow, for which only conduction perpendicular to the flow direction plays a part. Transient flow switches between the other two types. To summarize:

	Flow		
Driving force	Laminar	Transient	Turbulent
Density differences	X	X	x
Density and pressures	X	X	x
Pressures	X	X	X

1.4.3 Quantifying the convective surface film coefficient

1.4.3.1 Analytically

The few solvable cases underline that the convective surface film coefficient changes from spot to spot along a surface. When the heat flow is the subject of interest, using a surface-averaged value circumvents that complication:

$$h_c = \frac{1}{A} \int h_{cA} dA$$

The standard convective surface film coefficients are such averages, unless local surface temperature values are the subject of interest. Then spot-by-spot values should be used.

1.4.3.2 Numerically

Thanks to CFD, simulations of convection have improved a lot, albeit the velocity profiles in the boundary layer near surfaces still have to be assumed. Table 1.1 and Figure 1.21 list some CFD-related results.

Table 1.1 Convective surface film coefficient (h_e) along a rectangular, detached building: CFD-based correlations for wind speeds (v_w) below 15 m/s and a 10 °C surface-to-air temperature difference (Emmel et al., 2007)

	Surface to wind angle, °	h_e, W/(m².K)
Outer walls	0	$5.15 v_w^{0.31}$
	45	$3.34 v_w^{0.34}$
	90	$4.78 v_w^{0.71}$
	135	$4.05 v_w^{0.77}$
	180	$3.54 v_w^{0.76}$
Roofs	0	$5.11 v_w^{0.78}$
	45	$4.6 v_w^{0.79}$
	90	$3.76 v_w^{0.85}$

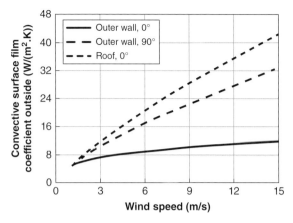

Fig. 1.21 Convective surface film coefficient outside, rectangular building, CFD-based correlations

1.4.3.3 Dimensionally

Most information about convection is still obtained from experiment and dimensional analysis. This latter technique determines which dimensionless ratios amongst the fluid properties, geometric parameters and kinematic parameters that define the movement, need the same value in experiment as in reality to allow extrapolation to real-world situations. These ratios either follow from the differential equations or from Buckingham's π-theorem, which states that if a problem depends on n single-valued physical properties involving p basic dimensions, then $n-p$ dimensionless ratios define the solution. The description of forced convection, for example, needs seven physical properties (L, λ_{fl}, v_{fl}, ρ_{fl}, μ_{fl}, c_{fl}, h_c: see above) with four basic dimensions: L (representative length) t (time) M (mass) and θ (temperature). So, three dimensionless ratios (π_1, π_2, π_3) are required, expressed as $\pi_1 = f(\pi_2, \pi_3)$. These follow from rewriting the π-functions as:

$$\pi = L^a \lambda_{fl}^b v_{fl}^c \rho_{fl}^d \mu_{fl}^e c_{fl}^f h_c^g$$

an equation only containing the basic dimensions, or:

$$\pi = [L]^a \left[\frac{ML}{t^3\theta}\right]^b \left[\frac{L}{t}\right]^c \left[\frac{M}{L^3}\right]^d \left[\frac{M}{Lt}\right]^e \left[\frac{L^2}{t^2\theta}\right]^f \left[\frac{M}{t^3\theta}\right]^g$$

Because the three ratios must be dimensionless, the sum of the exponents per dimension should be zero, or:

For M			b				+	d	+	e			+	g	=	0
For L	a	+	b	+	c	−	3d		−	e	+	2f			=	0
For t		−	3b	−	c				−	e	−	2f	−	3g	=	0
For θ		−	b								−	f	−	g	=	0

The three so follow from recalculating the π-function three times, first for g=1, c=0, d=0, then for g=0, a=1, f=0, and finally for g=0, e=1, c=0:

Solution 1	a=1, b=−1, e=0, f=0	or	$\pi_1 = \dfrac{h_c L}{\lambda_{fl}}$
Solution 2	b=0, c=1, d=1, e=−1	or	$\pi_2 = \dfrac{\rho_{fl} v_{fl} L}{\mu_{fl}}$
Solution 3	a=0, b=−1, d=0, f=1	or	$\pi_3 = \dfrac{c_{fl} \mu_{fl}}{\lambda_{fl}}$

Natural convection instead demands four dimensionless ratios, of which two combine into one. The three for forced convection are called the Reynolds, Nusselt and Prandtl numbers, and the three for natural convection are the Grasshof, Nusselt and Prandtl numbers.

Reynolds number is given by:

$$\mathrm{Re} = v_{fl} L / \nu_{fl} \qquad (= \pi_2) \tag{1.51}$$

with v_{fl} being velocity, ν_{fl} the kinematic viscosity of the fluid and L a characteristic length representing the geometry, such as the hydraulic diameter for pipes, the dimension in the flow direction for walls, or a calculated length for more complex cases. The Reynolds number gives the ratio between the inertia force and viscous friction. If the number is small, friction gains and the result is laminar flow. If large, the inertia force wins and turbulent flow occurs. So the number determines the flow type: laminar for $\mathrm{Re} \leq 2000$, turbulent for $\mathrm{Re} \geq 20\,000$, and transient in between.

The Nusselt number is calculated as:

$$\mathrm{Nu} = h_c L / \lambda_{fl} \qquad (= \pi_1) \tag{1.52}$$

where λ_{fl} is the thermal conductivity of the fluid. Multiplying the left- and right-hand sides of the convective surface film coefficient equation (see above) by the characteristic length gives:

$$h_c L / \lambda_{fl} = (\mathrm{grad}\,\theta)_s / \left[(\theta_{fl} - \theta_s)/L\right]$$

meaning that the Nusselt number represents the ratio between the temperature gradient in the fluid at the surface and the mean temperature gradient along the characteristic length. A large value indicates a significant gradient at the surface and a small gradient along the characteristic length, as is the case for high fluid velocities.

Physically, the number says that conduction governs heat transfer at a surface. Even in a turbulent flow regime, a laminar boundary layer remains whose thickness reduces with increasing fluid velocity but never becomes zero. The number underlines the importance of convection compared with conduction.

The Prandtl number is:

$$\text{Pr} = \nu_{fl}/a_{fl} \qquad (=\pi_3) \tag{1.53}$$

This number combines heat and mass transfer by rating two analogous quantities: the thermal diffusivity, which determines how easily a local temperature change spreads into the fluid, and the kinematic viscosity ν, which indicates how easily a local velocity change affects the fluid.

The Grasshof number in turn is given by:

$$\text{Gr} = \beta_{fl} g L^3 \Delta\theta / \nu_{fl}^2 \tag{1.54}$$

with β_{fl} the volumetric expansion coefficient, g the acceleration due to gravity and $\Delta\theta$ the representative temperature difference. Grasshof replaces Reynolds in the case of natural convection. Velocity then is indeed mainly the result of temperature-related differences in density (βg). As velocity influences temperature, the terms L^3 and ν_{fl}^2 replace L and ν_{fl} in the Reynolds number.

Finally, the Rayleigh number equals:

$$\text{Ra} = \text{Gr Pr} \tag{1.55}$$

This number has no physical meaning. It was introduced because in many formulae for natural convection, Grasshof and Prandtl appear as a product.

All experimental, numerical and analytical expressions for the convective surface film coefficient may be written as:

Natural convection: $\text{Nu} = c(\text{Ra})^n$

Mixed convection: $\text{Nu} = F(\text{Re}/\text{Gr}^{1/2}, \text{Pr})$

Forced convection: $\text{Nu} = F(\text{Re}, \text{Pr})$

where the coefficient c, the exponent n and the function $F()$ differ among geometries, the nature of the flow, and the flow direction for natural convection. A model and reality will coincide when in both cases the numbers have equal values.

1.4.4 Values for the convective surface film coefficient

1.4.4.1 Flat surfaces

In cases of natural convection, the characteristic length L (see the subscript) is the height for vertical surfaces, the side for square horizontal surfaces, and the average of length and width for rectangular horizontal surfaces. The mean temperature between wall surface and the undisturbed fluid fixes the property values. The relations are:

	Conditions	Functions
Vertical surfaces		
	$Ra_L \leq 10^9$	$Nu_L = 0.56\, Ra_L^{1/4}$
	$Ra_L > 10^9$	$Nu_L = 0.025\, Ra_L^{2/5}$
Horizontal surfaces		
Heat exchanged upwards	$10^5 < Ra_L \leq 2 \times 10^7$	$Nu_L = 0.56\, Ra_L^{1/4}$
	$2 \times 10^7 < Ra_L < 3 \times 10^{10}$	$Nu_L = 0.138\, Ra_L^{1/3}$
Heat exchanged downwards	$3 \times 10^5 < Ra_L < 10^{10}$	$Nu_L = 0.27\, Ra_L^{1/4}$

For forced convection, the relationships with L as the characteristic length are:

	Conditions	Functions
Laminar flow	$Pr > 0.1$ $Re_L < 5 \times 10^5$	$Nu_L = 0.644\, Re_L^{1/2} Pr^{1/3}$
Turbulent flow	$Pr > 0.5$ $Re_L > 5 \times 10^5$	$Nu_L = 0.036\, Pr^{1/3} \left(Re_L^{4/5} - 23{,}200 \right)$

In buildings with air-based HVAC systems, the convective surface film coefficients for mixed convection indoors are often related to the air change rate (n in air changes per hour, symbol ach), a number describing how many times an hour the air in a room is replaced by air delivered by the system. For a rectangular room (Fischer, 1995):

Configuration		Convective surface film coefficient W/(m².K)
Forced convection, air diffusers at the ceiling, room isothermal	Walls	$-0.199 + 0.18 n^{0.8}$
	Floor	$0.159 + 0.116 n^{0.8}$
	Ceiling	$-0.166 + 0.484 n^{0.8}$
Forced convection, air diffusers at the walls, room isothermal	Walls	$-0.110 + 0.132 n^{0.8}$
	Floor	$0.704 + 0.168 n^{0.8}$
	Ceiling	$0.064 + 0.00444 n^{0.8}$

1.4 Convection

In and around buildings, the fluid is air at atmospheric pressure and ambient temperature. This allows simplification of the given equations. For natural convection the temperature difference $\Delta\theta$ between the surface and the undisturbed air largely determines the surface film coefficient, a fact reflected by the following relationship:

$$h_c = a(\Delta\theta/L)^b$$

with as values for a, b and L:

	Conditions	a	b	L
Vertical surfaces				
	$10^{-4} < L^3\Delta T \leq 7$	1.4	¼	Height
	$7 < L^3\Delta T \leq 10^3$	1.3	⅓	1
Horizontal surfaces				
Heat flow upwards	$10^{-4} < L^3\Delta T \leq 0.14$	1.3	¼	Eq. side1
	$0.14 < L^3\Delta T \leq 200$	1.5	⅓	
Heat flow downwards	$2 \times 10^{-4} < L^3\Delta T \leq 200$	0.6	¼	1

For forced convection outdoors, wind is the main actor, which gives as relationships:

Wind speed	Relationship	Remarks
$v \leq 5$ m/s	$h_c = 5.6 + 3.9v$	For $v \leq 5$ m/s, natural convection still intervenes, hence the constant 5.6
$v > 5$ m/s	$h_c = 7.2\, v^{0.78}$	

The fact that the value rises with wind speed is a direct consequence of the reduction in boundary layer thickness.

In any case, all these simple relationships only apply for air flowing along freestanding flat surfaces. Angles between two surfaces and corners between three surfaces have a disturbing effect. Moreover, if the surfaces form a room, the overall flow pattern must satisfy the continuity equation. All this makes convection so complex that, for the sake of simplicity, all standards advance constant average values:

European Normalization (EN) standard	Heat loss	Surface temperatures
Natural convection (= inside)		
Vertical surfaces	3.5	2.5
Horizontal surfaces:		
Heat upwards	5.5	2.5
Heat downwards	1.2	1.2
Forced convection (= outside)	19.0	19.0

Reference temperatures in rooms use the air temperature at a point 1.7 m above the floor's centre. Outdoors, the reference is the air temperature measured by the nearest weather station. When calculating surface temperatures using local convective surface film coefficients, the reference moves to the air temperature just outside the boundary layer.

In case of large temperature differences, a complex geometry or surfaces screened by furniture where we require a more correct calculation, the more complete formulae given in the tables above or formulae mentioned in the literature must be used.

1.4.4.2 Cavities

The word cavity refers to an air or gas layer with a small width compared with either the length or height. At the warm face the convective heat flux equals:

$$q_{c1} = h_{c1}(\theta_{s1} - \theta_c)$$

At the cold face, it is:

$$q_{c2} = h_{c2}(\theta_c - \theta_{s2})$$

In both relations, θ_c is the gas temperature in the middle of the cavity. If the cavity remains unvented, the two must on average be equal, giving as mean flux:

$$q_c = \frac{h_{c1} h_{c2}}{h_{c1} + h_{c2}} (\theta_{s1} - \theta_{s2}) \tag{1.56}$$

Replacing the two surface film coefficients by a common value h_c simplifies this formula to:

$$q_c = (h_c/2)(\theta_{s1} - \theta_{s2})$$

In reality, convection in a cavity is partly conduction, so the expression above is mostly rewritten as if conduction dominates, though with the thermal conductivity of the gas (λ_{fl}) multiplied by the Nusselt number:

$$q + q_c = (\lambda_{fl} \mathrm{Nu}) \Delta \theta_s / d = h'_c \Delta \theta_s \tag{1.57}$$

In this formula, d is the cavity width in m and $\Delta \theta_s$ the temperature difference between both cavity faces in °C. In a horizontal, infinitely extending cavity, circular eddies, called Bénard cells, develop. In vertical infinitely extending cavities, some air rotation intervenes. In both cases, the Nusselt number can be written as:

$$\mathrm{Nu}_d = \max \left[1; 1 + \frac{m \, \mathrm{Ra}_d^r}{\mathrm{Ra}_d + n} \right] \quad (10^2 \leq \mathrm{Ra}_d \leq 10^8) \tag{1.58}$$

1.4 Convection

with:

	m	n	r
Horizontal cavity			
Heat transfer downwards	0		
Heat transfer upwards	0.07	3200	1.33
Vertical cavity, or tilted cavity with slope above 45°	0.024	10 100	1.39
Tilted cavity with slope below 45°			
Heat transfer upwards	0.043	4100	1.36
Heat transfer downwards	0.025	13 000	1.36

For the Raleigh number, the temperature difference between the two faces figures as the reference, while the cavity width (d) acts as characteristic length. A Raleigh number below 100 means conduction with $Nu_d = 1$.

Convection in finite cavities diverges strongly from infinite cavities. With d, H and L the cavity width, height and length in m, ^{lam} a superscript for laminar, ^{turb} a superscript for turbulent, and ^{transient} a superscript for transient flow, the Nusselt numbers become:

	Nu_d
Vertical cavity	
Raleigh number upper limit value (Ra_{max}) for the applicability of Nu_d depending on the ratio H/d:	$\max(Nu_d^{lam}, Nu_d^{turb}, Nu_d^{transient})$, with $Nu_d^{lam} = 0.242 \left(\frac{Ra_d d}{H} \right)^{0.273}$
H/d: 5 20 40 80 100 Ra_{max}: 10^8 2×10^6 2×10^5 3×10^4 1.2×10^4	$Nu_d^{turb} = 0.0.0605\, Ra_d^{0.33}$
	$Nu_d^{transient} = \left[1 + \left(\frac{0.104 Ra_d^{0.293}}{1 + (6310/Ra_d)^{1.36}} \right)^3 \right]^{0.33}$
Horizontal cavity	
Heat transfer upwards	$Ra_d \leq 1708$: 1 $Ra_d > 1708$: $\max\left[1, 1537 d^2 \left(\frac{\Delta\theta}{L} \right)^{1/4} \right]$
Heat transfer downwards	1
Tilted cavity: see literature	

1.4.4.3 Pipes

Experimental and semi-experimental work on convection between pipes and the ambient fluid has resulted in a series of formulae. For natural convection:

Vertical pipe	$Ra_d \leq 10^9$	$Nu_L = 0.555\, Ra_d^{1/4}$
	$Ra_d > 10^9$	$Nu_L = 0.021\, Ra_d^{2/5}$
Horizontal pipe	$Ra_d \leq 10^9$	$Nu_L = 0.530\, Ra_d^{1/4}$

For forced convection:

$Re_d < 500$	$Nu_d = 0.43 + 0.48\, Re_d^{1/2}$
$Re_d > 500$	$Nu_L = 0.46 + 0.00128\, Re_d$

In all equations, the characteristic length (d) relates to the outer diameter of the pipe, while all properties link to the average (θ_{conv}) between the temperature in the undisturbed fluid (θ_{fl}) and the temperature of the pipe's outside surface (θ_s).

1.5 Radiation

1.5.1 In general

Thermal radiation differs fundamentally from conduction and convection. Radiation involves electromagnetic waves in the heat exchange. Any surface warmer than 0 K emits electromagnetic waves, while their absorption by surfaces agitates atoms and electrons, which is effectively heating. Electromagnetic waves span an impressive interval of wavelengths (λ), but only the 10^{-7} to 10^{-3} m range with ultraviolet (UV), visible light (L) and infrared (IR) are quoted as being thermal (Table 1.2).

Table 1.2 Categorization of electromagnetic radiation by wavelength

Wavelength	Radiation type
$\lambda \leq 10^{-6}\,\mu m$	Cosmic radiation
$10^{-6} < \lambda \leq 10^{-4}\,\mu m$	Gamma rays
$10^{-4} < \lambda \leq 10^{-2}\,\mu m$	X-rays
$10^{-2} < \lambda \leq 0.38\,\mu m$	Ultraviolet
$0.38 < \lambda \leq 0.76\,\mu m$	Visible light
$0.76 < \lambda \leq 10^3\,\mu m$	Infrared
$10^3 < \lambda$	Radio waves

Due to its electromagnetic nature, thermal radiation does not require a medium. On the contrary, only in a vacuum, where the photons move at a speed of 299 792.5 km/s, is transfer unhindered. How much will be emitted depends on the nature of a surface and its temperature, while net heat exchanges only happen among surfaces at different temperatures. Aside, the wavelength is given by the ratio between the propagation speed in m/s and the frequency in Hz.

1.5.2 Definitions

Table 1.3 outlines how thermal radiation is quantified, with the spectral values standing for 'deduced with respect to wavelength'. A single wavelength gives monochromatic radiation, while several wavelengths together give coloured radiation.

1.5.3 Reflection, absorption and transmission

When a radiant flux (q_{uid}) emitted by a surface at temperature T touches another surface, a part is absorbed (b_{ra}), a part reflected (e_{rr}) and, if transparent, a part

Table 1.3 Variables of radiant heat transfer

Variable	Definition, units	Equations
Radiant heat Q_R	The heat emitted or received in the form of electromagnetic waves. Scalar, units J	
Radiant heat flow Φ_R	The radiant heat per unit of time. Scalar, units W	$\dfrac{dQ_R}{dt}$
Radiant heat flux q_R	The radiant heat flow per unit of surface. As a surface emits radiation and receives it from all directions, the flux, units W/m², is a scalar. The term *irradiation*, symbol E, is used for the incoming radiation, the term *emittance*, symbol M, for the emitted radiant heat flux.	$\dfrac{d^2 Q_R}{dA\, dt}$
Radiation intensity I	The radiant energy emitted in a specific direction. The intensity is a vector, units W/(m².rad) with $d\omega$ the elementary angle in the direction considered.	$\dfrac{dq_R}{d\omega}$ or $\dfrac{d^2 \Phi_R}{dA\, d\omega}$
Luminosity L	The ratio between the radiant heat flow rate in a direction ϕ and the apparent surface, seen from that direction. The luminosity is a vector, units W/(m².rad). It describes how a receiving surface sees an emitting one.	$\dfrac{d^2 \Phi_R}{\cos\phi\, dA\, d\omega}$

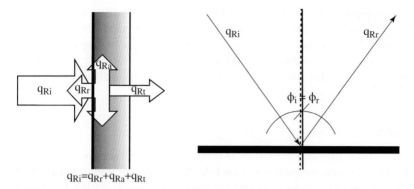

Fig. 1.22 Reflection, absorption and transmission at a surface, specular reflection

transmitted (a_{rt}):

$$\alpha = q_{Ra}/q_{Ri} \qquad \rho = q_{Rr}/q_{Ri} \qquad \tau = q_{Rt}/q_{Ri} \qquad (1.59)$$

where α, ρ and τ are the average absorptivity, reflectivity and transmissivity at a temperature T of the receiving surface. Conservation of energy now imposes that the sum of these three must be 1, or:

$$\alpha + \rho + \tau = 1$$

This does not hold if the three are at different temperatures – never add in such a case. A difference also exists between diffuse and specular reflection. The latter obeys the laws of optics, with the incident and reflected beam in the same plane, and the angles with the normal to the surface (ϕ_r and ϕ_i) being equal (Figure 1.22). Most surfaces, however, show diffuse reflection, meaning that the reflected radiation scatters in all directions.

Reflectivity in a given direction (α) can be defined in relation to the radiation intensity incident under an angle ϕ on the surface:

$$\rho_\phi = I_{Rr\alpha}/I_{Ri\phi}$$

with $I_{Rr\alpha}$ the reflected intensity in direction α. For a specularly reflecting surface, the reflectivity becomes:

$$\rho_\phi = (I_{Rr}/I_{Ri})_\phi$$

with ρ_ϕ being a function of the angle of incidence.

Most building and insulation materials are opaque for thermal radiation ($\tau = 0$). What arrives is absorbed in a thin surface layer, 10^{-6} m thick for metals and 10^{-4} m for other materials. Therefore the terms 'absorbing surface' or 'absorbing body' are

1.5 Radiation

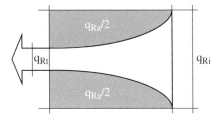

Fig. 1.23 Absorption of radiation passing through transparent materials

often used. Instead, most gases, fluids and solids such as glass and a number of synthetics are selectively transparent, although they show selective mass absorption depending on their extinction coefficient (a) (Figure 1.23):

$$\frac{dq_R}{q_R} = -a\,dx \tag{1.60}$$

For a layer with thickness d, the transmitted and absorbed radiant heat flux then equals:

$$q_{Rt} = q_{Ri}\exp(-ad) \qquad q_{Ra} = q_{Ri} - q_{Rt} = q_{Ri}[1 - \exp(-ad)]$$

with q_{Ri} the incoming flux. Absorptivity and transmissivity so become:

$$\alpha = 1 - \exp(-ad) \qquad \tau = \exp(-ad)$$

Specular reflectivity at an irradiated interface separating two media is given by:

$$\rho = \frac{I_r}{I_i} = \left[\frac{n_1\cos(\phi_i) - n_2\cos(\phi_t)}{n_1\cos(\phi_i) + n_2\cos(\phi_t)}\right]^2$$

where n_1 and n_2 are the refractive indexes of the media at either side, for example air ($n = 1$) and glass, ϕ_i is the angle of incidence in the first medium, and ϕ_t is the angle of transmittance to the second medium.

As stated, absorptivity, reflectivity, and transmissivity vary with temperature, and thus with wavelength, although the angle of incidence also matters. The impact can be impressive. Take glass, whose transmissivity for visible light is large, whereas for UV and IR it approaches zero with the absorptivity then exceeding a value of 0.9. Those differences explain the greenhouse effect. The short-wave, high-temperature solar radiation transmitted by the glass is absorbed by all surfaces indoors and re-emitted as low-temperature IR radiation, which the glass absorbs, leaving conduction as the only way to get rid of the heat. At the same time, the radiant bodies indoors slowly release the absorbed solar heat. The combination can make the indoors uncomfortably warm. Transparent synthetic materials act analogously, although some also transmit IR.

1.5.4 Radiant bodies

Ideal black surfaces absorb all incident radiation ($\alpha=1, \rho=0, \tau=0, \alpha \neq f(\lambda,\phi)$). Their study is enlightening for grey bodies, which have a constant absorptivity, blank bodies, which have an absorptivity of zero, and coloured bodies, whose absorptivity depends on the temperature and the direction of the incident radiation. Although blank and grey bodies are ideal, and do not exist in reality, most real surfaces are assumed to behave as grey bodies. A distinction is made between short solar (subscript S) and long-wave ambient radiation (subscript L). Both stand for a different absorptivity and reflectivity.

1.5.4.1 Black

Of all surfaces, black bodies (subscript b) emit most radiant energy, independent of temperature. Their emissivity is 1. In fact, according to the second law of thermodynamics, in closed systems black bodies at different temperatures must evolve irrevocably towards temperature equilibrium. Because at equilibrium each of them emits and absorbs the same amount of radiation, the emissivity must equal the absorptivity, which is 1. With respect to direction, a black body obeys Lambert's law: luminosity constant. Hence, the related radiant intensity must obey (Figure 1.24):

$$I_{b\phi} = L_b \cos \phi \tag{1.61}$$

This equation, known as the cosine law, offers a simple relationship between emittance and luminosity. From the definitions in Table 1.3:

$$M_b = L_b \int_\omega \cos \phi \, d\omega$$

where the integral covers the hemisphere. The angle $d\omega$ is calculated assuming a hemisphere with radius r_o surrounds the surface dA with emittance M_b (Figure 1.25). The angle $d\omega$ thus equals:

$$d\omega = r_0^2 \sin \phi \, d\phi \, d\vartheta$$

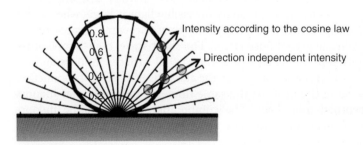

Fig. 1.24 The thick line representing the effect of the cosine law on radiation intensity

1.5 Radiation

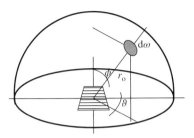

Fig. 1.25 Proving the cosine law

while on the hemisphere, the intensity drops to $I_{b\phi}/r_0^2$, or:

$$M_b = L_b \int_0^{2\pi} \int_0^{\pi/2} \frac{\cos \phi}{r_0^2} r_0^2 \sin \phi \, d\phi \, d\vartheta = \left[-\pi L_b \cos^2 \phi \right]_0^{\pi/2} = \pi L_b \tag{1.62}$$

Planck's law gives the spectral density of the emittance:

$$M_{b\lambda} = \frac{2\pi c^2 h \lambda^{-5}}{\exp\left(\dfrac{ch}{k\lambda T}\right) - 1} \tag{1.63}$$

with constants: c as the speed of light in m/s, h as Planck's constant (6.624×10^{-34} J.s), and k as Boltzmann's constant (1.38047×10^{-23} J/K). The products $2\pi c^2 h$ and ch/k are called the radiation constants for a black body, symbols C_1 (3.7415×10^{-16} W.m^2) and C_2 (1.4388×10^{-2} m.K). Figure 1.26 shows the spectral density of the emittance for different values of the absolute temperature.

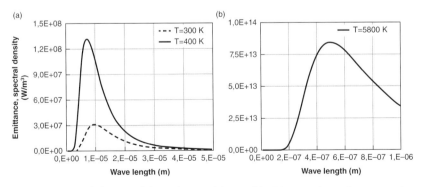

Fig. 1.26 Spectral density of the emittance (a) at ambient temperature, at a warmer temperature and (b) for the sun as a black body

The emittance, equal to the surface under the curve, increases quickly with temperature, while the maxima occur at ever-shorter wavelengths. In the $[\lambda, M_{B\lambda}]$ plane, these maxima have as a geometric locus a fifth order hyperbole, while their wavelengths obey Wien's law:

$$\lambda_M T = 2898 \qquad (\lambda_M \text{ in } \mu m) \tag{1.64}$$

At 20 °C, with $\lambda_M = 2898/293.15 = 9.9\,\mu m$, the maximum lies in the infrared interval. For the sun, with a radiant temperature of 5800 K, with $\lambda_M = 2898/5800 = 0.5\,\mu m$, it sits in the middle of the visible light interval. The emittance M_b itself follows from integrating Planck's law with respect to the wavelength:

$$M_b = \int_0^\infty M_{b\lambda}\, d\lambda = \frac{2\pi^5 k^4}{15 c^2 h^3} T^4 = \sigma T^4 \tag{1.65}$$

This equation is known as the Stefan-Boltzmann law, with σ as Stefan's constant, $5.67 \times 10^{-8}\,W/(m^2.K^4)$. This and Wien's law preceded Planck's, for which quantum mechanics had to come first. The Stefan-Boltzmann law is mostly written as:

$$M_b = C_b \left(\frac{T}{100}\right)^4 \tag{1.66}$$

with C_b the black body constant, $5.67\,W/(m^2.K^4)$, and $T/100$ the reduced radiant temperature. Luminosity and radiation intensity then become:

$$L_b = \frac{M_b}{\pi} = \frac{C_b}{\pi}\left(\frac{T}{100}\right)^4 \qquad I_{b\phi} = \frac{d^2 \Phi_{Rb}}{dA\,d\omega} = L_b \cos\phi = \frac{C_b}{\pi}\left(\frac{T}{100}\right)^4 \cos\phi$$

When two black bodies 1 and 2 with surfaces A_1 and A_2 and no medium in between are positioned as shown in Figure 1.27, the elementary radiant heat flow going from 1 to 2 equals:

$$d^2 \Phi_{R,1\to 2} = I_{b1} dA_1\, d\omega_1 = \frac{M_{b1}}{\pi} \cos\phi_1\, dA_1\, d\omega_1$$

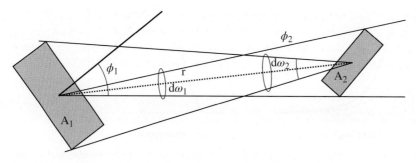

Fig. 1.27 Calculation of the view factor between the surfaces A_1 and A_2

1.5 Radiation

with $d\omega_1$ the angle at which A_1 sees A_2:

$$d\omega_1 = dA_2 \cos\phi_2 / r^2$$

This redresses that flow interchanged to:

$$d^2\Phi_{R,1\to 2} = \frac{M_{b1}}{\pi} \cos\phi_1 \cos\phi_2 \, dA_1 \frac{dA_2}{r^2}$$

From surface A_2 to surface A_1 the flow is:

$$d^2\Phi_{R,2\to 1} = \frac{M_{b2}}{\pi} \cos\phi_1 \cos\phi_2 \, dA_1 \frac{dA_2}{r^2}$$

The resulting flow between the two becomes:

From body 1 to body 2: $\quad d^2\Phi_{R,12} = d^2\Phi_{R,1\to 2} - d^2\Phi_{R,2\to 1}$

$$= \frac{(M_{b1} - M_{b2})\cos\phi_1 \cos\phi_2 \, dA_1 \, dA_2}{\pi r^2}$$

From body 2 to body 1: $\quad d^2\Phi_{R,21} = d^2\Phi_{R,2\to 1} - d^2\Phi_{R,1\to 2}$

$$= \frac{(M_{b2} - M_{b1})\cos\phi_1 \cos\phi_2 \, dA_1 \, dA_2}{\pi r^2}$$

If both are finite in shape, then that flow becomes:

From body 1 to body 2: $\quad \Phi_{R,12} = (M_{b1} - M_{b2})A_1 \left[\frac{1}{\pi A_1} \int_{A_1}\int_{A_2} \frac{\cos\phi_1 \cos\phi_2 \, dA_2 \, dA_1}{r^2} \right]$

(1.67)

From body 2 to body 1: $\quad \Phi_{R,21} = (M_{b2} - M_{b1})A_2 \left[\frac{1}{\pi A_2} \int_{A_2}\int_{A_1} \frac{\cos\phi_1 \cos\phi_2 \, dA_1 \, dA_2}{r^2} \right]$

(1.68)

The term between square brackets in both formulae is called the view factor, symbol F. Other names are the angle, shape or configuration factor. If A_1 is considered as emitting, the view factor is written F_{12}. Vice versa, F_{21} is used. View factors are geometric quantities indicating what fraction of the radiant flow emitted by the one reaches the other. The size of each body, their form, the distance, and the angle at which they see each other all define the value, which equals 1 when all emitted radiation touches the other.

Concerning the view factor properties, firstly, reciprocity exists in the sense that $A_1 F_1 = A_2 F_2$. This relationship follows from the definition. Further, when a surface A_2 surrounds a surface A_1, the view factor from A_1 to A_2 must be 1 (Figure 1.28), a

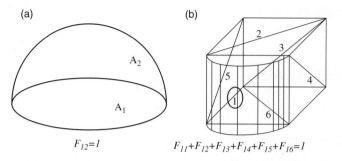

Fig. 1.28 View factor between (a) surface 1 completely surrounded by surface 2 or (b) the surfaces 2 to 6. In the second case, surface 1 also radiates to itself

result that applies for each surface surrounded by $n-1$ others that together form a closed volume:

$$\sum_{j=2}^{n} F_{1j} = 1$$

Finally, two infinitely parallel surfaces, such as the bounding faces of a cavity, also have a view factor of 1.

Some simple configurations allow an analytical calculation. For a point sitting at a distance D from a rectangle with sides L_1 and L_2, the view factor calculates as the ratio between the angle at which the point sees the rectangle with surface $L_1 \times L_2 = A_2$, and its whole view angle 4π:

$$F_{12} = \frac{1}{4\pi} \int_{A_2} \frac{\cos \phi}{r^2} dA_2$$

For a point above a corner, the formula becomes $(\cos \phi = D/r,\ r^2 = D^2 + x^2 + y^2)$:

$$F_{12} = \frac{1}{4\pi} \int_0^{L_1} \int_0^{L_2} \frac{D}{(D^2 + x^2 + y^2)^{3/2}} dy\, dx$$

Integration gives:

$$F_{12} = \frac{1}{8} - \frac{1}{4\pi} a\tan\left(\frac{D\sqrt{L_1^2 + L_2^2}}{L_1 L_2}\right)$$

Other positions of the point convert to the corner case by the construction of Figure 1.29. The resulting view factor is:

$$F_{12} = F_{1a} + F_{1b} + F_{1c} + F_{1d}$$

1.5 Radiation

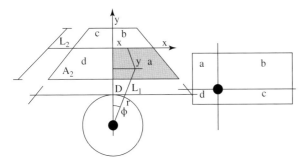

Fig. 1.29 View factor, point at distance D of surface A_2

Radiation between the human head and the ceiling is an example of a point-to-surface situation.

Another case concerns an infinitesimal surface dA_1 at an orthogonal distance D from a rectangle with sides L_1 and L_2 and surface A_2 in parallel. The formula for the view factor then is:

$$F_{12} = \frac{1}{dA_1} \int_{dA_1} \int_{A_2} \frac{\cos\phi_1 \cos\phi_2 \, dA_2 \, dA_1}{\pi r^2}$$

The way dA_1 sees each infinitesimal surface dA_2 on A_2 is independent of its position, or:

$$F_{12} = \frac{1}{\pi} \int_{A_2 \text{ seen by } dA_1} \frac{\cos\phi_1 \cos\phi_2 \, dA_2}{r^2}$$

Suppose now that surface dA_1 lies at a distance D above the corner (0, 0) of a rectangle A_2 (Figure 1.30). The view factor then is:

$$F_{12} = \frac{1}{\pi} \int_0^{L_1} \int_0^{L_2} \frac{\cos\phi_1 \cos\phi_2}{r^2} \, dx \, dy$$

an equation that can be simplified to $\left(\cos\phi_1 = \cos\phi_2 = D/r, \; r^2 = D^2 + x^2 + y^2\right)$:

$$F_{12} = \frac{1}{\pi} \int_0^{L_1} \int_0^{L_2} \frac{D^2 \, dy \, dx}{\left(D^2 + x^2 + y^2\right)^2}$$

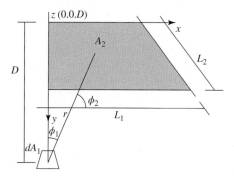

Fig. 1.30 View factor, infinitesimal surface dA_1 parallel to surface A_2 at an orthogonal distance D from the corner (0, 0)

with as a solution:

$$F_{12} = \frac{1}{2\pi}\left[\frac{L_1}{\sqrt{D^2+L_1^2}}\operatorname{atan}\left(\frac{L_2}{\sqrt{D^2+L_1^2}}\right) + \frac{L_2}{\sqrt{D^2+L_2^2}}\operatorname{atan}\left(\frac{L_1}{\sqrt{D^2+L_2^2}}\right)\right]$$

Other configurations have to be solved numerically. Consider an infinitesimally small perpendicular surface dA_1 at a distance D from a rectangular surface A_2 that has sides L_1 and L_2. dA_1 does not see that part of A_2 that lies behind the intersection of the plane containing dA_1 with A_2. The numerical formula for the view factor thus becomes:

$$F_{12} = \frac{D\Delta x \Delta y}{2\pi}\left[\sum_{x=\Delta x/2 \text{ to } L_1-\Delta x/2 \text{ step } \Delta x}\sum_{y=\Delta y/2 \text{ to } L_2-\Delta y/2 \text{ step } \Delta y}\frac{\sqrt{x^2+y^2}}{(x^2+y^2+D^2)^2}\right]$$

To give a practical example, the surface configurations in beam-shaped rooms include three pairs of two identical parallel walls, while all are perpendicular to each other and have common edges and corners. An analytical calculation of all view factors is not doable but related numerics are easily programmed on a spreadsheet.

With the view factor included, the radiant heat flows and fluxes are written as:

$$\phi_{R,12} = A_1 F_{12}(M_{b1} - M_{b2}) \quad q_{R,12} = F_{12}(M_{b1} - M_{b2})$$
$$\phi_{R,21} = A_2 F_{21}(M_{b2} - M_{b1}) \quad q_{R,21} = F_{21}(M_{b2} - M_{b1})$$
(1.69)

For a number of black bodies radiating to each other, the flow and flux per body are:

$$\phi_{R1n} = A_1 \sum_{j=2}^{n}\left[F_{1j}(M_{b1} - M_{bj})\right] \quad q_{R1n} = \sum_{j=2}^{n}\left[F_{1j}(M_{b1} - M_{bj})\right] \quad (1.70)$$

with 1 being the body considered, and 2 to n the $n-1$ others.

1.5.4.2 Grey

Related laws are similar to those for black bodies. Only the radiant exchange differs. For each wavelength and direction, a grey body emits a constant fraction compared with a black body. The ratio between both is called the emissivity (e). Conservation of energy tells us that the absorptivity (α) must equal that emissivity, giving as reflectivity (ρ):

$$\rho = 1 - \alpha = 1 - e$$

Grey bodies with reflectivity of 1 are blank. Lambert's law holds ($L = C^t$). The radiant heat flux obeys the cosine law, while the emittance is:

$$M = \pi L$$

The spectral emittance obeys Planck's law, but multiplied by the emissivity e. The total emittance thus is:

$$M = e\, C_b \left(\frac{T}{100}\right)^4 \tag{1.71}$$

Each grey body reflects radiation. If eM_b is the emittance of one and E the irradiation by all others, the radiosity of that one equals:

$$M' = e\, M_b + \rho E \tag{1.72}$$

The difference between radiosity and irradiation defines the emitted flux:

$$q_R = M' - E \tag{1.73}$$

Eliminating the unknown irradiation between the two results in:

$$q_R = M' - \frac{M' - e M_b}{\rho} = -\frac{e}{\rho}(M' - M_b) \tag{1.74}$$

Or, the radiant heat flux received equals:

$$q_R = \frac{e}{\rho}(M' - M_b) \tag{1.75}$$

Otherwise said, a grey surface looks black but with a grey filter in front, coupled to the black by a radiant resistance, equal to the ratio between the grey reflectivity and its emissivity (ρ/e). The black body has an emittance M_b, and the grey filter a radiosity M'. For two grey bodies 1 and 2, separated by a transparent medium, the resulting radiant flow interchanged is:

From black body 1 to grey filter 1: $\Phi_{R,11} = \frac{e_1}{\rho_1}(M_{b1} - M'_1) A_1$
From grey filter 1 to grey filter 2: $\Phi_{R,12} = F_{12}(M'_1 - M'_2) A_1$
From grey filter 2 to black body 2: $\Phi_{R,22} = \frac{e_2}{\rho_2}(M'_2 - M_{b2}) A_2$

The second equation is identical to the radiant heat flow between two black bodies. Indeed, as well the emittance of a black body as the radiosity of a grey body obeys Lambert's law for diffuse radiation. In the three equations, the flows $\Phi_{R,11}$, $\Phi_{R,12}$ and

$\Phi_{R,22}$ must be identical, or, elimination of the unknown radiosities M'_1 and M'_2 gives as flows from 1 to 2 and 2 to 1:

$$\Phi_{R,12} = \left[\frac{1}{\dfrac{\rho_1}{e_1} + \dfrac{1}{F_{12}} + \dfrac{\rho_2 A_1}{e_2 A_2}}\right] (M_{b1} - M_{b2}) A_1 \tag{1.76}$$

$$\Phi_{R,21} = \left[\frac{1}{\dfrac{\rho_2}{e_2} + \dfrac{1}{F_{21}} + \dfrac{\rho_1 A_2}{e_1 A_1}}\right] (M_{b2} - M_{b1}) A_2 \tag{1.77}$$

The term between the brackets stands for the radiation factor, symbol F_R. If A_1 is seen as emitting, we write $F_{R,12}$, while if A_2 is the emitter, $F_{R,21}$. Dividing both equations by the surface seen as emitting gives the radiant heat fluxes.

A common configuration consists of two infinitely large, parallel isothermal surfaces. Then $F_{12} = F_{21} = 1$ and $A_1 = A_2$, giving as flux ($\rho = 1 - e$):

$$q_{R,12} = \frac{\Phi_{R,12}}{A_1} = \left[\frac{1}{\dfrac{1}{e_1} + \dfrac{1}{e_2} - 1}\right] (M_{b1} - M_{b2}) \tag{1.78}$$

In building construction, the term between the brackets represents the radiation factor in an infinite cavity. If one of the faces is blank, for example face 1, then $F_{R,12} = 1/(1/0 + 1/e_2 - 1) = 0$. If one is black, for example face 2, then $F_{R,12} = 1/(1/e_1 + 1/1 - 1) = e_1$

Another common configuration is an isothermal surface (1) surrounded by another isothermal surface (2). F_{12} now is 1 and the radiant flow becomes:

$$\Phi_{R,12} = \frac{e_1 e_2}{e_2 \rho_1 + e_1 e_2 + \dfrac{e_1 \rho_2 A_1}{A_2}} (M_{b1} - M_{b2}) A_1$$

If both surfaces are almost black ($e > 0.9$), the denominator nears 1, giving:

$$\Phi_{R,12} = e_1 e_2 (M_{b1} - M_{b2}) A_1$$

If, moreover, the surrounded surface (A_1) is very small compared with the surrounding one (A_2), which brings their ratio close to zero, the equation further simplifies to:

$$\Phi_{R,12} = e_1 (M_{b1} - M_{b2}) A_1 \tag{1.79}$$

1.5 Radiation

In such a case, the resulting radiant heat flow only depends on the emissivity of the surface (A_1) that is surrounded.

When multiple isothermal grey surfaces face each other, all at different temperatures, the radiant heat flow between one of them (A_1) and all others ($n-1$) is written as:

From the equivalent black body 1 to grey filter 1: $\Phi_{R,11} = \dfrac{e_1}{\rho_1}(M_{b1} - M'_1) A_1$

From grey filter 1 to the $n-1$ other grey filters: $\Phi_{R,1 \text{ to } j} = \sum_{j=2}^{n} F_{1j}(M'_1 - M'_j) A_1$

As both flows are equal, the black body emissivity is given by:

$$M_{b1} = \left(1 + \dfrac{\rho_1}{e_1}\sum_{j=2}^{n}F_{1j}\right) M'_1 - \dfrac{\rho_1}{e_1}\sum_{j=2}^{n}(F_{1j}M'_j) \tag{1.80}$$

For the 2 to n surrounding bodies, this equation converts into:

$$M_{b1} = \dfrac{M'_1}{e_1} - \dfrac{\rho_1}{e_1}\sum_{j=2}^{n}(F_{1j}M'_j) \tag{1.81}$$

Per grey body, the radiosities M'_j are unknown but the black body emittances M_{b1} are known. The result is a system of n equations with n unknowns:

$$\left[M_{bj}\right]_n = [F]_{n.n}\left[M'_j\right]_n \tag{1.82}$$

where $[F]_{n.n}$ is the radiation matrix for the n isothermal grey bodies. Solving it gives the radiosities M'_j as functions of the black body emittances M_{bj}. The radiant heat flows then follow from inserting M'_j in the equations given above for the heat exchange between the black body and its grey filter.

1.5.4.3 Coloured

For coloured bodies, emissivity, absorptivity and reflectivity depend on the wavelength, which changes with temperature and sometimes direction. Kirchhoff's law ($e = \alpha$) still applies, but Lambert's law does not, as it requires the same emission per direction. The spectral emittance per wavelength thus differs from black, although the ratio between the coloured and black emittance at the same temperature still gives the average emissivity for that temperature. To simplify things, coloured bodies are often considered grey, albeit with temperature-dependent emissivity. For the emittance and irradiance at strongly different temperatures, ambient and solar for example, Kirchoff's law no longer applies because the short-wave absorptivity (α_S) differs from the ambient long wave emissivity (e_L). To give an example, for polished aluminium, α_S equals 0.2 to 0.4, while e_L is 0.05.

1.5.5 Simple formulae

Thermal radiation seems quite straightforward to model. However, calculating all angle factors is cumbersome, while the system of equations for multiple grey bodies can be very large. A simpler approach, therefore, is welcomed. In a first step, reality is reduced to two radiant surfaces: the one considered (surface 1) and the remaining $n-1$ others shaping the environment. In a second step, the environment is supposed black at a radiant temperature θ_r, which is the temperature it should have as a black body to interchange the same radiant flow as in reality. Solving the system of equations for all surfaces present gives the radiosity for surface 1 (M'_1) as a linear combination of the (black body) emittances of all surfaces present:

$$M'_1 = \sum_{i=1}^{n} a_{ri} M_{bi}$$

Insertion into the equation for the grey body radiant heat flow received and equating with the radiant heat flow found for the surrounded surface 1, which is usually very small compared to the $n-1$ surfaces in the environment, fixes that radiant temperature:

$$\theta_r = \sqrt[4]{\frac{1}{\rho_1}\left(\sum_{i=1}^{n} a_{ri} T_i^4 - e_1 T_1^4\right)} - 273.15$$

Analagously with convection, the radiant heat flow and heat flux can then be written as:

$$q_r = h_r(\theta_{s1} - \theta_r) \qquad \Phi_r = h_r(\theta_{s1} - \theta_r)A \tag{1.83}$$

In both, h_r represents the surface film coefficient for radiation (W/(m².K)), while θ_{s1} is the temperature of surface 1. The surface film coefficient varies with the configuration considered. If the environment surrounds surface 1, the value follows from equalling the equation above to the one derived for where the surrounded surface is small compared with the one surrounding:

$$h_r = e\, C_b \frac{\left[\left(\dfrac{T_{s1}}{100}\right)^4 - \left(\dfrac{T_r}{100}\right)^4\right]}{\theta_{s1} - \theta_r} \tag{1.84}$$

The term between brackets is called the temperature ratio for radiation (F_T):

$$F_T = \frac{T_m}{5000}\left[\left(\frac{T_{s1}}{100}\right)^2 + \left(\frac{T_r}{100}\right)^2\right] \approx \frac{4}{100}\left(\frac{T_m}{100}\right)^3 \tag{1.85}$$

1.6 Building-related applications

As that ratio hardly varies for temperatures between −10 and 50 °C, the simple expression on the right-hand side usually suffices. It considers the flux equation as being more or less linear. Insertion in the surface film coefficient for radiation gives:

$$h_r = e_1 \, C_b \, F_T \tag{1.86}$$

Another case is parallel bodies. If the other surface is isothermal, temperature θ_{s2}, the detour via the radiant temperature becomes superfluous and the surface film coefficient can be directly written as:

$$q_r = h_r(\theta_{s1} - \theta_{s2})$$

with:

$$h_r = \frac{5.67 F_T}{\dfrac{1}{e_1} + \dfrac{1}{e_2} - 1}$$

When surfaces in the environment have the same temperature as the one surrounded, only part of that environment will participate in the radiant exchange, while the radiant temperatures should only include those at different temperature ($= \theta'_r$). In such a case, the surface film coefficient for radiation becomes:

$$q_r = h_r(\theta_{s1} - \theta'_r)$$

with:

$$h_r = \frac{e_1 \, C_b \, F_{12} \, F_T}{e_1 + \rho_1 F_{12}}$$

where F_{12} is the view factor between surface A_1 and those in the environment at different temperatures. If surface A_1 is almost black, the denominator tends to 1 and:

$$h_r = e_1 \, C_b \, F_{12} \, F_T = 5.67 e_1 \, F_{12} \, F_T \tag{1.87}$$

Consider, for example a corner formed by two identical outer walls. Both are equally warm. The result is a radiant exchange with half the space, containing surfaces at a different temperature. The view factor is 0.5. The surface film coefficient for radiation then equals:

$$h_r = e_1 \, C_b \, F_T / 2 = 2.84 \, e_1 \, F_T \tag{1.88}$$

Of course, surfaces at the same temperature can also be included in the radiant temperature. The view factor then remains 1, but the radiant temperature will change.

1.6 Building-related applications

1.6.1 Surface film coefficients and reference temperatures

In real buildings, conduction, convection and radiation combine. Consider an outside wall without a cavity. Heat is conducted across the wall. Between the inside surface and indoors, convection and radiation take over. The same happens between the outside surface and outdoors. Both can be considered the product of a temperature difference with a surface film coefficient (h_c, h_r). However, convection and radiation are so intertwined that, when possible, the related surface film coefficients are combined into one value inside (h_i) and another outside (h_e), both linked to a specific reference temperature. Of course, the two modes may also remain separated. Then, an air and a radiant temperature will characterize both ambients.

1.6.1.1 Indoors

Assume an isothermal surface at temperature θ_{si}. The convective heat flux exchanged with the air is:

$$[q_{ci}] = h_{ci,s}(\theta_{i,ob} - \theta_{si})$$

with $h_{ci,s}$ the average convective surface film coefficient and $\theta_{i,ob}$ the average air temperature directly outside the boundary layer. If not that temperature, but the air temperature in the centre of the room 1.7 m above the floor (θ_i) is taken as the reference, the flux changes to:

$$q_{ci} = h_{ci}(\theta_i - \theta_{si})$$

with h_{ci} the average convective surface film coefficient, now linked to that new reference temperature:

$$h_{ci} = h_{ci,s}\left(\frac{\theta_{i,ob} - \theta_{si}}{\theta_i - \theta_{si}}\right) \tag{1.89}$$

The radiant heat flux at the inside surface equals:

$$q_{ri} = h_{ri}(\theta_{ri} - \theta_{si})$$

with θ_{ri} the radiant temperature characterizing the indoor environment. The total heat flux at the surface thus becomes:

$$q_i = q_{ci} + q_{ri} = h_{ci}(\theta_i - \theta_{si}) + h_{ri}(\theta_{ri} - \theta_{si}) = (h_{ci} + h_{ri})\left[\left(\frac{h_{ci}\theta_i + h_{ri}\theta_{ri}}{h_{ci} + h_{ri}}\right) - \theta_{si}\right] \tag{1.90}$$

The sum $h_{ci} + h_{ri}$ stands for the inside surface film coefficient for heat, symbol h_i, units W/(m².K). The weighted average between the central air temperature and the radiant temperature, named $\theta_{ref,i}$, serves as reference temperature for the surface considered. Standard values for the convective part h_{ci} were given when discussing convection. The radiant part, which convenes with a surface facing a surrounding environment, equals:

$$h_{ri} = 5.67 e_L F_T$$

1.6 Building-related applications

with e_L the long wave emissivity of the surface and F_T the temperature ratio for radiation in the interval $\theta_{si} - \theta_{ri}$, mostly a value around 0.95. Since most finishes have a long wave emissivity of 0.8 to 0.9, the result is $4.3 \leq h_{ri} \leq 4.95$ W/(m².K)

The standardized inside surface film coefficients (W/(m².K)) thus become:

Vertical surfaces		Horizontal surfaces	
Any flow direction	$\theta \approx 7.7$	Heat flow upwards	10
		Heat flow downwards	6

The 2017 *ASHRAE Handbook of Fundamentals* gives a more complete set of values:

Position	Heat flow direction	h_i (W/(m².K)) for a surface emissivity		
		0.9	0.2	0.05
Horizontal	Upward	9.26	5.17	4.32
Sloping 45°	Upward	9.09	5.00	4.15
Vertical	Horizontal	8.29	4.20	3.35
Sloping 45°	Downward	7.50	3.41	2.56
Horizontal	Downward	6.13	2.10	1.25

None of these is accurate for cases that deviate substantially from the assumptions made. Back to the theory, then, to define case-relevant inside surface film coefficients. If necessary, radiation and convection must remain decoupled.

How do we determine the reference temperature? Calculating the radiant temperature is quite complex, which is why, provided the room is beam-shaped and all surfaces are grey with long-wave emissivity ≈0.9, the area-weighted average surface temperature is used as an acceptable estimate:

$$\theta_{ri} = \sum_{k=1}^{n} (A_k \theta_{sk}) / \sum_{k=1}^{n} A_k \tag{1.91}$$

If the surface considered is part of a vertical, sloped or horizontal envelope assembly, the last with upward heat flow, and if it behaves as a grey body, the reference temperature becomes:

$$\theta_{\text{ref},i} = 0.44\,\theta_i + 0.56\,\theta_{ri}$$

a result close to the average between the central air and the radiant temperature, a value governing thermal comfort in buildings and called the operative temperature, symbol θ_o:

$$\theta_{\text{ref},i} = \theta_o = (\theta_i + \theta_{ri})/2 \tag{1.92}$$

For reflective surfaces, the reference temperature nears the central air temperature: $\theta_o \approx \theta_i$, as convection then dominates. For grey horizontal inside partition and envelope assemblies where the heat flows downward, the reference turns into:

$$\theta_{\text{ref},i} = 0.2\,\theta_i + 0.8\,\theta_{ri} \tag{1.93}$$

The larger the impact envelope assemblies with really cold inside surfaces have on the radiant temperature vertical, sloped and horizontal inside partitions face, the less evident is the use of the inside reference temperatures just defined.

1.6.1.2 Outdoors

Outdoors, three heat fluxes strike the surface. The first is convection to the outside air:

$$q_{ce} = h_{ce,j}(\overline{\theta}_{e,j} - \theta_{se})$$

with $h_{ce,j}$ the average convective surface film coefficient and $\overline{\theta}_{e,j}$ the average temperature of the air outside the boundary layer, which is usually replaced by the outside temperature measured under a thermometer hut, 1.7 m above grade, in the nearest weather station (θ_e). The average flux thus changes to:

$$q_{ce} = h_{ce}(\theta_e - \theta_{se})$$

where h_{ce} is the average convective surface film coefficient for that weather station value:

$$h_{ce} = h_{ce,j}\frac{\overline{\theta}_{e,j} - \theta_{se}}{\theta_e - \theta_{se}}$$

A second heat flux comes from long-wave radiation between the surface, the terrestrial environment and the sky, which is assumed black. The black body emittance from the surface (s) to the other two (e and sk) equals:

$$M_{bs} = \left[1 + \frac{\rho_{Ls}}{e_{Ls}}(F_{se} + F_{ssk})\right]M'_s - \frac{\rho_{Ls}}{e_{Ls}}(F_{se}M'_e + F_{ssk}M_{bsk})$$

From the environment (e) to the surface (s) and the sky (sk), it is:

$$M_{be} = \left[1 + \frac{\rho_{Le}}{e_{Le}}(F_{es} + F_{esk})\right]M'_e - \frac{\rho_{Le}}{e_{Le}}(F_{es}M'_s + F_{esk}M_{bsk})$$

In both, e_{Ls} and ρ_{Ls} are the long-wave emissivity and reflectivity of the surface, e_{Le} and ρ_{Le} the average long wave emissivity and reflectivity of the terrestrial environment, F_{se} the view factor between the surface and the environment, F_{ssk} the view factor between the surface and the sky, F_{es} the view factor between the terrestrial environment and the surface, and F_{esk} the view factor between the terrestrial environment and the sky. M_{bsk} is the black body emittance of the sky, while M'_s and M'_e are the radiosities of the surface and the terrestrial environment. As the terrestrial

1.6 Building-related applications

environment and the sky surround the surface completely, the sum of the view factors F_{se} and F_{ssk} is 1, while the view factor F_{es} is close to 0 as the surface is infinitely small compared with the terrestrial environment. The view factor F_{esk} is close to 1 since nearly all radiation from the terrestrial environment reaches the sky. The two balances thus simplify to:

$$M_{bs} = \frac{1}{e_{Ls}}M'_s - \frac{\rho_{Ls}}{e_{Ls}}(F_{se}M'_e + F_{ssk}M_{bsk}) \qquad M_{be} = \frac{1}{e_{Le}}M'_e - \frac{\rho_{Le}}{e_{Le}}M_{bsk}$$

Solving both for M'_s and inserting the result in the equation $q_{rse} + q_{rssk} = e_{Ls}(M_{bs} - M'_s)/\rho_{Ls}$, knowing that $e_{Ls}(F_{se} + F_{ssk})M_{bs} = e_{Ls}M_{bs}$, gives:

$$q_{rse} + q_{rssk} = q_{re} = e_{Ls}F_{se}(M_{bs} - e_{Le}M_{be}) + e_{Ls}F_{ssk}\left[M_{bs} - \left(\rho_{Le}\frac{F_{se}}{F_{ssk}} + 1\right)M_{bsk}\right]$$

Presume now that the terrestrial environment is a black body at outdoor temperature. Using this assumption, and the experimental fact that during clear nights the sky temperature drops some 21 °C below the air temperature in the atmospheric boundary layer, simplifies the radiant heat flux between the surface and the overall environment outdoors to:

$$q_{rs} = e_{Ls}C_b[(F_{se}F_{Tse} + F_{ssk}F_{Tssk})(\theta_e - \theta_{se}) - 21F_{ssk}F_{Tssk}(1 - 0.87c)]$$

where F_{Tse} is the temperature ratio for radiation between the surface and the terrestrial environment, F_{Tws} is the temperature ratio for radiation between the surface and the sky, and c the cloudiness factor (0 for a clear sky and 1 for an overcast sky).

A third heat flux comes from the sun. Each square metre of exterior surface absorbs the solar beam, diffuse and reflected short-wave radiation (E_{ST}) proportionally to its short-wave absorptivity (α_S):

$$q_{se} = \alpha_S E_{ST}$$

Summing up the three heat fluxes gives:

$$q_e = h_{ce}(\theta_e - \theta_{se}) + 5.67e_{Ls}(F_{se}F_{Tse} + F_{ssk}F_{Tssk})(\theta_e - \theta_{se})$$
$$- 120\, e_{Ls}F_{ssk}F_{Tssk}(1 - f_c) + \alpha_K E_{ST}$$

With the outside surface film coefficient for radiation (h_{re}) equal to:

$$h_{re} = 5.76 e_{Ls}(F_{se}F_{Tse} + F_{ssk}F_{Tssk})$$

this equation rewrites as:

$$q_e = (h_{ce} + h_{re})\left\{\left[\theta_e + \frac{\alpha_K E_{ST} - e_{Ls}120\, F_{ssk}F_{Tssk}(1 - f_c)}{h_{ce} + h_{re}}\right] - \theta_{se}\right\} \qquad (1.94)$$

The term between brackets ([]) with units °C may act as the reference temperature and is called the (average) sol-air temperature, symbol θ_e^*, over a given time interval (1 hour, 1 day, 1 week, 1 month). Consider it as the fictive air temperature, which ensures that the heat exchanged with the outside surface equals the value obtained by solar irradiation, long-wave radiation and convection, provided that the outside convective surface film coefficient is $19\,W/(m^2.K)$. The sol-air temperature depends on the radiant properties of the surface, its inclination, its orientation, the weather, the time interval considered, and so on. Its value differs between applications. The sum $h_{ce} + h_{re}$ yet shapes the outside surface film coefficient h_e, units $W/(m^2.K)$.

The flux equation above then becomes:

$$q_e = h_e(\theta_e^* - \theta_{se}) \tag{1.95}$$

Replacement of the temperature ratios for radiation F_{Tse} and F_{Tssk} by one value F_T, and from the fact that $F_{se} + F_{ssk} = 1$, the surface film coefficient for radiation (h_{re}) simplifies to:

$$h_{re} = 5.67\, e_L F_T \tag{1.96}$$

As the temperature factor F_T ranges between 0.8 and 0.9, a probable interval for its value is $4.4e_L \leq h_{re} \leq 5.1e_L\, W/(m^2.K)$.

Provided that outside surfaces are grey with a long-wave emissivity of 0.9, h_{re} of 4–$4.6\,W/(m^2.K)$ looks most likely. Thus, the standard outside surface film coefficient for heat transfer could be some $23\,W/(m^2.K)$. The EN standard takes $25\,W/(m^2.K)$, while the 2017 *ASHRAE Handbook of Fundamentals* makes a distinction between winter and summer:

	Direction of heat flow	$h_e\,(W/(m^2.K)$
Winter (wind speed 6.7 m/s)	Any	34.0
Summer (wind speed 3.4 m/s)	Any	22.7

For more accurate numerics one should return to the complete heat balance, including a more precise evaluation of the mean wind velocity. Long-term measurements in the early 1980s by the Laboratory of Building Physics, KULeuven, at the leeward side of an existing building for example gave a much lower average than $25\,W/(m^2.K)$.

1.6.2 Steady state: flat assemblies

1.6.2.1 Thermal transmittance of envelope parts and partitions

The use of surface film coefficients simplifies the calculation of the steady-state heat flux ambient-to-ambient across flat assemblies. Consider the outer wall in Figure 1.31.

1.6 Building-related applications

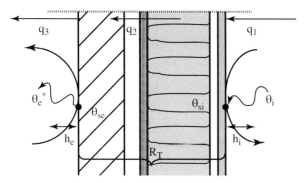

Fig. 1.31 Outer wall, thermal transmittance

Indoors, the reference temperature is θ_o, while outdoors it is θ_e^*. Assume a heated indoors and cold outdoors. From indoors to the inside surface the heat flux is:

$$q_1 = h_i(\theta_o - \theta_{si})$$

with θ_{si} the inside surface temperature. Across the assembly the heat flux equals:

$$q_2 = (\theta_{si} - \theta_{se})/R_T$$

with θ_{se} the outside surface temperature and R_T the total thermal resistance of the assembly. From the outside surface to outdoors, the heat flux is:

$$q_3 = h_e(\theta_{se} - \theta_e^*)$$

In steady state, the three must be equal, with a common value q. Rearrangement and addition gives:

$$q/h_i = \theta_o - \theta_{si}$$
$$qR_T = \theta_{si} - \theta_{se}$$
$$q/h_e = \theta_{se} - \theta_e^*$$

Sum: $q(1/h_i + R_T + 1/h_e) = \theta_o - \theta_e^*$

a result that is rewritten as $q = U(\theta_o - \theta_e^*)$ with:

$$U = \frac{1}{1/h_e + R + 1/h_i} \qquad (1.97)$$

The quantity U is called the thermal transmittance of the assembly, units W/(m².K). The lower that value, the less heat passes across, or the property reflects the insulation quality of a wall, roof or floor, separating the indoors from outdoors.

Fig. 1.32 Masonry, clearly not a layer where the heat flux develops one-dimensionally

Thus a calculation for ambient-to-ambient, accounting for radiation and convection at the inside and outside surfaces, looks simple. It suffices to add two surface resistances to the total resistance:

- indoors, a value $1/h_i$, denoted R_i, equal to $0.13\,\mathrm{m^2.K/W}$ for vertical surfaces, $0.1\,\mathrm{m^2.K/W}$ for sloped and horizontal surfaces if the heat flows upwards, and $0.17\,\mathrm{m^2.K/W}$ if downwards.
- outdoors, one value suffices for $1/h_e$, denoted R_e: $0.04\,\mathrm{m^2.K/W}$ independent of the slope and flow direction.

The inverse of the thermal transmittance is called the thermal resistance ambient-to-ambient, symbol R_a, units $\mathrm{m^2.K/W}$. The thermal transmittance as defined has the prefix 'clear wall', because possible two- and three-dimensional effects are not considered. However, due to voids in the mortar joints and vertical perforations in the bricks, heat transfer across a masonry wall for example already develops three-dimensionally (see Figure 1.32).

For inside partitions, the surface film coefficient at both sides is the value inside (h_i), giving, as thermal transmittance:

$$U = 1/(R + 2/h_i) \tag{1.98}$$

Reference temperatures are operative on both sides of the partition.

1.6.2.2 Average thermal transmittance of parts in parallel

Consider an assembly with surface A_T, composed of n parts in parallel with surfaces A_i that all face the outdoors (Figure 1.33).

Each part is different. If lateral conduction between parts is negligible and if all face the same reference temperature indoors, the heat flow across each is:

$$\Phi_i = U_i\, A_i\, \Delta\theta$$

1.6 Building-related applications

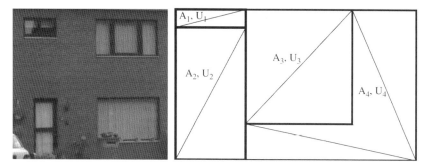

Fig. 1.33 Assembly composed of *n* parallel parts with hardly any lateral heat exchange

In case they face different operative temperatures indoors ($\theta_{o,j}$), that equation converts to:

$$\Phi_i = a_i U_i A_i \Delta\theta$$

with a_i a reduction factor equal to:

$$a_i = \frac{\theta_{o,j} - \theta_e}{\theta_{o,ref} - \theta_e}$$

with $\theta_{o,ref}$ the indoor operative temperature taken as the reference. The heat flow across the whole now equals the sum of the flows across each, or:

$$\Phi_T = \sum_{i=1}^{n} \Phi_i = \Delta\theta \sum_{i=1}^{n} (a_i U_i A_i) \tag{1.99}$$

Rewriting gives:

$$\Phi_T = \Delta\theta \, U_m \sum_{i=1}^{n} A_i \tag{1.100}$$

where U_m is the average clear-wall thermal transmittance of *n* parts in parallel, a value equal to the surface weighted average of their (clear wall) thermal transmittances:

$$U_m = \sum_{i=1}^{n} (a_i U_i A_i) / \sum_{i=1}^{n} A_i = \sum_{i=1}^{n} (a_i U_i A_i)/A_T \tag{1.101}$$

Conversion to resistances gives:

$$R_{am} = A_T / \sum_{i=1}^{n} \left(\frac{a_i A_i}{R_{ai}}\right) \tag{1.102}$$

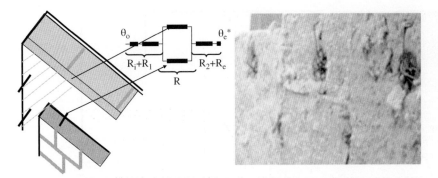

Fig. 1.34 Cavity wall, with an electrical analogy accounting for the ties that perforate the fill

1.6.2.3 Electrical analogy

As long as lateral conduction between parts is negligible, an electrical analogy allows us to solve quite complex cases. Consider a cavity wall. The ties perforate the fill. If A_t is the tie section, R_t their thermal resistance, A_{is} is the wall area and R_{is} is the thermal resistance of the insulation, then the overall thermal resistance (R) of the insulation equals (Figure 1.34):

$$R = \frac{A_{is}}{\dfrac{A_{is} - A_t}{R_{is}} + \dfrac{A_t}{R_t}}$$

With $R_1 + R_i$ the thermal resistance from the insulation to indoors and $R_2 + R_e$ the thermal resistance from the insulation to outdoors, the value using the two ambients becomes:

$$R_T = (R_i + R_1) + R + (R_2 + R_e) = (R_i + R_1) + \frac{A_i}{\dfrac{A_i - A_t}{R_{is}} + \dfrac{A_t}{R_t}} + (R_2 + R_e)$$

1.6.2.4 Thermal resistance of an unvented cavity

In an infinitely extending unvented cavity, the nature, distance and temperature difference between the bounding surfaces, the slope, the heat flow direction, and the mean temperature of the gas fill will all affect conduction, convection and radiation, giving as the total heat flux:

$$q_T = \left(\frac{\lambda_g \, \text{Nu}}{d} + \frac{C_b \, F_T}{1/e_{L1} + 1/e_{L2} - 1} \right)(\theta_c - \theta_{c2})$$

with λ_g the thermal conductivity of the gas, Nu the case-specific Nusselt number, e_{L1} and e_{L2} the long-wave emissivities and θ_{c1} and θ_{c2} the temperatures of the bounding

1.6 Building-related applications

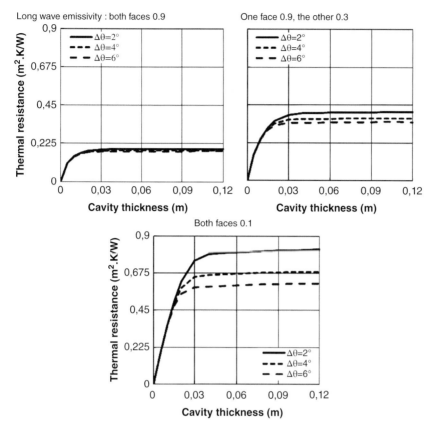

Fig. 1.35 Thermal resistance of an infinite vertical cavity for three different sets of parameters

surfaces. The thermal resistance then is:

$$R_c = \left(\frac{\lambda_g \, \text{Nu}}{d} + \frac{C_b F_T}{1/e_{L1} + 1/e_{L2} - 1}\right)^{-1} \tag{1.103}$$

Figure 1.35 gives values for a vertical air-filled cavity wherein the air temperature is 10 °C, depending on the width, the temperature difference across and the long-wave emissivity of the bounding surfaces.

The thermal resistance increases considerably with lower long-wave emissivity of the bounding surfaces, while at low emissivity the temperature difference across gains influence, underlining the dominance of radiation. The absence, in turn, of any additional gain once the cavity width passes 20–30 mm illustrates that radiation does not depend on this variable, while more convection gradually compensates for the drop in conduction. For non-vented cavities with limited dimensions, the values in Table 1.4 allow first-order calculations.

Table 1.4 Thermal resistance of cavities

Thickness (mm)	Vertical cavity		Horizontal cavity	
	R_c (m² K/W), both surfaces grey	R_c (m² K/W), one surface reflecting	R_c (m² K/W), heat flow up	R_c (m² K/W), heat flow down
$0 < d < 5$	0.00		0.00	0.00
$5 \leq d < 7$	0.11		0.11	0.11
$7 \leq d < 10$	0.13		0.13	0.13
$10 \leq d < 15$	0.15		0.15	0.15
$15 \leq d < 25$	0.16	0.35	0.17	0.17
$25 \leq d < 50$	0.16	0.35	0.17	0.19
$50 \leq d < 100$	0.16	0.35	0.18	0.21
$100 \leq d < 300$	0.16	0.35	0.18	0.22
$d > 300$	0.16	0.35	0.18	0.23

1.6.2.5 Interface temperatures

Both surface resistances R_i and R_e resemble the thermal resistance of a 1 m thick air layer with thermal conductivity h_i or h_e, whereby the reference temperatures act as fictitious 'surface temperatures' that activate the steady-state heat transfer. So for any composite envelope assembly, temperatures in the [R, θ] plane lie on a straight line linking $[0, \theta_e^*]$ to $[R_a, \theta_o]$ or, for inside partitions, linking $[0, \theta_{o,1}]$ to $[R_a, \theta_{o,2}]$ with both slopes, giving the heat flux When tracing the layer sequence, both surface resistances must be respected (see Figure 1.36).

The temperature on the inside surface is:

Envelope assembly: $\quad \theta_{si} = \theta_o - R_i \dfrac{\theta_o - \theta_e^*}{R_a} = \theta_o - \dfrac{U_{h_i}}{h_i}(\theta_o - \theta_e^*)$ (1.104)

Inside partition: $\quad \theta_{si} = \theta_{i.1} - R_i \dfrac{\theta_{o,1} - \theta_{o,2}}{R_a} = \theta_{o,1} - \dfrac{U_{h_i}}{h_i}(\theta_{o,1} - \theta_{o,2})$ (1.105)

The suffix h_i underlines that the clear wall thermal transmittance must be calculated using the surface film coefficient in the denominator. The temperatures at the interfaces equal:

$$\theta_x = \theta_i - q(R_i + R_{si}^x)$$

1.6 Building-related applications

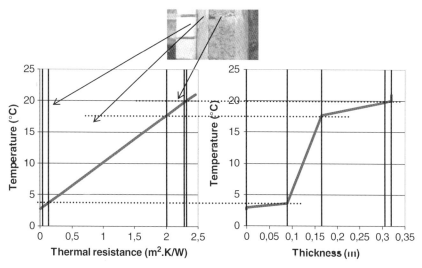

Fig. 1.36 Composite envelope assembly (filled cavity wall): temperature curve

1.6.2.6 Solar transmittance

The solar heat flux across any envelope part can be written as:

$$q_S = gE_{ST} \tag{1.106}$$

with E_{ST} the total incident solar radiation on the outside face and q_S the heat flux transmitted, both in W/m². The factor g, called the solar transmittance, encompasses the direct and indirect solar gains. The direct ones are:

$$q_{Sd} = \tau_S E_{ST}$$

with τ_S the total shortwave transmissivity of the part. For opaque parts, transmissivity and the direct gains are zero, but not so for transparent parts.

Indirect gains occur because opaque and transparent parts absorb a fraction of the solar flux impinging. They so warm up and conduct absorbed heat to the inside, where convection and long wave radiation dissipates it. For single glass with shortwave absorptivity α_S, estimating the indirect gains is easy as they are part of the heat flux dissipated by convection and radiation from the inside surface to the indoors:

$$q_{Si} = h_i(\theta_{si} - \theta_o) \tag{1.107}$$

where θ_{si} is the unknown inside surface temperature. Assuming that the glass is equally warm gives $\theta_{si} = \theta_x = \theta_{se}$ with θ_{se} the outside surface temperature and θ_x the glass temperature. The thermal balance for 1 m² of glass thus becomes (sum of the absorbed solar radiation, the heat flux outdoors and the heat flux indoors to the glass being zero):

$$\alpha_S E_{ST} + h_e(\theta_e - \theta_x) + h_i(\theta_o - \theta_x) = 0$$

with θ_e the outdoor air and θ_o the indoor operative temperature. The glass temperature thus is:

$$\theta_x = \theta_{si} = \frac{\alpha_S E_{ST}}{h_i + h_e} + \frac{h_e \theta_e + h_i \theta_i}{h_i + h_e}$$

The second term on the right-hand side stands for that temperature if single glass could not absorb solar radiation, and the first for the increase due to the fact that it does. Combining this with the equation for the heat flux to the indoors gives:

$$q_{Si} = \frac{h_i \alpha_S E_{ST}}{h_i + h_e} + \frac{h_i h_e (\theta_e - \theta_i)}{h_i + h_e} \tag{1.108}$$

Only the first term on the right is linked to the sun, thus representing the indirect gains, or:

$$q_{Si} = \frac{h_i \alpha_S E_{ST}}{h_i + h_e}$$

The solar transmittance for single glass thus equals:

$$g = \frac{q_{Sd} + q_{Si}}{E_{ST}} = \tau_S + \frac{\alpha_S}{1 + h_e/h_i} \tag{1.109}$$

The gains not only depend on the short-wave transmissivity but also on the short-wave absorptivity of the glass and the ease by which the absorbed heat dissipates to the indoors. In fact, a lower ratio between the outside and inside surface film coefficient increases the transmittance.

For double glass, calculating the solar transmittance is more demanding. Let τ_{S1}, ρ_{S1} and α_{S1} be the transmissivity, reflectivity and absorptivity, all shortwave, of one pane, and τ_{S2}, ρ_{S2} and α_{S2} for the other pane. Reflection in the cavity breaks the transmission of solar radiation (Figure 1.37) into a geometric series with ratio $\rho_{S1}\rho_{S2}$, whose sum gives:

$$q_{Sd} = \frac{\tau_{S1} \tau_{S2}}{1 - \rho_{S1}\rho_{S2}} E_{ST} \tag{1.110}$$

In general, the denominator $1 - \rho_{S1}\rho_{S2}$ nears 1. So the guide that the product of the transmissivities of both panes fixes the transmissivity of double glass is quite correct as a rule of thumb.

1.6 Building-related applications

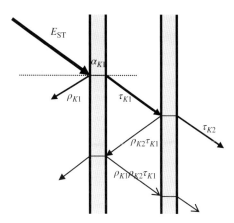

Fig. 1.37 Double glass, solar transmittance

Assuming $\theta_e = \theta_o = 0\,°C$ and both panes are isothermal, the indirect gains can be written as:

$$q_{Si} = h_i \theta_{x2}$$

with θ_{x2} the sun-induced temperature of the inside pane, a value ensuing from the heat balance per pane (1 is outside, 2 inside):

Pane 1: $\alpha_{S1} \underbrace{\dfrac{1 - \rho_{S1}\rho_{S2} + \tau_{S1}\rho_{S2}}{1 - \rho_{S1}\rho_{S2}}}_{f_1} E_{ST} - h_e \theta_{x1} + \dfrac{\theta_{x2} - \theta_{x1}}{R_c} = 0$

Pane 2: $\alpha_{S2} \underbrace{\dfrac{\tau_{S1}}{1 - \rho_{S1}\rho_{S2}}}_{f_2} E_{ST} + \dfrac{\theta_{x1} - \theta_{x2}}{R_c} - h_i \theta_{x2} = 0$

This system has as a solution:

$$\theta_{x2} = \dfrac{\dfrac{\alpha_{S1} f_1}{R_c} + \alpha_{S2} f_2 \left(h_e + \dfrac{1}{R_c}\right)}{\left(h_i + \dfrac{1}{R_c}\right)\left(h_e + \dfrac{1}{R_c}\right) - \dfrac{1}{R_c^2}} E_{ST} \qquad (1.111)$$

Inserting the outcome into the equation for the indirect gains gives, as solar transmissivity:

$$g = \tau_{S1}\tau_{S2} + h_i \dfrac{\dfrac{\alpha_{S1} f_1}{R_c} + \alpha_{S2} f_2 \left(h_e + \dfrac{1}{R_c}\right)}{\left(h_i + \dfrac{1}{R_c}\right)\left(h_e + \dfrac{1}{R_c}\right) - \dfrac{1}{R_c^2}} \qquad (1.112)$$

The result shows how to decrease the gains across double glazing: limit the direct transmission and lower either the inside surface film coefficient or the short-wave absorptivity of both panes. For multiple glazing including a shade, a same approach applies, with the short-wave transmissivity equal to the product of the short-wave transmissivities of all panes and the shade, and the indirect gains derived from a system of heat balances per pane and the shade.

1.6.3 Local inside surface film coefficients

Average surface film coefficients are usable when area-average phenomena such as the heat lost or gained are at stake. To quantify temperatures or heat fluxes at specific spots, local values are preferred. In general, the following holds:

$$h_{ix}(\theta_{ref,i} - \theta_{six}) = h_{cix}(\theta_{ix} - \theta_{six}) + h_{rix}(\theta_{rix} - \theta_{six}) \tag{1.113}$$

with h_{ix} the local surface film coefficient linked to a reference temperature $\theta_{ref,i}$, h_{cix} the local convective surface film coefficient, θ_{ix} the local air temperature outside the boundary layer, θ_{six} the local surface temperature, h_{rix} the local radiant surface film coefficient, and θ_{rix} the radiant temperature the spot faces. If R' is an equivalent thermal resistance that links the inside surface of that spot to the environment on the other side, then:

$$h_{ix}(\theta_{ref,i} - \theta_{six}) = (\theta_{six} - \theta_j)/R' \tag{1.114}$$

This equation is an approximation. In fact, the equivalent thermal resistance depends on the distribution of the local inside surface film coefficients (h_{ix}) over the whole surface. For the envelope, θ_j is the temperature outdoors (j = e), while for inside partitions, it is the reference temperature in the neighbouring space. Eliminating the local surface temperature θ_{six} from both equations and solving for the local surface film coefficient gives:

$$h_{ix} = \frac{h_{cix} + h_{rix} - p_T}{1 + R'p_T} \tag{1.115}$$

where:

$$p_T = \frac{h_{cix}(\theta_{ref,i} - \theta_{ix}) + h_{rix}(\theta_{ref,i} - \theta_{rix})}{\theta_{ref,i} - \theta_j} \tag{1.116}$$

If the reference temperature indoors ($\theta_{ref,i}$), the relationship with the local air temperature just outside the boundary layer (θ_{ix}), the relationship with the radiant temperature facing the spot (θ_{rix}) and the local inside surface film coefficients h_{cix} and h_{rix} are known, then combining the equations above gives the heat flux at the spot considered, on condition that the equivalent thermal resistance is known. The questions left are how to link the local reference temperature indoors to the overall

1.6 Building-related applications

reference indoors ($\theta_{\text{ref},i}$), and what values to use for the local surface film coefficients.

As overall reference indoors, the air temperature in the room's centre, 1.7 m above floor level, is chosen (θ_i). Assuming the local air temperature increases linearly along the room's height with little slope if well insulated and less convective heated, then the relation with the overall reference could be:

$$\frac{\theta_{ix} - \theta_j}{\theta_i - \theta_j} = 1 + 0.2 p_c U_m (y - 1.7) \tag{1.117}$$

with y the height ordinate, θ_j the reference temperature in the neighbouring room or outdoors, p_c a convection factor (1 for air heating, 0.9 for convectors, 0.4 to 0.8 for radiators, 0.4 for floor heating), and U_m the weighted average thermal transmittance of all walls in the room. The relationship reflects the outcome of a series of measurements in a test room, where the heating system and the insulation of the outside walls could be varied.

For the local radiant temperature (θ_{rix}) a uniform value is assumed, proportional to the overall reference value, with a gradient depending on the local convective surface film coefficient, the convection factor and the weighted average thermal transmittance of all walls in the room:

$$\frac{\theta_{rix} - \theta_i}{\theta_i - \theta_j} = \frac{h_{cix}}{h_{cix} + \dfrac{(p_c - 0.4) U_m}{0.6}} \tag{1.118}$$

This equation followed from computer simulations of the radiant heat exchange in rooms of different shapes.

The local convective surface film coefficient (h_{cix}) is set at 2.5 W/(m².K), while the following values are used for the local radiant surface film coefficient (h_{rix}):

Corner between three envelope assemblies or two and an inside partition:
    5.5 e_L Surfaces more than 0.5 m from the edge lines
    3.4 e_L Surfaces less than 0.5 m from the edge lines but more than 0.5 m from the corner
    2.2 e_L Surfaces less than 0.5 m from the corner

Edge between two envelope assemblies or one and an inside partition:
    5.5 e_L Surfaces more than 0.5 m from the edge line
    3.4 e_L Surfaces less than 0.5 m from the edge line

Envelope assembly or inside partition:
    5.5 e_L –

Where furniture hides a wall, a combined inside surface film coefficient of 2 W/(m².K) applies.

1.6.4 Steady state: two and three dimensions

1.6.4.1 Pipes

At the outside surface of a pipe, the heat flow equals:

$$\Phi_{n+1} = 2\pi R_{n+1} h_2 (\theta_{s,2} - \theta_{ref,2}) \tag{1.119}$$

with h_2 the surface film coefficient outside the pipe, $\theta_{ref,2}$ the reference temperature there, $\theta_{s,2}$ the outside surface temperature and R_{n+1} the outside radius. At the inside surface the flow is:

$$\Phi_1 = 2\pi R_1 h_1 (\theta_{ref,1} - \theta_{s,1}) \tag{1.120}$$

with h_1 the surface film coefficient between fluid and pipe, $\theta_{ref,1}$ the temperature of the fluid, $\theta_{s,1}$ the inside surface temperature and R_1 the inside radius. Across the pipe, the flow numbers:

$$\Phi_{1,n+1} = \frac{\theta_{s,1} - \theta_{s,2}}{\sum_{i=1}^{n} \left[\frac{\ln(R_{i+1}/R_i)}{2\pi \lambda_i} \right]}$$

with Σ indicating that the pipe wall may consist of several layers. In steady state, the three heat flows are equal. Rearrangement and addition gives:

$$\frac{\Phi_{n+1}}{2\pi R_{n+1} h_2} = \theta_{s,2} - \theta_{ref,2}$$

$$\Phi_{1,n+1} \sum_{i=1}^{n} \left[\frac{\ln(R_{i+1}/R_i)}{2\pi \lambda_i} \right] = \theta_{s,1} - \theta_{s,2}$$

$$\frac{\Phi_1}{\theta_{ref,1} - \theta_{s,1}} = 2\pi R_1 h_1$$

Sum: $\quad \Phi_{1,n+1} \left\{ \dfrac{1}{2\pi R_{n+1} h_2} + \sum_{i=1}^{n} \left[\dfrac{\ln(R_{i+1}/R_i)}{2\pi \lambda_i} \right] + \dfrac{1}{2\pi R_1 h_1} \right\} = \theta_{ref,1} - \theta_{ref,2}$

For flat assemblies, this sum is rewritten as:

$$\Phi_{1,n+1} = U_{pipe} (\theta_{ref,1} - \theta_{ref,2})$$

where U_{pipe} stands for the thermal transmittance, now per metre run, of the pipe:

$$U_{\text{pipe}} = \frac{1}{\dfrac{1}{2\pi R_{n+1} h_2} + \sum_{i=1}^{n}\left[\dfrac{\ln(R_{i+1}/R_i)}{2\pi \lambda_i}\right] + \dfrac{1}{2\pi R_1 h_1}} \quad (\text{W}/(\text{m.K})) \quad (1.121)$$

Insulation will lower the heat loss of pipes transporting warm fluids, or the gains of pipes transporting cold fluids. A difference is that the additional benefit of thicker insulation drops off more rapidly than for flat assemblies.

1.6.4.2 Floors on grade

Calculation of the thermal transmittance of a floor on grade is a typical example of a three-dimensional heat flow problem solved using a simplified method. The thermal transmittance is written as:

$$U = aU_{\text{o,floor}} \quad (1.122)$$

with a being a reduction factor and $U_{\text{o,floor}}$ the thermal transmittance of the floor as if it were a flat assembly facing the outside. Valuing the reduction factor starts with fixing what is called the characteristic floor dimension:

$$B' = 2A_{\text{fl}}/P \quad (\text{m}) \quad (1.123)$$

with A_{fl} the floor's surface area and P that part of the floor's perimeter touching the outdoors and called the free perimeter (see Figure 1.38).

Then the equivalent soil thickness (d_t) of the floor that replaces its thermal resistance is fixed:

$$d_t = d_{\text{fw}} + \lambda_{\text{gr}}\left(\frac{1}{h_e} + R_{\text{T,fl}} + \frac{1}{h_i}\right) \quad (\text{m}) \quad (1.124)$$

Fig. 1.38 The arrow shows a part of the free perimeter

with d_{fw} the average thickness of the foundation walls along the free perimeter in m, λ_{gr} the thermal conductivity of the soil, $R_{T,fl}$ the thermal resistance of the floor, h_i the surface film coefficient indoors, 6 W/(m².K), and h_e the surface film coefficient outdoors, 25 W/(m².K).

Finally, the reduction factor, which depends on the ratio between the equivalent soil thickness and the characteristic floor dimension, follows from:

$$\text{For } d_t < B': \quad a = \frac{1}{U_{o,floor}} \left(\frac{2\lambda_{gr}}{\pi B' + d_t} \right) \ln\left(\frac{\pi B'}{d_t} + 1 \right)$$

$$\text{For } d_t \geq B': \quad a = \frac{1}{U_{o,floor}} \left(\frac{\lambda_{gr}}{0.457 B' + d_t} \right)$$

(1.125)

1.6.4.3 Thermal bridges

The term thermal bridge applies to all spots in the envelope where heat flows two- or three-dimensionally. The term may be taken literally: not only is more heat lost or gained than across neighbouring flat parts, except for single glazing, but during the heating season the inside surface will also stay colder there. CVM is used to calculate the heat exchange with the ambient using surface film resistances, through which the heat moves normally to the end faces. Where local surface temperatures are of interest, local values apply. When it is the overall heat loss or gain that matters, the standard values are used. The energy balance for a control volume with a centre point on an end face combines six heat flows: four from the neighbouring centre points on the face, one from the neighbouring control volume in the material, and one across the surface film resistance with the ambient reference temperature as source. Consider Figure 1.39.

The surface of a control volume touching the end face extends parallel to the $[y, z]$ plane. The heat flow from the ambient temperature θ_1 in (i, m, n) to its centre point (s, m, n) equals:

$$\Phi_{s,m,n}^{i,m,n} = h_i\left(\theta_{i,m,n} - \theta_{s,m,n}\right)a^2$$

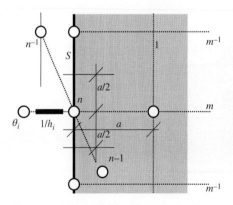

Fig. 1.39 CVM method, with control volumes at the inside or outside surface

1.6 Building-related applications

The heat flows from the four neighbouring centre points on the end face to that point are:

$$\Phi^{s,m,n}_{s,m+1,n} = \lambda_1(\theta_{s,m+1,n} - \theta_{s,m,n})\frac{a}{2} \quad \Phi^{s,m,n}_{s,m-1,n} = \lambda_1(\theta_{s,m-1,n} - \theta_{s,m,n})\frac{a}{2}$$

$$\Phi^{s,m,n}_{s,m,n+1} = \lambda_1(\theta_{s,m,n+1} - \theta_{s,m,n})\frac{a}{2} \quad \Phi^{s,m,n}_{s,m,n-1} = \lambda_1(\theta_{s,m,n-1} - \theta_{s,m,n})\frac{a}{2}$$

The heat flow from the neighbouring control volume in the material to that point is:

$$\Phi^{s,m,n}_{l,m,n} = \lambda_1(\theta_{1,m,n} - \theta_{s,m,n})a$$

Setting to zero gives:

$$a\,h_i\theta_{i,m,n} + \lambda_1\frac{\theta_{s,m+1,n} + \theta_{s,m-1,n} + \theta_{s,m,n+1} + \theta_{s,m,n-1}}{2} + \lambda_1\theta_{l,m,n} - (a\,h_i + 3\lambda_1)\theta_{s,l,m,n} = 0$$

with $a\,h_i\theta_{i,m,n}$ the known term. Mesh points in the corners and others give analogous equations. Figure 1.40 shows the results of a CVM calculation for a wall with a load-bearing outer leaf, the cavity closed at the window, and a non-load-bearing inner leaf.

In practice, a distinction is made between geometric and structural thermal bridges. The former follow from the three-dimensional nature of building enclosures. The latter are a consequence of structural decisions, such as concrete girders and columns penetrating the envelope, or discontinuities in the thermal insulation. Structural integrity often explains their existence. For example, take a balcony – continuity with the floor slab is needed to balance the cantilever moment (Figure 1.41).

Neutralizing both thermal bridge types demands continuity of the thermal insulation. Ideally, the insulation should be traceable on the drawings without crossing parts that create easy heat flow paths. Complete avoidance is often not possible, although the impact must remain manageable.

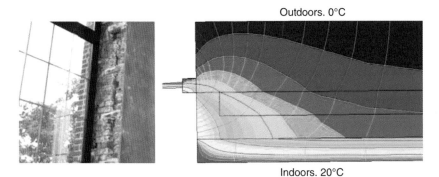

Fig. 1.40 CVM calculation result

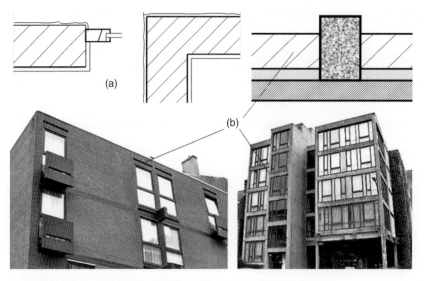

Fig. 1.41 (a) Geometric thermal bridges; (b) structural thermal bridges

Using CVM at the design stage is demanding. Therefore the concepts of a linear and a local thermal transmittance have been introduced. The first, symbol ψ, units W/(m.K), stands for the extra heat flow a two-dimensional thermal bridge gives per metre run and per kelvin temperature difference between the ambient on both sides. The second, symbol χ, units W/K, quantifies the extra heat flow a three-dimensional thermal bridge induces per kelvin temperature difference between the ambient on both sides. Calculating the first demands a well-defined one-dimensional reference and agreement on what surface to consider, inside or outside, with a preference for outside as it allows use of the facade drawings (Figure 1.42). The detail that acts as a thermal bridge is ignored at first. Otherwise,

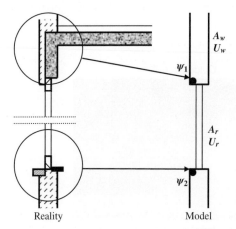

Fig. 1.42 Linear thermal transmittances. The dummy consists of flat parts with lines perpendicular to the section representing the linear thermal bridges

1.6 Building-related applications

a flat dummy replaces reality, with a dot where the thermal bridge sits, and the one-dimensional heat loss is calculated.

Then, using the correct structural drawings the real two-dimensional heat transfer and inside surface temperatures are calculated. When Φ_{2D} is the two-dimensional heat flow and Φ_o the heat flow across the flat dummy, the linear thermal transmittance equals:

$$\psi = \frac{\Phi_{2D} - \Phi_o}{L \Delta\theta} \qquad (1.126)$$

with L the length of the thermal bridge. If an assembly only contains a local thermal bridge, the local thermal transmittance (χ) becomes:

$$\psi = \frac{\Phi_{3D} - \Phi_o}{L \Delta\theta} \qquad (1.127)$$

Often, a local thermal bridge emerges where linear ones cross. If so, two references must be calculated: first the one-dimensional, and then including the linear thermal bridges. Then the local transmittance is extracted from a three-dimensional calculation:

$$\chi = \frac{\Phi_{3D} - \Phi_{2D}}{\Delta\theta} \qquad (1.128)$$

Once all linear and local thermal transmittances are known, the whole-wall thermal transmittance of a flat assembly with thermal bridges follows from:

$$U = U_o + \frac{\sum_{i=1}^{n}(\psi_i L_i) + \sum_{j=1}^{m} \chi_j}{A} \qquad (1.129)$$

where U_o is its clear wall thermal transmittance, A the surface area considered, n the number of linear thermal bridges, L_i their length and m the number of local thermal bridges.

The lowest inside surface temperature ($\theta_{s,min}$) that an envelope-related thermal bridge calculation gives is mostly transposed into a non-dimensional temperature factor:

$$f_{h_i} = \frac{\theta_{s,min} - \theta_e}{\theta_o - \theta_e} \qquad (1.130)$$

with θ_o the reference temperature indoors and θ_e the one outdoors. The suffix h_i is a reminder that the local surface film coefficient must be used when calculating surface temperatures. A CVM calculation with a 1 K temperature difference between the environments on either side directly gives the temperature factor.

The higher the linear or local thermal transmittance and the lower the temperature factor, the more problematic is a thermal bridge. The inside surface will collect more dirt, have a greater mould risk, may become a preferred spot for surface condensation and see crack sensitivity increase, while taking a disproportionate share in the heat loss or gain.

Thermal bridge catalogues have been published that contain the linear thermal transmittances, local thermal transmittances and temperature factors for reveals, lintels, dormer windows, balconies, and so forth, considering various designs, material combinations and layer thicknesses. Interactive CD-ROMs and software tools to calculate two- and three-dimensional heat flows and temperature fields are also available.

1.6.4.4 Windows

Windows transfer heat three-dimensionally, as the IR picture in Figure 1.43 shows.

Calculating the thermal transmittance of a window (U_{window}) thus requires appropriate software tools. However, as this is not really practical, frames are characterized by an equivalent thermal transmittance ($U_{eq,frame}$), multi-pane glass by a central thermal transmittance ($U_{o,glass}$), and the glazing/spacer/frame combination by a linear thermal transmittance (ψ_{spacer}). This allows us to write:

$$U_{window} = \frac{A_{glass} U_{o,glass} + A_{frame} U_{eq,frame} + \psi_{spacer} L_{spacer}}{A_{window}} \tag{1.131}$$

The surface taken by the frame (A_{frame}) coincides with its normal projection onto an outside plane parallel to the window. The visible glass surface (A_{glass}) is defined the same way, while the length of the spacer (L_{spacer}) equals the total perimeter of all

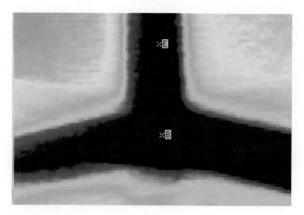

Fig. 1.43 Window: IR picture of the frame and the double glass

1.6 Building-related applications

Fig. 1.44 Window, thermal transmittance

glazing parts, measured out to out (see Figure 1.44). Table 1.5 lists approximate thermal transmittances and linear thermal transmittances for different types of frames, glazing and spacers.

1.6.4.5 Building envelopes

Building envelopes, also called building enclosures, shield the indoors from the outdoors, from unheated neighbouring spaces, sometimes from water volumes and from the soil. The assemblies forming the enclosure encompass low-slope and sloped roofs, outside walls, walls separating heated from unheated spaces, glazed surfaces, floors on grade, floors above crawlspaces, floors above unheated basements, floors separating heated from unheated spaces and floors separating the indoors from the outdoors (see Figure 1.45). For party walls, the assumption typically but not necessarily is that both buildings are at a same temperature.

Envelopes are by definition three-dimensional. Quantifying the time-averaged heat flow for a 1 °C difference with the outdoors is done by decomposing the envelope

Table 1.5 Frames, glass and spacers, thermal transmittances and linear thermal transmittances

Window frames	U_{frame} W/(m².K)	Glazing	U_{glass} W/(m².K)
Hardwood, $d = 70$ mm	2.08	Double	2.8
Aluminium, 20 mm thermal cut	2.75	Double, low-e, argon-filled	1.1
PVC, three room frame	2.00	Triple, low-e, argon-filled	0.6
Spacers			
Metal	ψ W/(m.K)	Insulating	ψ W/(m.K)
$U_{\text{frame}} < 5.9$, $U_{\text{glass}} > 2.0$ W/(m².K)	0.06	$U_{\text{frame}} < 5.9$, $U_{\text{glass}} > 2.0$ W/(m².K)	0.05
$U_{\text{frame}} < 5.9$, $U_{\text{glass}} < 2.0$ W/(m².K)	0.11	$U_{\text{frame}} < 5.9$, $U_{\text{glass}} < 2.0$ W/(m².K)	0.07

Fig. 1.45 Building envelope

into flat and curved parts with area A_j and clear thermal transmittance $U_{o,j}$ coupled in parallel. The contact lines and linear details, each with length L_k, are represented by linear thermal transmittances, while all spots where heat flows three-dimensionally get local thermal transmittances. This gives, as average thermal transmittance of an envelope:

$$U_m = \frac{\sum_{j=1}^{n}(a_j A_j U_{o,j}) + \sum_{k=1}^{m}(a_k L_k \psi_k) + \sum_{l=1}^{p}(a_l \chi_l)}{A_T} \quad (1.132)$$

In this formula, a_j, a_k and a_l are reduction factors. A value of 1 stands for parts separating the indoors from the outdoors, a value below 1 for parts separating the building from unheated neighbouring spaces, for floors on grade (see above), floors above unheated basements, floors above crawlspaces and vertical walls contacting the soil. For party walls the value is typically 0. Contact with water gives:

$$a = \frac{1}{1 - 0.04 U_o}$$

Still, how to measure surfaces and lengths has to be decided upon. The outside dimensions are handy, as these are available from the facade drawings. When for reasons of simplicity, linear and local thermal transmittances are overlooked, then out-to-out gives the smallest error, although it is best not to overlook thermal bridging.

Bad workmanship can seriously degrade the real average thermal transmittance, compared with what was calculated. A formula that reflects this has the form:

$$U_m = \frac{\sum_{j=1}^{n}(a_j A_j U_{o,j}/\eta_{\text{ins},j}) + \sum_{k=1}^{m}(a_k L_k \psi_k/\eta_{\text{ins},k}) + \sum_{l=1}^{p}(a_l \chi_l/\eta_{\text{ins},l})}{A_T} \quad (1.133)$$

1.6 Building-related applications

with $\eta_{ins,j}$, $\eta_{ins,k}$ and $\eta_{ins,l}$ the insulation efficiencies, with values of 1 for perfect workmanship, but far below 1 for poor workmanship allowing air looping around, wind washing behind and indoor air washing in front of the insulation.

1.6.5 Heat balances

The use of surface film coefficients does not reflect reality in an exact way. In case the concepts do not work, a return to and solution of the separate heat balances is preferred. First, the surfaces or interfaces where the temperature and heat flows or fluxes are the unknowns are selected. These form the calculation points, with their number defining the number of balances needed. Then, in each calculation point, conservation of energy applies: the sum of all heat flows coming from the environment or neighbouring volumes equals zero. In this way, each calculation point supplies an equation in which its temperature and some or all unknown temperatures of the neighbouring points feature as variables, and the known temperatures are given. Solving the system then gives the requested temperatures and heat flows or fluxes. The challenge lies in not overlooking any intervening heat flows or fluxes.

1.6.6 Transient

1.6.6.1 Periodic: flat assemblies

To calculate the periodic response from environment to environment, the surface film resistances are again assumed to represent 1 m thick air layers with thermal conductivity h_i and h_e, but a volumetric specific heat capacity of 0. The reference temperature of either thus becomes a fictitious surface temperature. For both, the following apply:

$$\omega_n = \sqrt{\frac{2in\pi\rho c\lambda}{T}} = 0 \qquad \cosh(\omega_n R) = 1$$

$$\omega_n \sinh(\omega_n R) = 0 \qquad \frac{\sinh(\omega_n R)}{\omega_n} = \frac{0}{0} = \lim_{n \to \infty}\left(\frac{\sinh(\omega_n R)}{\omega_n}\right) = R$$

turning the complex surface matrixes into:

$$W_i = \begin{bmatrix} 1 & 1/h_i \\ 0 & 1 \end{bmatrix} \qquad W_e = \begin{bmatrix} 1 & 1/h_e \\ 0 & 1 \end{bmatrix}$$

Transposition into real matrices gives:

$$W_i = \begin{bmatrix} 1 & 0 & 1/h_i & 0 \\ 0 & 1 & 0 & 1/h_i \\ 0 & 0 & 1 & 0 \\ 0 & 0 & 0 & 1 \end{bmatrix} \qquad W_e = \begin{bmatrix} 1 & 0 & 1/h_e & 0 \\ 0 & 1 & 0 & 1/h_e \\ 0 & 0 & 1 & 0 \\ 0 & 0 & 0 & 1 \end{bmatrix} \qquad (1.134)$$

For an envelope assembly, the system matrix environment to environment thus becomes:

$$W_{na} = W_i \, W_{n1} \, W_{n2} \, W_{n3} \, \ldots \, W_{nn} \, W_e$$

For an inside partition, it changes to:

$$W_{na} = W_i \, W_{n1} \, W_{n2} \, W_{n3} \, \ldots \, W_{nn} \, W_i$$

For single-layer assemblies, these products reduce to $W_{na} = W_i \, W_n \, W_e$ when part of the envelope, and $W_{na} = W_i \, W_n \, W_i$ when an inside partition.

All this, of course, is a simplification. On the one hand, the radiant part of the surface film resistance involves all other faces seen by the surface; on the other hand, due to the volumetric specific heat capacity of air, limited air velocity and the interactions with other surfaces and furniture, some inertia is involved.

1.6.6.2 Periodic: spaces

Assume that the envelope and partitions enclosing a space can be decomposed into parallel flat assemblies whereby the windows lack thermal inertia. To simplify the calculations, ventilation accounts for a constant outside airflow, while air exchanges with neighbouring spaces are lacking and the solar and internal gains get injected in the space's centre. That centre's operative temperature θ_o is thermally linked to all assemblies by surface film coefficients h_i, which combine convection and radiation (Figure 1.46).

The response to a periodic heat input consists of a zero harmonic, which equals the average response as steady-state reality, a first harmonic with as period (T) the time span of, for example, 1 day, and higher harmonics with periods $T/2$, $T/3$, and so on.

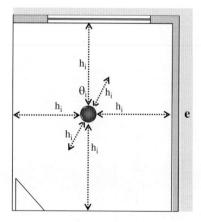

Fig. 1.46 Replacing a space by its centre point

1.6 Building-related applications

The zero harmonic with the operative temperature in the centre (θ_o) as an unknown equals:

$$\sum_{j=1}^{n}\left[a_{e,j}U_{e,j}A_{e,j}\left(\theta_{e,j}^{*}-\theta_{o}\right)\right] + \sum_{k=1}^{m}\left[U_{w,k}A_{w,k}\left(\theta'_{e,k}-\theta_{o}\right)\right] + \sum_{l=1}^{p}\left[U_{i,l}A_{i,l}(\theta_{i}-\theta_{o})\right]$$

$$+\underbrace{0.34\,nV(\theta_{e}-\theta_{i})}_{(4)}+\underbrace{\sum_{k=1}^{m}\left(g_{w,k}f_{w,k}r_{w,k}E_{\text{sun},w,k}\right)+\overline{\Phi}_{\text{intern}}}_{(5)} = 0$$

(1.135)

The suffix e stands for all opaque envelope assemblies, the suffix w for the windows, and the suffix i for the inside partitions. $\theta_{e,j}^{*}$ is the sol-air temperature for envelope assembly j and $\theta'_{e,k}$ the specific sol-air temperature for windows:

$$\theta'_e = \theta_e - \frac{120\, e_L F_{w,sk}(1-f_c)}{h_e}$$

wherein θ_e is the outside air temperature, e_L the long-wave emissivity of the glass, $F_{w,sk}$ the view factor between window and sky, f_c the cloudiness factor and h_e the outside surface film coefficient. Term (5) in the balance equation gives the solar gains across the windows. θ_i in the ventilation term (4) is the air temperature in the space, assumed equal to the operative temperature θ_o. The θ_i's are the operative temperatures in the neighbouring spaces, the A's the surface areas and the U's the clear wall thermal admittances. V is the air volume in the space, n the ventilation rate (ach), g the solar transmittance of the windows including their shading devices, E_{sun} the solar irradiation on the windows, f the ratio between glass and total area per window, and r a shadow factor. The product $g_{w,k}f_{w,k}r_{w,k}E_{\text{sun},w,k}$ gives the average solar and $\overline{\Phi}_{\text{intern}}$ the average internal gains over the base period – all in SI units!

The harmonic reponses look like:

$$\sum_{j=1}^{n}\Phi_{e,j}^{n}+\sum_{k=1}^{m}\Phi_{w,k}^{n}+\sum_{l=1}^{p}\Phi_{i,l}^{n}+\Phi_{\text{vent}}^{n}+\sum_{k=1}^{m}\Phi_{\text{sun},w,k}^{n}+\Phi_{\text{intern}}^{n} = \left(\rho_{a}c_{a}+\frac{c_{f}M_{f}}{V}\right)V_{j}\frac{d\theta_{o}^{n}}{dt}$$

(1.136)

with $\Phi_{e,j}^{n}$ the nth harmonic of the heat flow across the opaque envelope assemblies, $\Phi_{i,k}^{n}$ the nth harmonic of the heat flow across the inside partitions, $\Phi_{w,l}^{n}$ the nth harmonic of the heat flow across the windows, Φ_{vent}^{n} the nth harmonic of the enthalpy flow by ventilation, $\Phi_{\text{sun},w,k}^{n}$ the nth harmonic of the solar gains, Φ_{intern}^{n} the nth harmonic of the internal gains, θ_{o}^{n} the nth harmonic of the operative temperature, c_f specific heat capacity and M_f the weight of all furniture and furnishings. The operative temperature and the heat flows can now be written as:

Operative (and air) temperature	$\theta_o^n = \alpha_o^n \exp(2in\pi t/T)$
Transmission	$\Phi^n = \hat{\Phi}^n \exp(2in\pi t/T)$
Ventilation	$\Phi_{vent}^n = \hat{\Phi}_{vent}^n \exp(2in\pi t/T)$
Solar gains	$\Phi_{sun}^n = \hat{\Phi}_{sun}^n \exp(2in\pi t/T)$
Internal gains	$\Phi_{internal}^n = \hat{\Phi}_{internal}^n \exp(2in\pi t/T)$

In these formulae, α_o^n is the complex operative temperature, $\hat{\Phi}_x^n$ the complex heat flow, T the base period, n the order of the harmonic and i the imaginary unit. Entering these expressions in the harmonic heat balance gives:

$$\sum_{j=1}^{n} \hat{\Phi}_{e,j}^n + \sum_{k=1}^{m} \hat{\Phi}_{w,k}^n + \sum_{l=1}^{p} \hat{\Phi}_{i,l}^n + \hat{\Phi}_{V,j}^n + \sum_{k=1}^{m} \hat{\Phi}_{sun,w,k}^n + \hat{\Phi}_{intern}^n = i(\omega_n \rho_a cV)\alpha_o^n$$

with ω_n the pulsation of the nth harmonic and c the equivalent specific heat capacity in the space, often set as five times the specific heat capacity of air:

$$c = c_a + c_f M_f/(\rho_a V) \approx 5c_a \approx 5000 \quad (1.137)$$

If necessary, a more accurate value can be calculated, using the weight and specific heat capacity of the materials comprising the furniture and furnishings.

Applying the definitions of temperature damping, dynamic thermal resistance and admittance allows the rewriting of the separate complex heat flows. To keep it simple, the results are given for the first harmonic only. Higher harmonics give identical expressions, but with the transient properties, complex temperatures and complex heat flow rates for the harmonic considered. Assuming that heat goes from outside to inside, the heat flow across the opaque envelope assemblies can be written as:

$$\hat{\Phi}_{e,j}^n = \alpha'_{e,j} A_{e,j} = \left(\frac{1}{D_{q,e,j}} \alpha_{e,j}^* - \frac{D_{\theta,e,j}}{D_{q,e,j}} \alpha_o\right) A_{e,j} = \left(\frac{1}{D_{q,e,j}} \alpha_{e,j}^* - Ad_{e,j}\alpha_o\right) A$$

For windows the thermal transmittance remains the intervening property, giving as heat flow:

$$\hat{\Phi}_{w,k}^n = \alpha'_{w,k} A_{i,l} = \left[U_{w,k}(\alpha'_{e,k} - \alpha_o)\right] A_{w,k}$$

The heat flows across the opaque partitions with neighbouring spaces are written as:

$$\hat{\Phi}_{i,l}^n = \alpha'_{i,l} A_{i,l} = (\alpha_l/D_{q,i,l} - Ad_{i,l}\alpha_o) A_{i,l}$$

The constant ventilation rate gives:

$$\Phi_{vent} = 0.34 \, nV(\alpha_e - \alpha_o)$$

1.6 Building-related applications

If, besides the solar irradiation, the solar transmittance of the window with shading is also variable, the complex component of the solar gains equals:

$$\hat{\Phi}_{sun,w,k} = f_{w,k} A_{w,k} (\alpha'_{sun,w,k})$$

with:

$$\alpha'_{sun,w,k} = \text{Harm}(g_{w,k} f_{w,k} r_{w,k} q_{sun,w,k})$$

wherein $q_{sun,w,k}$ is the flux touching the outside face of the shading. Harm(...) indicates that the product between brackets forms a Fourier series. The complex components of the internal gains finally follow from a Fourier analysis:

$$\hat{\Phi}^n_{intern} = \text{Harm}(\Phi_{intern})$$

Transposing all these equations into the balance and solving for the complex operative temperature gives:

$$\alpha_o = \frac{\sum_{j=1}^{n}\left(\frac{A_{e,j}}{D_{q,e,j}}\alpha^*_{e,j}\right) + \sum_{k=1}^{m}(U_{w,k}A_{w,k}\alpha'_{e,k}) + \sum_{l=1}^{p}\left(\frac{A_{i,l}}{D_{q,i,l}}\alpha_l\right) + 0.34nV\alpha_e + \sum_{k=1}^{m} f_{w,k}A_{w,k}\text{Harm}(g_{w,k}r_{w,k}q_{sun,w,k}) + \text{Harm}(\hat{\Phi}_{intern})}{\sum_{j=1}^{n}(A_{e,j}Ad_{e,j}) + \sum_{k=1}^{m}(U_{w,k}A_{w,k}) + \sum_{l=1}^{p}(A_{i,l}Ad_{i,l}) + 0.34nV + i(6000\omega V)}$$

The solution presumes a transposition of complex to real numbers.

If the sol-air and specific sol-air temperatures for glazing are assumed equal to the outdoor temperature ($\theta'_e = \theta^*_e = \theta_e$ and $\alpha'_e = \alpha^*_e = \alpha_e$), which means neither solar radiation nor under-cooling, then, for a ventilation rate of zero and all neighbouring spaces at the same operative temperature as the one considered, that formula simplifies to:

$$\alpha_o = \left\{ \frac{\sum_{j=1}^{n}\left(\frac{A_{e,j}}{D_{q,e,j}}\right) + \sum_{k=1}^{m}(U_{w,k}A_{w,k})}{\sum_{j=1}^{n}(A_{e,j}Ad_{e,j}) + \sum_{k=1}^{m}(U_{w,k}A_{w,k}) + \sum_{l=1}^{q}\left[A_{i,l}\left(Ad_{i,l} - \frac{1}{D_{i,l}}\right)\right] + i(6000\,\omega V)} \right\} \alpha_e$$

(1.138)

The term between the large brackets contains only construction-related characteristics: surface areas and the inverse of the dynamic thermal resistances and the admittances of all opaque envelope parts fixing the thermal inertia and storage capacity, surface area and thermal transmittance of the windows, for all inside partitions the surface area and thermal storage capacity, and this last also for the air, the furniture and furnishings. The inverse stands for the ratio between the complex outdoor air and complex indoor operative temperatures:

$$D_{\theta,\text{space}} = \left\{ \frac{\sum_{j=1}^{n}(A_{e,j}Ad_{e,j}) + \sum_{k=1}^{m}(U_{w,k}A_{w,k}) + \sum_{l=1}^{q}\left[A_{i,l}\left(Ad_{i,l} - \frac{1}{D_{i,l}}\right)\right] + i(6000\,\omega V)}{\sum_{j=1}^{n}\left(\frac{A_{e,j}}{D_{q,e,j}}\right) + \sum_{k=1}^{m}(U_{w,k}A_{w,k})} \right\}$$

(1.139)

That inverse is called the room damping for the harmonic considered, and reflects how well a space dampens the temperature swings outdoors. The first harmonic usually suffices to classify a building space as dampening well, or dampening poorly.

1.6.6.3 Thermal bridges

For the consideration of thermal bridges, see Section 1.2.4 on transient conduction. Combine what is advanced there with the surface resistance approach explained above, under steady-state thermal bridges. As a reminder, surface resistances lack capacitance.

1.7 Problems and solutions

Problem 1.1

Calculate the thermal transmittance of an outside wall, inside to outside, assembled as follows ($h_i = 7.7$ W/(m².K), $h_e = 25$ W/(m².K)):

Layer	Thickness, cm	Thermal conductivity, W/(m.K)	Thermal resistance, m².K/W
Plaster	1	0.3	
Inside leaf	14	0.5	
Cavity fill	8	0.04	
Unvented air cavity	4		0.17
Brick veneer	9	0.9	

Solution

All quantities must be expressed in SI units. So, metres (m), not centimetres (cm):

$$U_o = \frac{1}{1/h_i + \sum R_j + 1/h_e}$$

$$= \frac{1}{1/8 + 0.01/0.3 + 0.14/0.5 + 0.08/0.04 + 0.17 + 0.09/0.9 + 1/25}$$
$$= 0.36 \text{ W}/(\text{m}^2.\text{K})$$

Never give more than two digits.

1.7 Problems and solutions

Problem 1.2

Calculate the thermal transmittance of a low-slope roof, inside to outside, assembled as follows ($h_i = 10$ W/(m².K), $h_e = 25$ W/(m².K)):

Layer	Thickness, cm	Thermal conductivity, W/(m.K)
Plaster	1	0.3
Concrete floor	14	2.5
Screed	10	0.6
Vapour barrier	1	0.2
Thermal insulation	12	0.028
Membrane	1	0.2

Solution

$U_o = 0.21$ W/(m².K).

Problem 1.3

Calculate the clear wall thermal transmittance of a timber frame outer wall, inside to outside, assembled as follows ($h_i = 7.7$ W/(m².K), $h_e = 25$ W/(m².K), studs not considered):

Layer	Thickness (d) cm	Thermal conductivity (λ), W/(m.K)	Thermal resistance (R), m².K/W
Gypsum board	1.2	0.2	
Air space	2		0.17
Airflow retarder	0.02	0.2	
Thermal insulation	20	0.04	
Outside sheathing	2	0.14	
Unvented air cavity	2		0.17
Brick veneer	9	0.9	

Solution

$U_o = 0.17$ W/(m².K).

Problem 1.4

Calculate the sol-air temperature for a horizontal surface subjected to a solar irradiation of 750 W/m². The outdoor air temperature is 30 °C, the outside surface film coefficient 12 W/(m².K). The surface has a short-wave absorptivity of 0.9 and a long-wave emissivity of 0.8. The long-wave losses to the clear sky reach 100 W/m² (low-slope roof).

Repeat the calculation for a daily mean outdoor air temperature of 24 °C, a daily mean solar irradiation of 169 W/m² and a daily mean long-wave loss to the clear sky of 50 W/m². This is representative of a south-oriented vertical wall during a hot summer's day in a temperate climate. Redo the exercise for a cold winter's day with a daily mean outdoor air temperature of −15 °C, a daily mean solar irradiation of 109 W/m² and a daily mean long-wave loss to the clear sky of 50 W/m². The short-wave absorptivity and long-wave emissivity of the outside wall face are 0.5 and 0.8, respectively, while the outside surface film coefficient reaches 16 W/(m².K) and the cloudiness reaches 0.8.

Solution

The equivalent temperature in the first situation is:

$$\theta_e^* = \theta_e^* + \frac{a_K E_S - e_L q_L}{h_e} = 30 + \frac{0.9 \times 750 - 0.8 \times 100}{12} = 79.6°C$$

which is high. The mean equivalent temperature for the south-oriented wall during the hot summer's day touches:

$$\theta_e^* = \theta_e^* + \frac{a_K E_S - e_L q_L}{h_e} = 24 + \frac{0.5 \times 169 - 0.8 \times 50}{16} = 26.8°C$$

During the cold winter's day, one has:

$$\theta_e^* = \theta_e^* + \frac{a_K E_S - e_L q_L}{h_e} = -15 + \frac{0.5 \times 109 - 0.8 \times 50}{16} = -13.6°C$$

Problem 1.5

Return to Problem 1.1. Calculate the highest and lowest daily mean temperatures for all interfaces, knowing that the equivalent outdoor temperature has the same value as in the repeat part of Problem 1.4. The operative temperature indoors is 21 °C in winter and 25 °C in summer. The surface film coefficient outside is 16 W/(m².K), and the surface film resistance inside 0.13 m².K/W. Draw the result.

Solution

The temperatures are given by $\theta_j = \theta_i - (\theta_i - \theta_e^*) \sum_{i=1}^{j} R/R_a$. As a table and a figure:

Layer	ΣR, m².K/W	Temp, cold winter's day, °C	Temp, warm summer's day, °C
	0	21.0	25.0
	0.13	19.5	25.3
1	0.16	19.1	25.4
2	0.44	15.9	26.0
3	2.44	−7.1	30.3

1.7 Problems and solutions

4	2.61	−9.0	30.6
5	2.71	−10.2	30.9
	2.78	−10.9	31.0

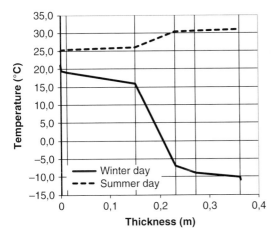

The insulation backs the temperature difference. It is as if the wall splits into a part leaning to the indoors and a part leaning to the outdoors which experiences the greatest temperature change.

Problem 1.6

Now repeat Problem 1.5 for the timber frame wall of Problem 1.3. The outdoor sol-air and air temperatures in winter and summer, the operative temperatures indoors in winter and summer and the inside and outside surface film coefficients are as given above. Draw the result.

Solution

As a table and a figure:

Interface	ΣR_j, m².K/W	Temp, cold winter's day, °C	Temp, warm summer's day, °C
$1/h_i$	0	21.0	24.0
Gypsum board	0.13	20.3	24.2
Air space	0.19	20.0	24.2
Airflow retarder	0.36	19.0	24.4
Thermal insulation	0.36	19.0	24.4
Outside sheathing	5.36	−8.3	30.4
Air cavity	5.50	−9.1	30.6
Brick veneer	5.67	−10.0	30.8
$1/h_e$	5.77	−10.6	30.9

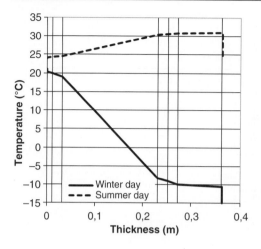

Problem 1.7

Take the low-sloped roof of Problem 1.2. Calculate the highest and lowest daily mean temperatures for all interfaces, knowing that the daily mean outdoor equivalent temperature in summer reaches 40 °C for a daily mean air temperature of 24 °C, while in winter these values are −19.5 °C and −15 °C respectively. The average surface film coefficient outside during windless days is 12 W/(m².K). Inside, the operative temperature is 21 °C in winter and 25 °C in summer. The inside surface film coefficient equals 6 W/(m².K) in summer and 10 W/(m².K) in winter. Draw the result.

Solution

As a table and a figure:

Interface	Winter		Summer	
	ΣR_j, m².K/W	Temp, °C	ΣR_j, m².K/W	Temp, °C
$1/h_i$	0	21.0	0	25.0
Render	0.17	19.6	0.10	25.3
Concrete floor	0.20	19.3	0.13	25.4
Screed	0.26	18.9	0.19	25.6
Vapour barrier	0.42	17.5	0.36	26.1
Thermal insulation	0.47	17.1	0.41	26.3
Membrane	4.76	−18.4	4.69	39.6
$1/h_e$	4.81	−18.8	4.74	39.7

1.7 Problems and solutions

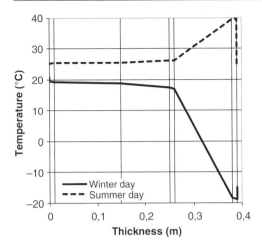

Problem 1.8

A manufacturer introduces a new sandwich panel with, as a section, inside to outside:

Layer	Thickness (d), cm	Thermal conductivity (λ), W/(m.K)	Thermal resistance (R), m².K/W
Aluminium	0.2	230	
VIP (vacuum insulation)	2	0.006	
Air cavity	2		0.15
Glass pane	1	Assume ∞	

The panel fills a curtain wall. Assume temperatures of 35 °C outdoors and 24 °C indoors. Solar irradiation on the glass pane reaches 500 W/m². No long-wave radiation needs to be considered. The surface film coefficient outdoors is 15 W/(m².K), and indoors is 7.7 W/(m².K). Short-wave radiant properties of the glass are: $a_S = 0.05$, $r_S = 0.20$, $\tau_S = 0.75$. The cavity face of the VIP has a short-wave absorptivity of 1. What will be the temperature in the glass pane? How large will the heat flux be entering the building across the panel?

Solution

The problem is solved by writing two heat balances: one for the glass and one for the VIP cavity face:

Glass (temperature θ_1): $h_e(\theta_e - \theta_1) + a_S E_S + \dfrac{\theta_{s2} - \theta_1}{R_{cav}} = 0$

VIP (temperature θ_{s2}): $\dfrac{\theta_1 - \theta_{s2}}{R_{cav}} + \tau_S E_S + \dfrac{\theta_i - \theta_{s2}}{R_{VIP} + R_{alu} + 1/h_i} = 0$

or:

$$\begin{cases} -(15 + 1/0.15)\theta_1 + \dfrac{\theta_{s2}}{0.15} = -0.05 \times 500 - 15 \times 35 \\ \dfrac{\theta_1}{0.15} - \theta_{s2}\left(\dfrac{1}{0.15} + \dfrac{1}{0.02/0.005 + 0.002/230 + 1/7.7}\right) \\ \quad = -0.75 \times 500 - \dfrac{1}{0.02/0.005 + 0.002/230 + 1/7.7} \times 24 \end{cases}$$

Solving this system of two equations gives $\theta_1 = 60.2\,°C$, $\theta_{s2} = 113.2\,°C$. The heat flow rate to the inside is $21.6\,W/m^2$. The high temperatures underline that the panel acts as a solar collector. The heat flux to the inside equals that transferred in the absence of solar irradiation by an assembly with thermal transmittance of $1.96\,W/(m^2.K)$, whereas the clear wall thermal transmittance of the manufactured panel is only $0.23\,W/(m^2.K)$. What measures could lower the temperatures within and heat flux across the panel?

Problem 1.9

Solve Problem 1.8 for the case when heat-absorbing glass, $a_S = 0.3$, $r_S = 0.19$, $\tau_S = 0.51$, is used, and the short-wave absorptivity and reflectivity of the VIP's cavity face is 0.5.

Solution

The temperature of the glass is $52.8\,°C$, and the temperature at the cavity side of the VIP, $70.2\,°C$. The heat flux to the inside equals $11.2\,W/m^2$, which corresponds to a U value of $1.02\,W/(m^2.K)$. The real U value remains $0.23\,W/(m^2.K)$

Problem 1.10

The roof of a mountain chalet is covered with 40 cm of snow ($\lambda = 0.07\,W/(m.K)$, $a_S = 0.15$). The outdoor temperature is $-15\,°C$, while indoors it is $22\,°C$. Solar irradiation reaches $600\,W/m^2$. The surface film coefficients are $15\,W/(m^2.K)$ outside and $10\,W/(m^2.K)$ inside. What insulation thickness is needed to stop the snow from melting in contact with the membrane, for an insulation material with apparent thermal conductivity of $0.023\,W/(m.K)$? What heat flux will be noted across the roof? The thermal resistance face-to-face of the roof without insulation is $0.5\,m^2.K/W$.

Solution

The thickness needed is 21 cm. The heat flux across the roof equals $2.24\,W/m^2$.

Problem 1.11

An intensely ventilated attic receives an insulated ceiling, composed of metal girders, mounted 60 cm centre-to-centre with a 120 mm thick thermal insulation in between. The girder section is given in the figure below. Suppose the insulation has a thermal conductivity of $0\,W/(m.K)$, while for the metal the value is $\infty\,W/(m.K)$. The surface film coefficients are $25\,W/(m^2.K)$ at the attic side and $6\,W/(m^2.K)$ inside. The attic temperature is $-10\,°C$, and the temperature indoors $20\,°C$. Does the heat loss differ between the profile mounted with the broader flange indoors (a) or vice versa (b)? What is the metal temperature in both cases? Calculate the U value of the ceiling.

1.7 Problems and solutions

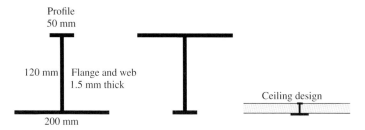

Solution

The heat balance for the profile in case (a) is: $0.2 \times 6 \times (20 - \theta_x) + 0.05 \times 25 \times (-10 - \theta_x) = 0$. In case (b) it is: $0.05 \times 6 \times (20 - \theta_x) + 0.2 \times 25 \times (-10 - \theta_x) = 0$. So yes, the heat losses do differ. In case (a) the metal temperature is 4.7 °C, while in case (b) it is −8.3 °C. For 1/0.6 girders per metre run, in case (a), $U = 1.02$ W/(m².K), while in case (b) it is 0.47 W/(m².K)

Problem 1.12

Solve Problem 1.11 for a metal profile with flanges of 100 mm each.

Solution

The temperature of the steel profiles is 4.2 °C, and the U value of the ceiling is 0.53 W/(m².K).

Problem 1.13

A reinforced concrete column with sides of 0.4 m is positioned between two glass panels in such a way that the glass lines up with the column's inside face. The glass is considered a surface with thickness zero. The temperature indoors is 21 °C, and the temperature outdoors 0 °C. The inside surface film coefficient equals 8 W/(m².K), and the outside surface film coefficient 25 W/(m².K). Calculate the temperature field in and the heat loss across the column.

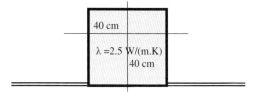

Solution

Assume the column reacts as a flat wall. The U value then is:

$$\frac{1}{1/25 + 0.4/2.5 + 1/8} = 3.1 \text{ W}/(\text{m}^2\text{K})$$

The temperature at the inside surface thus equals $21 - 3.1 \times (21 - 0)/8 = 12.9$ °C, giving a temperature factor of 0.65. The heat loss becomes $3.1 \times 0.4 \times 21 = 25.8$ W/m. Or, the thermal transmittance and temperature factor are close to those for double glass. In the column's centre, the temperature is 7.8 °C.

A first upgrade consists of applying a very simple CVM grid with the centre of the column as the calculation point (point 1, 2). The heat balance there is:

$$\frac{0.2(21 - \theta_x)}{1/8 + 0.2/2.5} + 3\frac{0.2(0 - \theta_x)}{1/25 + 0.2/2.5} = 0$$

giving as the central temperature 3.4 °C, and as the inside surface temperature, 10.3 °C, which gives a temperature factor of 0.49, a value 24.6% lower than just calculated. The heat loss now is 34.3 W/m – 33.7% higher than with the flat wall assumption. In a second upgrade, the grid over half the column is refined to 6 calculation points, of which 5 lie on the perimeter.

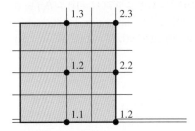

The heat balances are:

Point 1,1
$$8 \times 0.1 \times (21 - \theta_{1,1}) + \frac{2.5 \times 0.1}{0.2}(\theta_{1,2} - \theta_{1,1}) + \frac{2.5 \times 0.1}{0.2}(\theta_{2,1} - \theta_{1,1}) = 0$$

Point 2,1
$$8 \times 0.1 \times (21 - \theta_{2,1}) + \frac{2.5 \times 0.1}{0.2}(\theta_{1,1} - \theta_{2,1}) + \frac{2.5 \times 0.1}{0.2}(\theta_{2,2} - \theta_{2,1})$$
$$+ 25 \times 0.1 \times (0 - \theta_{2,1}) = 0$$

Point 1,2
$$\frac{2.5 \times 0.1}{0.2}(\theta_{1,1} - \theta_{1,2}) + \frac{2.5 \times 0.1}{0.2}(\theta_{1,3} - \theta_{1,2}) + \frac{2.5 \times 0.2}{0.2}(\theta_{2,2} - \theta_{1,2}) = 0$$

Point 2,2
$$\frac{2.5 \times 0.1}{0.2}(\theta_{2,1} - \theta_{2,2}) + \frac{2.5 \times 0.2}{0.2}(\theta_{1,2} - \theta_{2,2}) + \frac{2.5 \times 0.1}{0.2}(\theta_{2,3} - \theta_{2,2})$$
$$+ 25 \times 0.2 \times (0 - \theta_{2,2}) = 0$$

Point 1,3
$$\frac{2.5 \times 0.1}{0.2}(\theta_{1,2} - \theta_{1,3}) + 25 \times 0.1 \times (0 - \theta_{1,3}) + \frac{2.5 \times 0.1}{0.2}(\theta_{2,3} - \theta_{1,3}) = 0$$

Point 2,3
$$2 \times 25 \times 0.1 \times (0 - \theta_{2,3}) + \frac{2.5 \times 0.1}{0.2}(\theta_{2,2} - \theta_{2,3}) + \frac{2.5 \times 0.1}{0.2}(\theta_{1,3} - \theta_{2,3}) = 0$$

Solving this system gives as temperatures in the column:

0.4 0.9 0.4
1.4 2.9 1.4
4.9 8.1 4.9

The lowest temperature factor inside now sits at the corners, 0.25 – as poor as single glass. The heat loss equals:

$$\Phi = 2 \times 0.1 \times 8 \times (21 - 4.9) + 0.2 \times 8 \times (21 - 8.1) = 46.4 \text{ W/m}$$

that is 80% higher than given by the flat wall assumption. The last upgrade comes by including more control volumes, and using software for two-dimensional heat transport. The figure below shows the nearly correct answer in terms of temperatures:

Red: 19 to 20°C
Dark blue: 0 to 1°C
Each colour in between is a step of 1°C

Problem 1.14

Aerated concrete is chosen as the envelope material. The manufacturer promises a very good transient thermal response, which is formulated in terms of a much higher effective thermal resistance than calculated steady state. Is this true? The material properties are:

Situation	Density, kg/m³	Thermal conductivity (λ), W/(m.K)	Specific heat capacity, J/(kg.K)
Just applied (humid)	450	0.30	2700
After some years (air-dry)	450	0.13	1120

The wall thicknesses to consider are 10, 20 and 30 cm. The outside surface film coefficient is 25 W/(m².K), and the inside surface film coefficient 8 W/(m².K).

Solution

A way to evaluate the claim is through calculating the harmonic properties. A high dynamic thermal resistance seems to confirm it, but a low admittance indicates that this will not suffice to stabilize the indoor climate in the case of significant solar and internal gains.

To show how the harmonic calculation works, temperature damping, the dynamic thermal resistance and the admittance are calculated for the initially humid aerated concrete wall with $d = 10$ cm and $\lambda = 0.3$ W/(m.K).

Thermal diffusivity: $a = \frac{\lambda}{\rho c} = 2.469 \times 10^{-7}$ m²/s

X_n value: $X_n = d\sqrt{\frac{n\pi}{aT}} = 0.1\sqrt{\frac{3.1418}{2.469 \times 10^{-7} \times 3600 \times 24}} = 1.2135$

Functions $G_{n1}(X_n)$ to $G_{n6}(X_n)$:

G_{n1}	0.640
G_{n2}	1.437
G_{n3}	0.928
G_{n4}	0.486
G_{n5}	−1.431
G_{n6}	2.733

Layer matrices:

$$\text{Inside surface} \begin{bmatrix} 1 & 0 & 0.125 & 0 \\ 0 & 1 & 0 & 0.125 \\ 0 & 0 & 1 & 0 \\ 0 & 0 & 0 & 1 \end{bmatrix} \quad |1|$$

$$\text{Layer} \begin{bmatrix} 0.640429 & 1.437199 & 0.309307 & 0.161937 \\ -1.437199 & 0.640429 & -0.161937 & 0.309307 \\ -4.29249 & 8.198851 & 0.640429 & 1.437199 \\ -8.198851 & -4.29249 & -1.437199 & 0.640429 \end{bmatrix} \quad |2|$$

$$\text{Outside surface} \begin{bmatrix} 1 & 0 & 0.04 & 0 \\ 0 & 1 & 0 & 0.04 \\ 0 & 0 & 1 & 0 \\ 0 & 0 & 0 & 1 \end{bmatrix} \quad |3|$$

Matrix multiplication:

$$|2| \times |1| = \begin{bmatrix} 0.640429 & 1.437199 & 0.38936 & 0.341587 \\ -1.437199 & 0.640429 & -0.341587 & 0.38936 \\ -4.29249 & 8.198851 & 0.103868 & 2.462055 \\ -8.198851 & -4.29249 & -2.462055 & 0.103868 \end{bmatrix} \quad |4|$$

$$|3| \times |4| = \begin{bmatrix} 0.46873 & 1.765153 & 0.3993515 & 0.440069 \\ -1.765153 & 0.46873 & -0.440069 & 0.3993515 \\ -4.29249 & 8.198851 & 0.103868 & 2.462055 \\ -8.198851 & -4.29249 & -2.462055 & 0.103868 \end{bmatrix}$$

1.7 Problems and solutions

Harmonic properties:

Temperature damping $\quad |D_\theta| = \sqrt{0.46873^2 + 1.765133^2} = 1.83$

$$\phi_\theta = a\tan\left(\frac{1.765133}{0.46873}\right)\frac{12}{\pi} = 5\,\text{h}$$

Dynamic thermal resistance $\quad |D_q| = \sqrt{0.3993515^2 + 0.38936^2} = 0.59\,\text{m}^2\text{K/W}$

$$\phi_q = a\tan\left(\frac{0.38936}{0.3993515}\right)\frac{12}{\pi} = 3.2\,\text{h}$$

Admittance $\quad |Ad| = \dfrac{|D_\theta|}{|D_q|} = 3.09\,\text{W}/(\text{m}^2\text{K})$

$$\phi_{Ad} = \phi_\theta - \phi_q = 1.8\,\text{h}$$

It is for the reader to calculate the other cases, using a spreadsheet or program. The results are:

	Case 1 $d=10$ cm $\lambda=0.3$ W/(m.K)	Case 2 $d=20$ cm $\lambda=0.3$ W/(m.K)	Case 3 $d=30$ cm $\lambda=0.3$ W/(m.K)	Case 4 $d=10$ cm $\lambda=0.13$ W/(m.K)	Case 5 $d=20$ cm $\lambda=0.13$ W/(m.K)	Case 6 $d=30$ cm $\lambda=0.13$ W/(m.K)
$D\theta$	1.83	6.56	23.1	1.63	5.72	19.1
ϕ_θ (h)	5 h 00'	9 h 48'	14 h 36'	4 h 33'	9 h 19'	13 h 57'
D_q (m².K/W)	0.59	1.93	6.56	1.02	3.16	10.4
ϕ_q (h)	3 h 18'	7 h 54'	12 h 42'	2 h 24'	6 h 53'	11 h 28'
Ad (W/(m².K))	3.09	3.39	3.53	1.60	1.81	1.84
ϕ_{Ad} (h)	1 h 48'	1 h 52'	1 h 57'	2 h 09'	2 h 26'	2 h 29'

So aerated concrete is clearly not the wonder promised. To get sufficient temperature damping air-dry ($D_\theta > 15$), a thickness beyond 20 cm is needed. The same holds for the dynamic thermal resistance if a value beyond 4 m².K/W is the target. The admittance is low, surely when the aerated concrete is airdry. Therefore the material does not function in the way the manufacturers claim.

Problem 1.15

Take a living room with surface 4×6.5 m and ceiling height 2.5 m. The room has two exterior walls, one of 4×2.5 m² and the other 6.5×2.5 m², completely glazed with gas-filled, low-e double glazing, U value 1.3 W/(m².K) for $h_i = 7.7$ W/(m².K) and $h_e = 25$ W/(m².K). Both partition walls and the ceiling have a thermal resistance of 0.505 m².K/W from the surface in the living room to the ambient in the neighbouring space. The living room has floor heating with the network of pipes covered by a screed having a thermal resistance of 0.1 m².K/W. Walls, floor and ceiling are grey bodies with emissivity 0.9. The glass has a grey body emissivity of 0.92. The ventilation rate in the room is 1 ach and the surface film coefficient for convection reaches 3.5 W/(m².K). Calculate the glass, wall and ceiling temperatures, knowing that the indoor and outdoor air temperatures are 21 °C and −8 °C respectively.

Solution

The room is considered as a system with six grey surfaces: the windows with surface A_1 and A_2 and temperatures T_{s1} and T_{s2}, the two inside walls with surface A_3 and A_4 and surface temperatures T_{s3} and T_{s4}, the ceiling with surface A_5 and surface temperature T_{s5} and, the floor with surface A_6 and surface temperature T_{s6}. The floor heating has a temperature T_{fl}. Seven heat balances so are needed, convective for the room and one per wall surface.

Room balance: $Q_v + \sum_{j=1}^{6} h_c A_j (21 - \theta_{sj}) = 0$

or, with $Q_v = \rho_a c_a V(\theta_e - \theta_i)$, $\theta_i = 21\,°C$, $\theta_e = -8\,°C$, $V = 65\,m^3$, $A_1 = A_3 = 10\,m^2$, $A_2 = A_4 = 16.25\,m^2$, $A_5 = A_6 = 26\,m^2$, $c_a = 1008\,J/(kg.K)$, ρ_a and $h_c = 3.5\,W/(m^2.K)$:

$$-633.36 + 35T_{s1} + 56.875T_{s2} + 35T_{s3} + 56.875T_{s4} + 91T_{s5} + 91T_{s6} - 7680.75 = 0$$

Surface balances: the radiant heat flow rate at each surface is written as $q_R = e_L(M' - M_b)/\rho_L$. Linearization of the black body emittance M_b in a temperature interval 10–25 °C gives:

$$M_b = 307.75 + 5.57\theta_s, \quad r^2 = 0.999$$

The radiosity M' equals:

$$M'_j = \frac{M_{bj}}{e_j} - \frac{\rho_j}{e_j} \sum_{i=2}^{6} F_{ji} M'_j$$

with F_{ji} the view factor between each surface and the other five:

	Surface 1	Surface 2	Surface 3	Surface 4	Surface 5	Surface 6
Surface 1	–	0.187	0.070	0.187	0.278	0.278
Surface 2	0.115	–	0.115	0.210	0.280	0.280
Surface 3	0.070	0.187	–	0.187	0.278	0.278
Surface 4	0.115	0.210	0.115	–	0.280	0.280
Surface 5	0.107	0.175	0.107	0.175	–	0.436
Surface 6	0.107	0.175	0.107	0.175	0.436	–

1.7 Problems and solutions

The black body emittance of each surface thus becomes:

s1 $M_{b1} = \frac{1}{0.92}M'_{s1} - \frac{0.08}{0.92}(0.187M'_{s2} + 0.07M'_{s3} + 0.187M'_{s4} + 0.278M'_{s5} + 0.278M'_{s6})$

s2 $M_{b2} = \frac{1}{0.92}M'_{s2} - \frac{0.08}{0.92}(0.115M'_{s1} + 0.115M'_{s3} + 0.21M'_{s4} + 0.28M'_{s5} + 0.28M'_{s6})$

s3 $M_{b3} = \frac{1}{0.9}M'_{s3} - \frac{0.1}{0.9}(0.07M'_{s1} + 0.187M'_{s2} + 0.187M'_{s4} + 0.278M'_{s5} + 0.278M'_{s6})$

s4 $M_{b4} = \frac{1}{0.9}M'_{s4} - \frac{0.1}{0.9}(0.115M'_{s1} + 0.210M'_{s2} + 0.115M'_{s3} + 0.28M'_{s5} + 0.28M'_{s6})$

s5 $M_{b5} = \frac{1}{0.9}M'_{s5} - \frac{0.1}{0.9}(0.107M'_{s1} + 0.175M'_{s2} + 0.107M'_{s3} + 0.175M'_{s4} + 0.436M'_{s6})$

s6 $M_{b6} = \frac{1}{0.9}M'_{s6} - \frac{0.1}{0.9}(0.107M'_{s1} + 0.175M'_{s2} + 0.107M'_{s3} + 0.175M'_{s4} + 0.436M'_{s5})$

Inverting the matrix of this system of six equations gives the radiosities of the six as functions of the black body emittances:

Matrix:
$$\begin{bmatrix} 1.0870 & -0.0163 & -0.0061 & -0.0163 & -0.0242 & -0.0242 \\ -0.0100 & 1.0870 & -0.0100 & -0.0183 & -0.0243 & -0.0243 \\ -0.0078 & -0.0208 & 1.1111 & -0.0208 & -0.0309 & -0.0309 \\ -0.0128 & -0.0233 & -0.0128 & 1.1111 & -0.0311 & -0.0311 \\ -0.0119 & -0.0194 & -0.0119 & -0.0194 & 1.1111 & -0.0485 \\ -0.0119 & -0.0194 & -0.0119 & -0.0194 & -0.0485 & 1.1111 \end{bmatrix}$$

Inverted:
$$\begin{bmatrix} H'_1 \\ H'_2 \\ H'_3 \\ H'_4 \\ H'_5 \\ H'_6 \end{bmatrix} = \begin{bmatrix} 0.9208 & 0.0150 & 0.0058 & 0.0146 & 0.0219 & 0.0219 \\ 0.0092 & 0.9214 & 0.0090 & 0.0162 & 0.0221 & 0.0221 \\ 0.0074 & 0.0187 & 0.9010 & 0.0182 & 0.0273 & 0.0273 \\ 0.0115 & 0.0207 & 0.0112 & 0.9017 & 0.0275 & 0.0275 \\ 0.0108 & 0.0176 & 0.0105 & 0.0172 & 0.9032 & 0.0408 \\ 0.0108 & 0.0176 & 0.0105 & 0.0172 & 0.0408 & 0.9032 \end{bmatrix}$$

$$x \begin{bmatrix} 307.75 + 5.57\theta_{s1} \\ 307.75 + 5.57\theta_{s2} \\ 307.75 + 5.57\theta_{s3} \\ 307.75 + 5.57\theta_{s4} \\ 307.75 + 5.57\theta_{s5} \\ 307.75 + 5.57\theta_{s6} \end{bmatrix}$$

Introducing this result into the radiant heat flow rate equation allows the elimination of the constant 307.75. The combined heat balance per surface, including radiation, convection and conduction, is now

$$q_R + q_C + q_{cond} = 0$$

or:

$$-10.1367\theta_{s1} + 0.9599\theta_{s2} + 0.3726\theta_{s3} + 0.9352\theta_{s4} + 1.4023\theta_{s5} + 1.4023\theta_{s6} + 0\theta_{fl}$$
$$= 60.985$$

$$0.5907\theta_{s1} - 10.0972\theta_{s2} + 0.5771\theta_{s3} + 1.0391\theta_{s4} + 1.4129\theta_{s5} + 1.4129\theta_{s6} + 0\theta_{fl}$$
$$= 60.985$$

$$0.3726\theta_{s1} + 0.9378\theta_{s2} - 10.444\theta_{s3} + 0.9136\theta_{s4} + 1.3699\theta_{s5} + 1.3699\theta_{s6} + 0\theta_{fl}$$
$$= 115.084$$

$$0.5755\theta_{s1} + 1.0391\theta_{s2} + 0.5622\theta_{s3} - 10.410\theta_{s4} + 1.3765\theta_{s5} + 1.3765\theta_{s6} + 0\theta_{fl}$$
$$= 115.084$$

$$0.5393\theta_{s1} + 0.8831\theta_{s2} + 0.5269\theta_{s3} + 0.8603\theta_{s4} - 10.335\theta_{s5} + 2.0454\theta_{s6} + 0\theta_{fl}$$
$$= 115.084$$

$$0.5393\theta_{s1} + 0.8831\theta_{s2} + 0.5269\theta_{s3} + 0.8603\theta_{s4} + 2.0454\theta_{s5} - 18.355\theta_{s6} + 10\theta_{fl}$$
$$= -73.5$$

In these equations, the diagonal terms consist of:

$$\theta_{s1}, \theta_{s2} - \left[\frac{1}{(1/1.3 - 0.13)} + 3.5 + 0.9208 \left(\frac{0.92}{0.08}\right)\right]$$

$$\theta_{s3}, \theta_{s4}, \theta_{s5} - \left[\frac{1}{0.505} + 3.5 + (0.901, 0.9017, 0.9032)\left(\frac{0.9}{0.1}\right)\right]$$

$$\theta_{s6} - \left[\frac{1}{0.1} + 3.5 + 0.9032\left(\frac{0.9}{0.1}\right)\right]$$

Solving this system of six surface and one room balance equation gives temperatures:

Window 4×2.5 m²	$\theta_{s1} = 17.6$ °C
Window 6.5×2.5 m²	$\theta_{s2} = 17.8$ °C
Wall 4×2.5 m²	$\theta_{s3} = 21.9$ °C
Wall 6.5×2.5 m²	$\theta_{s4} = 21.8$ °C
Ceiling	$\theta_{s5} = 22.3$ °C
Floor	$\theta_{s6} = 29.2$ °C
Floor heating	$\theta_{fl} = 36.1$ °C

Problem 1.16

Repeat Problem 1.15 assuming normal double glass with $U = 2.9$ W/(m².K) for $h_i = 7.7$ W/(m².K) and $h_e = 25$ W/(m².K), while all other data remain the same.

Solution

With normal double glazing, the temperatures become:

Window $4 \times 2.5\,\text{m}^2$	$\theta_{s1} = 11.8\,°\text{C}$
Window $6.5 \times 2.5\,\text{m}^2$	$\theta_{s2} = 12.1\,°\text{C}$
Wall $4 \times 2.5\,\text{m}^2$	$\theta_{s3} = 21.9\,°\text{C}$
Wall $6.5 \times 2.5\,\text{m}^2$	$\theta_{s4} = 21.7\,°\text{C}$
Ceiling	$\theta_{s5} = 22.6\,°\text{C}$
Floor	$\theta_{s6} = 34.7\,°\text{C}$
Floor heating	$\theta_{fl} = 47.0\,°\text{C}$

The floor is much warmer now than acceptable for feet comfort (28 °C), or, heat loss is too high to only install floor heating. The room also needs a radiator or a convector.

Further reading

ASHRAE (2001) *ASHRAE Handbook of Fundamentals*, SI edn, Tullie Circle, Atlanta, GA.

ASHRAE (2005) *ASHRAE Handbook of Fundamentals*, SI edn, Tullie Circle, Atlanta, GA.

ASHRAE (2013) *ASHRAE Handbook of Fundamentals*, SI edn, Tullie Circle, Atlanta, GA.

ASHRAE (2017) *ASHRAE Handbook of Fundamentals*, SI edn, Tullie Circle, Atlanta, GA.

Blomberg, T. (1996) Heat Conduction in Two and in Three Dimensions, Computer Modelling of Building Physics Applications. Report TVBH-1008, Lund University of Technology.

Cammerer, J.S. (1962) *Wärme- und Kälteschutz in der Industrie*, Springer Verlag, Berlin [in German].

Carslaw, H.S. and Jaeger, J.C. (1986) *Conduction of Heat in Solids*, Oxford Science Publications, Oxford.

CSTC, Règles Th (1975) Règles de calcul des charactéristiques thermiques utiles des parois de construction de base des bâtiments et du coefficient G des logement et autres locaux d'habitation. DTU [in French].

Defraeye, T., Blocken, B. and Carmeliet, J. (2011) An adjusted temperature wall function for turbulent forced convective heat transfer for bluff bodies in the atmospheric boundary layer. *Building and Environment*, **46**, 2130–2141.

De Grave, A. (1957) *Bouwfysica 1*, Uitgeverij SIC, Brussels, [in Dutch].

de Wit, M.H. (1995) *Warmte en Vocht in Constructies (Heat and Moisture in Building Constructies)*, TU-Eindhoven [in Dutch].

DIN 4701 (1983) *Regeln für die Berechnung des Wärmebedarfs von Gebäuden*, German standard, DNA [in German].

Dragan, C. and Goss, W. (1995) Two-dimensional forced convection perpendicular to the outdoor fenestration surface – FEM solution. *ASHRAE Transactions*, **101** (Pt 1), 201–209.

El Sherbiny, S., Raithby, G. and Hollands, K. (1982) Heat transfer by natural convection across vertical and inclined air layers. *Journal of Heat Transfer*, **104**, 96–102.

Emmel, M., Abadie, O. and Mendes, N. (2007) New external convective heat transfer coefficient correlations for isolated low-rise buildings. IEA-ECBCS Annex 41, paper A41-T3-Br-07-2.

Feynman, R., Leighton, R. and Sands, M. (1977) *Lectures on Physics*, vol. **1**, Addison-Wesley Publishing Company, Reading, MA.

Fischer, D. (1995) An experimental investigation of mixed convection heat transfer in a rectangular enclosure. PhD thesis, University of Illinois, Urbana-Champaign.

Haferland, F. (1970) *Das Wärmetechnische Verhalten mehrschichtiger Aussenwände*, Bauverlag Gmbh, Wiesbaden-Berlin [in German].

Hagentoft, C.-E. (2001) *Introduction in Building Physics*, Studentlitteratur, Lund.

Hauser, G. and Stiegel, H. (1990) *Wärmebrücken Atlas für den Mauerwerksbau*, Bauverlag Gmbh, Wiesbaden Berlin [in German].

Hens, H. (1978, 1981) *Bouwfysica, Warmte en Vocht, Theoretische grondslagen, 1^e en 2^e uitgave*, ACCO, Leuven, [in Dutch].

Hens, H. (1991) *Bouwfysica 1, Warmte en Massatransport*, ACCO, Leuven [in Dutch].

Hens, H. (2007) *Building Physics – Heat, Air and Moisture, Fundamentals and Engineering Methods with Examples and Exercises*, 1st edn, Ernst & Sohn (A Wiley Company), Berlin.

IEA-Annex 14 (1990) *Condensation and Energy: Guidelines and Practice*, ACCO, Leuven.

Judkoff, R. and Neymark, J. (1995) *Building Energy Simulation Test (BESTEST) and Diagnostic Method*, NREL/TP-472-6231, Colorado National Renewable Energy Laboratory, Golden.

Kreith, F. (1976) *Principles of Heat Transfer*, Harper & Row Publishers, New York.

Kumaran, K. and Sanders, C. (2008) Boundary Conditions and Whole Building HAM Analysis, Final Report IEA-ECBCS. Annex 41, Whole Building Heat, Air, Moisture Response, pp. 65–110.

Lecompte, J. (1989) De invloed van natuurlijke convectie op de thermische kwaliteit van geïoleerde spouwconstructies (Impact of natural convection on the thermal quality of insulated cavity walls). Dissertation, KU-Leuven [in Dutch].

Lutz, P., Jenisch, R., Klopfer, H., Freymuth, H. and Krampf, L. (1989) *Lehrbuch der Bauphysik*, B.G. Teubner Verlag, Stuttgart [in German].

Mainka, G.W. and Paschen, H. (1986) *Wärmebrückenkatalog*, B.G. Teubner Verlag, Stuttgart [in German].

Murakami, S., Mochida, A., Ooka, R. and Kato, S. (1996) Numerical prediction of flow around a building with various turbulence models: Comparison of k-ε, EVM, ASM, DSM and LES with wind tunnel tests. *ASHRAE Transactions*, **102** (Pt 1), 979–990.

NBN B62-003 (1987) Berekening van de warmtedoorgangscoëfficiënt van wanden. Belgian standard, BIN [in Dutch].

NEN 1068 (1981) Thermische isolatie van gebouwen. Dutch standard, NNI [in Dutch].

Physibel, C.V. (1996) *Kobra Koudebrugatlas*, Maldcgam, Belgium [edited in several languages].

PREN 31077 (1993) *Windows, Doors and Shutters, Thermal Transmittance, Calculation Method*. European Standard, CEN.

Rietschel, R. (1970) *Heiz- und Klimatechnik, 15. Auflage*, Springer Verlag, Berlin [in German].

Roots, P. (1997) Heat transfer through a well insulated external wooden frame wall. Report TVBH-1009. Doctoral thesis, Lund, Sweden.

Saelens, D. (2002) Energy performance assessment of single storey multiple-skin facades. Doctoral thesis, KU-Leuven, Belgium.

Standaert, P. (1984) Twee- en driedimensionale warmteoverdracht: numerieke methoden, experimentele studie en bouwfysische toepassingen, (Two- and three-dimensional heat transfer: numerical models, experimental study and building physics related applications). Dissertation, KU-Leuven [in Dutch].

Taveirne, W. (1990) *Eenhedenstelsels en groothedenvergelijkingen: overgang naar het SI*, Pudoc, Wageningen [in Dutch].

Tavernier, E. (1985) De theoretische grondslagen van het warmtetransport. Kursus 'Thermische isolatie en vochtproblemen in gebouwen'. TI-KVIV [in Dutch], 36 pp.

TU-Delft (1975 –1985) Faculteit Civiele Techniek, Vakgroep Utiliteitsbouw-Bouwfysica. *Bouwfysica, naar de colleges van Prof A.C.Verhoeven*. Delft Technical University, the Netherlands [in Dutch].

Vogel, H. (1997) *Gerthsen Physik*, Springer Verlag, Berlin [in German].

Welty, J., Wicks, C. and Wilson, R. (1969) *Fundamentals of Momentum, Heat and Mass Transfer*, John Wiley & Sons, New York.

"Solar Heating and Cooling" is a research programme initiated by the International Energy Agency. The programme's work is accomplished through the international collaborative effort of experts from Member countries and the European Union. The results are published in a series with Ernst & Sohn and Wiley.

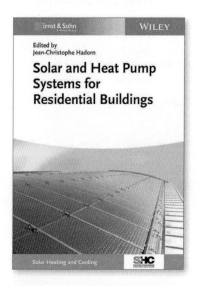

Most developed countries have adopted net-zero energy as a long term goal for new buildings. This book will aid designers in optimally using simulation tools for net-zero energy building design. It presents advanced modelling techniques as well as in-depth case studies. The strategies and technologies are also applicable for the design of energy-plus buildings.

Ed.: Jean-Christophe Hadorn
Solar and Heat Pump Systems for Residential Buildings
2015. 274 pages.
€ 79,–
ISBN 978-3-433-03040-0

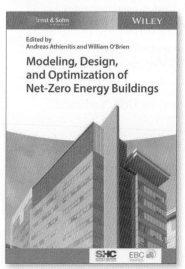

The combination of heat pumps and solar components has a great potential for improving the energy efficiency of house and hot water heating systems. This book compares different systems, analyses their performance, presents design methods and shows monitoring techniques.

Eds.: Andreas Athienitis, William O'Brien
Modeling, Design, and Optimization of Net-Zero Energy Buildings
2015. 396 pages.
€ 94,–
ISBN 978-3-433-03083-7

Order online:
www.ernst-und-sohn.de

Ernst & Sohn
Verlag für Architektur und technische
Wissenschaften GmbH & Co. KG

Customer Service: Wiley-VCH
Boschstraße 12
D-69469 Weinheim

Tel. +49 (0)6201 606-400
Fax +49 (0)6201 606-184
service@wiley-vch.de

* € Prices are valid in Germany, exclusively, and subject to alterations. Prices incl. VAT, excl. shipping. 1112126_dp

2 Mass transfer

2.1 Generalities

2.1.1 Quantities and definitions

The term 'mass transfer' points to the movement of air, vapour, water, other gases, other liquids and dissolved solids within and across materials, building assemblies and whole buildings. Examples are airflow in a room, vapour transport across a roof assembly, water and salt mitigation in bricks, blowing agents diffusing out of insulation materials, CO_2 absorption by fresh lime plaster, and so on. Mass flow can only develop in materials that are open-porous, that is, have accessible pores with an equivalent diameter larger than the diameter of the relevant molecules or molecule clusters. In materials without pores, with smaller pores than the said diameter, or with closed pores, mass transfer fails.

In buildings and building assemblies, air and moisture transport dominate. Moving air carries heat (enthalpy) and vapour, which has positive and negative effects. On the one hand, the passage of dry air increases the drying potential of an assembly and discharges water vapour before it may condense, while on the other hand, air looping in fibrous insulation layers and around carelessly mounted closed cell insulation will increase heat losses and gains. In turn, wetness is the most destructive of all climate-related building loads, and is the reason why assuring moisture tolerance is a main challenge for designers and builders.

Looking to open-porous materials, the word 'moisture' refers to water filling the pores in its two or three states, with various substances dissolved if liquid. So moisture includes ice, liquid plus vapour below $0\,°C$ and liquid plus vapour above $0\,°C$. In water vapour the molecules move separately, whereas in the liquid they form clusters of much larger diameter than the molecule's $0.28\,nm$. Therefore pores that allow vapour to pass may be unaccessible for liquid water. Some materials are thus waterproof but not vapour-proof. The solid state, ice, is crystalline with a 10% larger volume compared with liquid, which is the reason why frost can be so destructive.

The amounts of humid air, moisture and other fluids in a material depend upon:

Density (symbol ρ, units kg/m^3)	Mass per unit volume of material. A porous material has a lower density than its specific density.
Total porosity (symbol Ψ, units % m^3/m^3)	Volume of pores per unit volume of material.
Open porosity (symbol Ψ_o, units % m^3/m^3)	Volume of open pores per unit volume of material. The open pore fraction depends on the fluid. In general, the open porosity is lower than the total porosity ($\Psi_o \leq \Psi$).

With the pores filled with humid air, the following relationship exists between porosity and density:

$$\Psi = \frac{\rho_s - \rho}{\rho_s - \rho_a} \approx 1 - \frac{\rho}{\rho_s} \tag{2.1}$$

with ρ_s the specific density of the material and ρ_a the density of the humid air.

The air in a material is quantified as:

Air content (symbol w_a, units kg/m³)	Air mass per unit volume of material.
Air ratio (symbol X_a, units % kg/kg)	Air mass per unit mass of dry material.
Volumetric air ratio (symbol Ψ_a, units % m³/m³)	Volume of air per unit volume of material.
Air saturation degree (symbol S_a, units %)	Ratio between the current air content and the maximum possible. May also be defined as the fraction of open pores filled, as opposed to those accessible for air.

Moisture presence is quoted analogously:

Moisture content, symbol w, units kg/m³	Mass of moisture per unit volume of material
Moisture ratio, symbol X, units % kg/kg	Mass of moisture per unit mass of dry material
Volumetric moisture ratio, symbol Ψ, units % m³/m³	Volume of moisture per unit volume of material
Moisture saturation degree, symbol S, units: %	Ratio between the current moisture content and the maximum possible. May also be defined as the fraction of open pores filled against those accessible for moisture

These definitions can be extended to all kinds of fluids and gases. The first three average the presence over the material, although in reality its solid matrix does not contain air, moisture, dissolved salts, and so on. The real content in the pores is equal to ρS, with ρ the density of the fluid or gas and S its saturation degree. The following relationships hold between air content, air ratio, volumetric air ratio and density of the porous material:

$$X_a = 100 \frac{w_a}{\rho} \qquad \Psi_a = 100 \frac{w_a}{\rho_a} \tag{2.2}$$

Identical formulas link moisture content to moisture ratio and volumetric moisture ratio:

$$X = 100 \frac{w}{\rho} \qquad \Psi = 100 \frac{w}{1000} = \frac{w}{10} \tag{2.3}$$

2.1 Generalities

Air and moisture content are also coupled:

$$w_a = \rho_a \left(\frac{\Psi_o \text{ in \%}}{100} - \frac{w}{1000} \right) \qquad (2.4)$$

The saturation degree for air and moisture form a twin with sum 1, on condition that no other substances fill the pores ($S + S_a = 1$). This and the equations above indicates that no air will stay where liquid sits and vice versa. Vapour instead blends with air. For air, no rules impose which quantity to use (w_a, X_a of Ψ_a). Not so for moisture. Although all quantities relate to the same wetness, the numbers differ. For materials with density above 1000 kg/m³ moisture ratio gives the lowest number. For materials with density below 1000 kg/m³, this is true for the volumetric moisture ratio. As 'moisture' has a negative connotation, manufacturers often use the lowest and apparently best scoring number without mentioning the units. Therefore, the following rules prevail: use moisture content for stony materials, moisture ratio for wood-based materials, and volumetric moisture ratio for highly porous materials. Always give the units.

2.1.2 Saturation degrees

When water fills all open pores in a material, moisture content is maximal and air content zero, meaning the moisture saturation degree is 100% and for the air 0%. Conversely, when the pores contain no water, the material is dry and air content is maximal, with saturation degree 100%. The reality always lies between both extremes, whereby the values evolve in opposite directions. When the moisture saturation degree increases, that for the air drops, and vice versa. Consequently, per material, a saturation scale for both can be constructed, containing intervals and values that define the response (see Table 2.1).

For capillary-porous materials such as wood, brick, sand-lime stone, natural stone, concrete, and others, the following holds for the moisture saturation degree: $S_{m,H} < S_{m,cr} < S_{m,c} < S_{m,fr}$. For non capillary-porous materials, such as many insulation materials, no-fines concrete and others, the critical, capillary and frost saturation degrees lose their meaning.

2.1.3 Air and moisture transfer

Air and moisture transfer across porous materials could be accurately described if all related physical laws were known and the pore system was quantifiable as a hydraulic network. Neither is true. The transport laws are approximations, and quantifying the pore distribution using mercury porosity, electronic microscope images and X-ray tomography does not deliver an accurate three-dimensional topology of the network. Another complication resides in the changes of state from water to vapour or ice and vice versa, or from ice to vapour and vice versa, all the more because vapour blends with the air in the pores. Hence, only a phenomenological approach, with potentials and transport coefficients that engage the pores wherein flow occurs, remains.

Table 2.1 The air and moisture saturation scale

Saturation degree (%)		Meaning
Moisture	Air	
$S_m = 0$	$S_a = 100$	**Dry material.** All open pores contain air. This is never purely the case, although a moisture saturation degree can be close to 0.
$0 \leq S_m \leq S_{m,H}$	—	**Hygroscopic interval.** Relative humidity in the pores determines the moisture saturation degree, with $S_{m,H}$ reached at 98% relative humidity
$S_{m,cr}$	—	**Critical moisture saturation degree.** Below there is hardly any liquid transport in the material, but above this point there is mainly liquid transport. Airflow has little influence.
$S_{m,c}$	$S_{a,cr}$	**Capillary moisture saturation degree, critical air saturation degree.** At higher degrees of moisture saturation, airflow stops, which is why the value coincides with the critical air saturation degree. Below it, airflow is barely hindered. If moistening and drying are in balance, the water presence will rarely exceed that saturation degree under atmospheric conditions.
$S_{m,c} < S_m \leq 100\ S_{m,fr}$		To enter this interval, a porous material must be in contact with water for a long time without drying. Dissolution of the air left in the pores then pushes the moisture saturation degree beyond capillary. **The moisture saturation degree for frost $S_{m,fr}$** lies somewhere in that range, and stands for the value above which frost damage looms. For materials with high tensile strength, it can reach 90%. The closer the critical moisture saturation degree for frost to the capillary saturation degree, the poorer the frost resistance of a material.
$S_m = 100$	$S_a = 0$	**Moisture saturation.** All air has left the material. Requires water sorption under vacuum or boiling the material.

For air transport the drivers are external forces such as wind, water migrating across the pores and differences in air temperature and air composition. Moisture transport combines vapour and liquid movement, the last including the dissolved substances. Equivalent diffusion, air mitigation, capillary suction, gravity and external pressures all act as driving forces.

Mechanism	Driving force
Water vapour	
Equivalent diffusion	Differentials in vapour pressure in the pore air and the ambient air at both sides of an assembly.
Air transport	Gradients in total air pressure. Vapour migrates together with the air in and across assemblies.
Water	
Capillarity	Differentials in capillary suction connected to pore width. The larger a pore, the less it sucks but the more intense the flow.
Gravity	Differentials in weight between water heads. Gravity activates liquid flow in wider pores that hardly show capillary suction.
Pressure	Differentials in external total pressure. Air as well as water are responsible. Although pressure gradients are small compared with the water pressure gradients, both cause water flow across very large pores, cracks, open joints, and so on.

Which mechanisms intervene will depend on the nature of a material:

	Vapour		Water		
	Eq. diff.	Airflow	Capillarity	Gravity	Pressure
If non-capillary and non-hygroscopic	X	X		X	X
If non-capillary and hygroscopic	X	X		X	(X)
If capillary and non-hygroscopic	X		X	(X)	
If capillary and hygroscopic	X		X		

Non-capillary, non-hygroscopic materials have very large pores. Non-capillary, hygroscopic materials mix macro-pores too large to suck with sorption-active micro-pores that exert too much friction to allow capillary transport. Air flow is possible in the macro- but hardly in the micro-pores. The pores in capillary non-hygroscopic materials are small enough to be suction-active but too large to collect much sorption moisture. Capillary, hygroscopic materials, finally, have pores narrow enough to give both suction and sorption.

2.1.4 Moisture sources

The starting and boundary conditions that direct the moisture response depend on the sources involved. At any point, humid air fills the pores in the open-porous materials comprising the building parts. It also contacts the part's end faces, except where the part is below grade or touching water. Vapour in the air entrains hygroscopic moisture, although materials are termed air-dry as long as their moisture content does not pass the sorption equilibrium for the relative humidity present.

Looking to water first, leaking water and sewage pipes, leaky joints around shower basins and many others can form accidental water sources. Common is construction moisture entrained by precipitation, chemical reactions, and mixing with water during building. Another source is rising damp, which points to moisture being sucked up by wall assemblies. Rain is a leading source in wet climates. External water heads also act as such. They attack walls and floors below the water table and are a problem in water tanks and swimming pools.

Moving on to vapour, first comes relative humidity, which acts as the motor behind sorption and desorption. Another is surface condensation on the inside or outside faces of assemblies. Indoors it is seen as harmful, not so outdoors, where the amounts deposited may compare with wind-driven rain. Interstitial condensation within assemblies is feared as it may remain unnoticed for a long time, only emerging when rot or corrosion becomes visible or moisture starts dripping out (Figure 2.1).

Albeit unjustified in rainy climates, of the nine sources mentioned, the three related to water vapour are most feared. Their thorough understanding appears more scientific than for the five listed sources of water. The professional literature still deals extensively with vapour, but too often not critically and exhaustively. Craftsmen instead are more concerned about rain and rising damp.

2.1.5 Air and moisture in relation to durability

The presence of stagnant humid air in a material or assembly is not troublesome as such. Only when the air migrates do the effects turn negative. In fact, related flow entrains vapour and increases heat transmission. Vapour can condense, while the enthalpy flow will lower the insulation efficiency. Also, moisture does not always cause harm. Nobody complains about brick veneers turning wet by wind-driven rain. Things change when moisture stains appear on the inside surface of the inside leaf. Also, nobody minds about surface condensation on single glass. Only when the condensate wets the wooden window frame may damage occur. In any case, moisture is definitely a major cause of degradation. Studies show that up to 70% of the building fabric deficiencies are wetness-related. The reason is the strong polarity of the water molecule, which makes it an excellent solvent, an efficient chemical catalyst and a prerequisite for biological activity.

Invisible as damage is the increase in thermal conductivity, a problem for insulation materials that can absorb water or become wet by interstitial condensation. Coupled latent heat transfer enlarges the effect. Wetness further decreases strength and rigidity, an effect that can be very pronounced for resin-bound, wood-based materials. Many materials swell with increasing moisture content, and shrink when it decreases. Visible damage includes hydrolysis of the binder resin in particle board, strawboard, plywood and oriented strand board (OSB). Related losses in strength and stiffness may result in cracking even when carrying only a fraction of the design load. The same composites suffer from mould, mildew, fungae and bacteria once the relative humidity at their surface passes given thresholds. Fungae are particularly feared as they digest cellulose and lignine while producing water, a

2.1 Generalities

Water as source

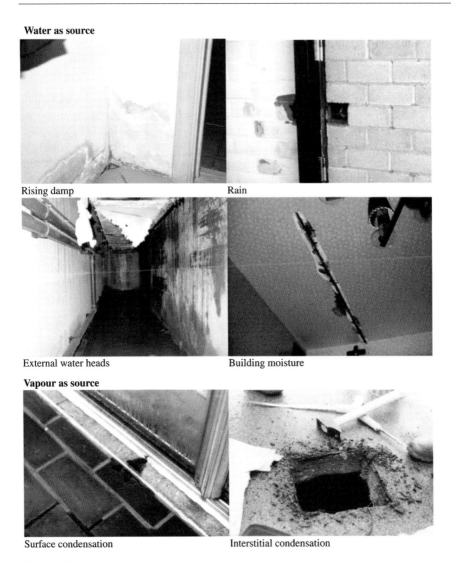

Fig. 2.1 Moisture sources

combination that disintegrates wood completely. Even stony materials can show visible damage. While moss and algae growth on their surface is mainly an aesthetic problem, moss roots may degrade the mortar joints in masonry, while the organic acids they produce can initiate metal corrosion. Salt attack with hydration and crystallization may pulverize bricks. Carbonatation or alkali silicate reaction can degrade concrete. Frost can spall stones and bricks. Steel corrodes visibly at high relative humidity, with the oxide swelling significantly compared with pristine metal. Thus corroding reinforcement bars can spall reinforced concrete. For synthetics, hydration is of most concern as it makes them weak and decreases cohesion over time. Temperature and vapour pressure fluctuations, for which interstitial condensation is often a sign, may irreversibly deform synthetic foams.

2.1.6 Links to energy transfer

Mass transfer means energy displacement. Indeed, the movement produces kinetic energy coupled to the velocity and potential energy as a result of gravity, suction and pressure. Most mass transfer phenomena in porous materials, however, give velocities far too low for kinetic energy to play any role, while differences in potential energy cause the displacement. Mass movement further activates heat transfer. Each gas, liquid and solid at temperature θ contains a quantity of heat per unit mass, given by:

$$Q = c(\theta - \theta_o) \tag{2.5}$$

where c is the specific heat capacity (J/(kg.K)) and θ_o a reference temperature, generally 0 °C. Each mass flow (G) thus is responsible for a sensible heat flow, equal to:

$$\Phi = Gc(\theta - \theta_o) \tag{2.6}$$

When phase changes occur in the moving mass, a latent heat flow equal to the heat of transformation, symbol l_o at reference temperature (units J/kg), is also carried along:

$$\Phi = G[c(\theta - \theta_o) + l_o] \tag{2.7}$$

In building physics most processes are more or less isobaric. In such a case, c is the specific heat capacity at constant pressure. The quantity of heat and heat flows just given are called the enthalpy of the migrating mass, symbol h if per kg, with units J/kg. It is this enthalpy transfer that lowers the insulation efficiency of an assembly subjected to air and moisture intrusion.

2.1.7 Conservation of mass

As stated, this chapter deals with air and moisture transfer. This does not mean that no other gases, such as radon, CO_2 or SO_2, could substitute in an open-porous material, nor that the moving liquid will not carry salts and other soluble substances, which crystallize where drying takes place and dissolve where moistening occurs. But first, some definitions:

Mass, symbol M, units kg	Quantifies the mass present or migrating. Is a scalar.
Mass flow, symbol G, units kg/s	Mass migrating per unit of time. Is also a scalar.
Mass flux, symbol g, units kg/(m^2.s)	Mass migrating per unit of time across a unit surface normal to the flow. The flux is a vector with the same direction as the surface vector. Components in Cartesian coordinates g_x, g_y, g_z; in polar coordinates g_R, g_ϕ, g_Θ.

2.2 Air

A suffix often characterizes the masses moving: a for air; da for dry air; v for water vapour, m for moisture and w for water.

Solving a mass transfer problem means determining two fields: that of the potentials (Po), which cause the transfer and that of the mass fluxes (**g**). The unknown quantities thus are Po(x,y,z,t) and **g**(x,y,z,t), the one scalar (Po), the other vectorial (**g**); computing requires a scalar and a vector equation. The scalar equation follows from mass conservation, stating that the mass flow exchanged between a system and its environment, plus the mass produced in or removed from the system per unit of time, called a source or a sink, equals the changing quantity of mass in the system per unit of time:

$$\text{div}(\mathbf{g}_x) \pm S_x = \frac{\partial w_x}{\partial t} \tag{2.8}$$

with S_x the source or sink in kg/(m³.s). Flows can be diffusive and convective, the first depending on the gradient and the second on the driving force itself. The relationships between fluxes and driving forces offer the vector equations needed.

2.2 Air

2.2.1 Overview

Dry air is a mixture of 21% by volume of oxygen (O_2), 78% nitrogen (N_2) and traces of other gases (CO_2, SO_2, Ar, Xe). Generally, dry air is assumed to behave as an ideal gas with equation of state:

$$p_{da} V = m_{da} R_{da} T \tag{2.9}$$

where p_{da} is the (partial) dry air pressure in Pa, T the temperature in K, m_{da} the mass of (dry) air in kg filling the volume V (m³), and R_{da} the gas constant for dry air, 287.055 J/(kg.K). In reality, dry air is non-ideal, obeying following equation of state:

$$\frac{p_{da} V}{n_{da} R_o T} = 1 + \frac{B_{aa}}{(V/n_1)} + \frac{C_{aaa}}{(V/n_1)^2}$$

with n_{da} the number of moles filling the volume V (m³), R_o the general gas constant, 8314.41 J/(mol.K), and B_{aa} and C_{aaa} viral coefficients, which at ambient temperature are almost zero:

$$B_{aa} = 0.349568 \times 10^{-4} - \frac{0.668772 \times 10^{-2}}{T} - \frac{2.10141}{T^2} + \frac{92.4746}{T^3} \quad (\text{m}^3/\text{mol})$$

$$C_{aaa} = 0.125975 \times 10^{-2} - \frac{0.190905}{T} + \frac{63.2467}{T^2} \quad (\text{m}^6/\text{mol}^2)$$

For the dry air concentration the ideal gas law gives:

$$\rho_{da} = \frac{m_{da}}{V} = p_{da}/(R_{da} T) \tag{2.10}$$

Moist air, a mixture of dry air and water vapour (suffix a), has as ideal gas equation:

$$P_a V = m_a R_a T$$

Concentration thus becomes:

$$\rho_a = P_a/(R_a T)$$

In both relations, P_a is the total air pressure in Pa, equal to the sum of the partial dry air and partial vapour pressure ($P_a = p_{da} + p_v$). Compared with dry air, the presence of water vapour modifies the air mass m_a, the gas constant R_a and the concentration. However, for temperatures below 50 °C the effect is so small that the following holds:

$$R_a \approx R_{da}, \quad P_a \approx p_{da}, \quad \rho_a \approx \rho_{da}$$

or:

$$P_a/(R_a T) \approx p_{da}/(R_{da} T)$$

Air transfer across assemblies occurs when wind, stack or fans create air pressure differentials between rooms and with the outdoors. For this to happen, the enclosure and building fabric must be air-permeable, which presumes the presence of air-permeable materials, cracks, overlaps, intended leaks, and so on.

2.2.2 Air pressure differentials

2.2.2.1 Wind

Wind pressures are given by:

$$P_w = C_p \frac{\rho_a v^2}{2} \approx 0.6 \, C_p v^2 \tag{2.11}$$

an equation that follows from Bernoulli's law applied to a horizontal wind whose velocity drops from a value v to 0 when running into an infinite large obstacle, multiplied by a pressure factor C_p that accounts for the difference between bumping against an infinite as opposed to a finite obstacle. Finiteness, in fact, deflects the air flow at the upper and side edges, where it forms vortexes, while a lee zone develops in front of and behind the obstacle.

The pressure factor couples the real pressure at a point on a surface to a reference wind velocity, in principle measured in an open field at a height of 10 m. Of course, other references can be used depending on the situation. For buildings, pressures on the façades often correlate to the velocity above the ridge. A change in reference alters the pressure factors. Also, wind direction, building location, building geometry and the façade spot considered will all affect the pressure factor. A positive C_p means overpressure, while negative C_p is underpressure. Overpressure is found at the wind side, and underpressure at the lee side and along faces more or less parallel to the wind (Figure 2.2). Wind pressure differences are extremely variable and may change sign regularly.

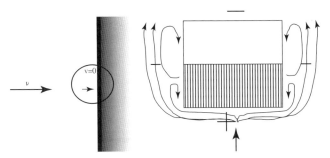

Fig. 2.2 Bernouilli's law and the wind pressure field around a building

2.2.2.2 Stack

Stack or buoyancy in gases and liquids has two causes: differences in temperature and differences in composition. Confined to the atmosphere, the decrease in air pressure with height (h) above sea-level is:

$$dP_a = -\rho_a g \, dh$$

where g is the acceleration due to gravity. Inserting the ideal gas law gives:

$$\frac{dP_a}{P_a} = -\frac{g \, dh}{R_a T} \tag{2.12}$$

With the temperature (T) constant, the solution of this differential equation is:

$$P_a = P_{a0} \exp\left(-\frac{gh}{R_a T}\right) \tag{2.13}$$

a result known as the barometric relation. Air pressure thus depends on height. However, temperature differences and another gas composition change the gas constant (R_a) into a variable inducing pressure gradients at the same height, which is called stack. The pressure difference between two points, one at height h_1, the other at height h_2, is:

$$\Delta P_a = g \, P_{a,h_o} (h_2 - h_1)/(R_a T)$$

with the suffix h_o denoting the average height. The stack potential at height h is then given by the pressure difference between that point in an air mass at variable temperature and composition, and a point at the same height in an air mass at constant temperature and composition:

$$P_{stack} = g \, P_{a,h_o} \left[\left(\int_{h=o}^{h} \frac{dh}{R_a(h)T(h)} \right) - \frac{h}{(R_a T)_o} \right] = g \, P_{a,h_o} h \left[\frac{1}{(R_a T)_m} - \frac{1}{(R_a T)_o} \right]$$

where $(R_a T)_m$ is the harmonic average of the product $R_a(h)T(h)$ at height h:

$$(R_a T)_m = h \left/ \int_{h=o}^{h} \frac{dh}{R_a(h)T(h)} \right.$$

The resulting stack between the two points becomes:

$$p_{T,1-2} = P_{stack2} - P_{stack1} = \mathbf{g}\, P_{a,h_o}\mathbf{h}\left[\frac{1}{(R_a T)_{m2}} - \frac{1}{(R_a T)_{m1}}\right]$$

If the temperatures T_1 and T_2 are different but constant between height zero and height h, while the composition of the air remains the same, stack simplifies to:

$$p_{T,1-2} \approx \frac{\mathbf{g}\, P_{a,h_o}(R_{a1}T_1 - R_{a2}T_2)\mathbf{h}}{R_{a1}T_1 R_{a2}T_2} \approx \frac{\rho_{a0}\,\mathbf{g}(R_{a1}\theta_1 - R_{a2}\theta_2)\mathbf{h}}{R_{am12}T_{m12}} \approx \rho_a \mathbf{g}\beta(\theta_1 - \theta_2)\mathbf{h} \tag{2.14}$$

with $\beta(=1/T_{m12})$ the compressibility of air. If, in contrast, the temperature is constant but the composition, in this case being the vapour concentration, differs, stack changes to:

$$p_{T,1-2} \approx \rho_a R_a \mathbf{g}\beta \mathbf{z}\left(\frac{1}{R_{a2}} - \frac{1}{R_{a1}}\right) \tag{2.15}$$

In case leaks connect air volumes, a neutral plane, stack zero, appears somewhere between the lowest and highest leak. In the absence of other pressure differentials, the air above the neutral plane moves from warm to cold or higher to lower vapour concentration, while the air below goes the other way. Where that neutral plane sits will depend on the leak distribution and size.

Except for high-rises, thermal stack air pressure differences are small. Vapour concentration-related stack is even negligible: 20 °C indoors and 0 °C outdoors in winter gives as thermal stack over a 2.5 m high room ($\rho_{a0} = 1.2\,\text{kg/m}^3$, $g = 9.81\,\text{m/s}^2$, $\beta = 1/273.15\,\text{K}^{-1}$, $\theta_{m12} - \theta_0 = 20\,°C, z = 2.5\,\text{m})\, p_T = 1.2 \times 9.81 \times 1/273.15 \times 20 \times 2.5 = 2.15$ Pa. Instead, for a 250 m high-rise, thermal stack between the lowest and highest floors will touch 215 Pa, a value as large as the pressure difference between wind- and leeside for a 62 km per hour wind. Thermal stack, however, is more stable over time than the wind pressure differences are.

2.2.2.3 Fans

Fans are part of any air heating, air conditioning or forced ventilation system. The pressure differentials they create are usually quite stable over time.

2.2.3 Air permeances

A distinction must be made between open porous materials, air-open layers, unintended and intended leaks. Open porous materials include no-fines concrete, mineral fibre, wood-wool cement and others. Air-open layers include sidings and finishes made of scaly or plate elements such as tiled, slated or corrugated sheet roof covers, lathed ceilings and layers composed of board or strip-like materials, such as plywood, OSB, insulation layers and underlays (Figure 2.3).

Unintended leaks include construction parts, boards and other elements with joints in between, cracks formed when tensile strength is exceeded, microcracks between

Fig. 2.3 Air open layers: tiled roof covers, slated roof covers and a lathed ceiling

mortar and bricks in masonry, nail holes, leaky joint fillers and loose connections. Intended leaks include the electricity boxes in walls, light spots in a ceiling, trickle vents above windows, and so on (Figure 2.4). Many outer wall, party wall and roof solutions have cavities and voids included.

In an open porous material, air displacement follows Poiseuille's law of proportionality between air flux and driving force, in this case being the air pressure gradient:

$$\mathbf{g}_a = -k_a \, \mathbf{grad}\, P_a \qquad (2.16)$$

The factor k_a is called the air permeability, with units of seconds. The minus sign indicates that air moves from points at higher to points at lower air pressure. The permeability is a scalar for isotropic materials, and a tensor with three main directions and a value differing per direction for anisotropic materials. Its value increases together with the open porosity and the number of macro-pores. For air permeable layers, leaks, cracks, joints, cavities and openings, the equations are:

Air permeable layers (per m²): $G_a = -K_a \Delta P_a$, with G_a in kg/(m².s), K_a in s/m.

Joints, cavities (per m): $\quad G_a = -K_a^\psi \Delta P_a$, with G_a in kg/(m.s), K_a^ψ in s.

Leaks, openings (per unit): $G_a = -K_a^\chi \Delta P_a$, with G_a in kg/s, K_a^χ in m.s.

$$(2.17)$$

with K_a, K_a^ψ and K_a^χ the air permeances and ΔP_a the air pressure differentials. Because the flow is not necessarily laminar, most permeances depend on the pressure

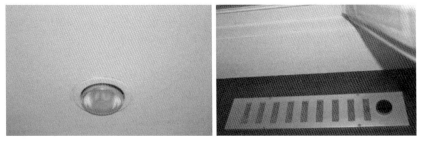

Fig. 2.4 Spotlight and trickle vent as intended leaks

Table 2.2 Friction factor f

Reynolds number $\mathrm{Re} = vd_H/\nu$	Flow	Friction factor f
$\mathrm{Re} \leq 2500$	Laminar	$96/\mathrm{Re}$
$2500 \leq \mathrm{Re} \leq 3500$	Critical	$\dfrac{0.038(3500 - \mathrm{Re}) + f_{T,\mathrm{Re}=2500}(\mathrm{Re} - 2500)}{1000}$
$\mathrm{Re} > 3500$	Turbulent	f_T
$\mathrm{Re} \gg 3500$	Stable turbulent	$f_T = C^{te}$, single function of roughness

differential:

$$K_a^x = a(\Delta P_a)^{b-1}$$

where a is the air permeance coefficient for a 1 Pa pressure differential and b the air permeance exponent. For laminar flow, b is 1, for turbulent flow 0.5, and for transition flow 0.5 to 1. Most air permeances are only quantifiable by experiment. For joints, leaks, cavities and openings with known geometry, using hydraulics helps. The pressure losses are first of all frictional:

$$\Delta P_a = f \frac{L}{d_H} \frac{\rho_a v^2}{2} \approx 0.42 f \frac{L}{d_H} g_a^2 \tag{2.18}$$

with f the friction factor, d_H the hydraulic diameter of the section, L its length and v the average transfer velocity. Bends, widenings, narrowings, entrances and exits add local losses:

$$\Delta P_a = \xi \frac{\rho_a v^2}{2} \approx 0.42 \, \xi \, g_a^2 \tag{2.19}$$

with ξ the local loss factor. Values for the friction factor are given in Table 2.2. Table 2.3 collects some local loss factors.

In the Reynolds number Re, v is the average flow velocity, ν the kinematic viscosity and d_H the hydraulic diameter, for a circular section the diameter of the circle; for a rectangular section, $d_H = 2ab/(a+b)$ where a and b are its sides, and for a cavity, $d_H = 2b$, where b is its width. The Reynolds number can be written as $\approx 56\,000\, \mathbf{g_a}\, d_H$, with $\mathbf{g_a}$ the air flux. The turbulent friction factor f is given by:

$$f_T = \left[2\log\left(-4.793\log\left(\frac{10}{\mathrm{Re}} + 0.2\,\varepsilon\right)\bigg/\mathrm{Re} + 0.2698\,\varepsilon\right)\right]^{-2}$$

where ε is the relative roughness (see Figure 2.5).

Calculation of the air permeances stems from continuity, stating that in each section the same flow must pass. Total pressure loss, the sum of the frictional and local

2.2 Air

Table 2.3 Local loss factors ξ

Local resistance		ξ			
Entering an opening:		0.5			
Leaving an opening:		1.0			
Widening: $\sigma = A_o/A_1$ A_o small section A_1 large section	Re $\leq$ 1000	$-0.036 + 9.6 \times 10^{-5}$Re $+ \Delta\xi$			
	1000 < Re $\leq$ 3000	1.28×10^{-5} Re$^{1.223}$ $+ \Delta\xi$			
	Re > 3000	0.21 Re$^{0.012}$ $+ \Delta\xi$			
	with $\sigma \leq 0.5$: $\Delta\xi = 0.78 - 1.56\sigma$				
	with $\sigma > 0.5$: $\Delta\xi = 0.48 - 0.96\sigma$				
Narrowing: $\sigma = A_o/A_1$ A_o small section A_1 large section	Re $\leq$ 1000	0.98 Re$^{-0.03}$ $+ A$			
	1000 < Re $\leq$ 3000	10.59 Re$^{-0.37}$ $+ A$			
	Re > 3000	0.57 Re$^{-0.01}$ $+ A$			
	where $A = 0.0373\, \sigma^2 - 0.067\, \sigma$				
Leak:		2.85			
Angle or curve: b_o width inlet channel b_1 width of channel after the curve refers to the inlet channel f_o friction factor in inlet channel	$\dfrac{k_{Re0}\xi_g k_0}{(d_H)_0}$ where $\xi_g = 0.885 \left(\dfrac{b_1}{b_0}\right)^{-0.86}$ and				
	Relative roughness ε	3000 $\leq$ Re < 40 000		Re > 40 000	
		k_{Re0}	$k_0/(d_H)_0$	k_{Re0}	$k_0/(d_H)_0$
	0	$45 f_o$	1	1.1	1
	0–0.001	$45 f_o$	1	1.0	$1 + 0.5 \times 10^3 a$
	> 0.001	$45 f_o$	1	1.1	1

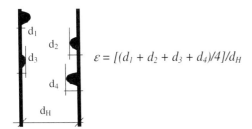

Fig. 2.5 Relative roughness

losses, equal to the driving force, allows calculation of the airflow then as a function of the pressure differential.

2.2.4 Airflow in open-porous materials

2.2.4.1 The conservation law adapted

With neither a source nor a sink present, the equation for conservation of mass simplifies to:

$$\text{div } \mathbf{g}_a = -\frac{\partial w_a}{\partial t} \tag{2.20}$$

The air content w_a depends on the total open porosity Ψ_0, the air saturation degree S_a, the air pressure P_a and temperature T, or:

$$w_a = \Psi_0 S_a \rho_a = \Psi_0 S_a P_a / (R_a T)$$

Whereas the gas constant R_a is more or less invariable and the open porosity Ψ_0 remains constant at least as long as no liquid fills the pores, the variables are the air saturation degree, the air pressure and the temperature. The differential of air content to time can thus be expressed as:

$$\frac{\partial w_a}{\partial t} = \frac{\Psi_0}{R_a T}\left(S_a \frac{\partial P_a}{\partial t} + P_a \frac{\partial S_a}{\partial t} - \frac{S_a P_a}{T}\frac{\partial T}{\partial t}\right)$$

With the saturation degree constant, this equation simplifies to:

$$\frac{\partial w_a}{\partial t} = \frac{\Psi_0 S_a}{R_a T}\left(\frac{\partial P_a}{\partial t} - \frac{P_a}{T}\frac{\partial T}{\partial t}\right)$$

In isothermal conditions, which excludes thermal stack, the derivative of temperature versus time becomes zero, or:

$$\text{div } \mathbf{g}_a = -\left(\frac{\Psi_0 S_a}{R_a T}\right)\frac{\partial P_a}{\partial t} \tag{2.21}$$

The ratio $\Psi_0 S_a / (R_a T)$ is called the isothermal volumetric specific air content, symbol c_a, units kg/(m^3.Pa). As the open porosity and saturation degree for air can never pass 1, replacing R_a by its value 287 J/(kg.K) gives:

$$c_a = \Psi_0 S_a / (R_a T) < 0.00348/T$$

meaning that the specific air content is a negligible quantity. Insertion of the air flux equation into the isothermal mass balance gives:

$$\text{div}(k_a \, \mathbf{grad}\, P_a) = c_a \frac{\partial P_a}{\partial t}$$

2.2 Air

If the air permeance (k_a) is constant, then:

$$\nabla^2 P_a = \frac{c_a}{k_a} \frac{\partial P_a}{\partial t} \tag{2.22}$$

The inverse of the ratio c_a/k_a in Eq. (2.22), k_a/c_a, stands for the isothermal air diffusivity of an open-porous material, symbol D_a (units m²/s). Its value is generally so large that each adjustment to air pressure happens within a few seconds. For pressures that fluctuate over extremely short time spans, such as wind gusts, the pressure response will show some damping and time shift, but usually both are absent. Consider, for example, mineral wool with density 40 kg/m³. The properties are:

Thermal	Air
$\lambda = 0.036$ W/(m.K)	$k_a = 8 \times 10^{-5}$ s
$\rho c = 33\,600$ J/(m³.K)	$c_a = 1.17 \times 10^{-5}$ kg/(m³.Pa)

The thermal diffusivity thus equals 1.1×10^{-6} m²/s and the isothermal air diffusivity 6.8 m²/s, 6 355 000 times larger than the thermal diffusivity. Or, when combining air with heat and moisture transfer, the isothermal air balance can remain steady state as $\nabla^2 P_a = 0$. The combination with the flux equation thus resembles steady-state heat conduction. Although in non-isothermal conditions (i.e. including stack) the isobaric volumetric specific air capacity increases ($\approx 3525/T^2$), steady state remains. The equations anyhow become:

$$\nabla^2(P_{ao} - \rho_a\, \mathbf{g}\, z) = 0 \qquad \mathbf{g}_a = -k_a\, \mathbf{grad}(P_{ao} - \rho_a\, \mathbf{g}\, z)$$

showing that a solution requires the heat balance. The same holds for non-isothermal flow through air permeable layers, joints, cavities, openings, and so on.

2.2.4.2 One dimension: flat assemblies

Assume a flat assembly made of air permeable materials. As the air flux and pressures then resemble steady-state heat conduction, the pressures form a straight line in a single-layer assembly (Figure 2.6), while the flux is:

$$g_a = k_a \frac{\Delta P_a}{d} = \frac{\Delta P_a}{d/k_a} \tag{2.23}$$

The ratio d/k_a is called the (specific) air resistance of the assembly: symbol W, units m/s. The inverse gives the air conductance, symbol K_a, units s/m.

For a composite assembly, the result is:

$$g_a = \Delta P_a \bigg/ \sum_{i=1}^{n} \frac{d_i}{k_{ai}} \tag{2.24}$$

with $\sum d_i/k_{ai}$ the total air resistance, symbol W_T, units m/s. The inverse $1/W_T$ is the total air conductance K_{aT}. The term d_i/k_{ai}, symbol W_i, represents the air resistance,

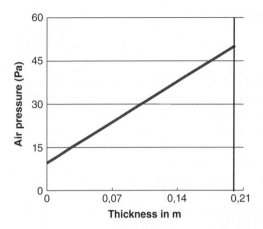

Fig. 2.6 Airflow across a single-layer flat assembly, air pressure line

and the inverse, $1/W_i = k_{ai}/d_i$, the air conductance K_{ai} of layer i. Air pressure in an interface is given by:

$$P_{aj} = P_{a1} + \frac{\sum_{i=1}^{j} W_i}{W_T}(P_{a2} - P_{a1}) = P_{a1} + g_a W_{1j} \qquad (2.25)$$

Graphically the assembly behaves as a single layer in the $[W, P_a]$ plane. Transferring the intersections with all interfaces in the correct order to the $[d, P_a]$ plane gives the air pressure course in the thicknesses (d) of the successive layers in the correctly drawn assembly (Figure 2.7).

The air pressure difference over a layer thus is:

$$\Delta P_{aj} = \frac{W_j}{W_T}(P_{a2} - P_{a1})$$

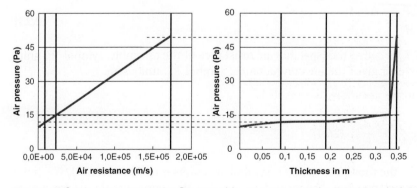

Fig. 2.7 Airflow across a composite flat assembly: air pressure in the $[W_a, P_a]$ and $[d, P_a]$ planes

2.2 Air

showing that the most air-tight layer buffers the greatest difference. Inclusion of such an 'air barrier' or 'air retarder' diminishes the total air permeance substantially and minimizes the air flux. Of course, air barriers must withstand the wind load.

Is this flat assembly approach of practical relevance? Not really. The prerequisite of having one-dimensional flow already excludes stack with its pressure profile along a wall's height. Also, wind is never uniform. Cracks, leaks and others also disrupt one-dimensionality.

2.2.4.3 Two and three dimensions

For isotropic open-porous materials, the calculation is a copy of two- and three-dimensional heat conduction. Of course, mass conservation substitutes energy conservation – the sum of air flows from every neighbour to each central control volume zero:

$$\sum G_{a,i+j} = 0$$

The algorithm thus becomes:

$$\sum_{\substack{i=l,m,n \\ j=\pm 1}} \left(K'_{a,i+j} P_{a,i+j} \right) - P_{a,l,m,n} \sum_{\substack{i=l,m,n \\ j=\pm 1}} K'_{a,i+j} = 0$$

with $K'_{a,i+j}$ the air permeance between every neighbouring and each central control volume. The value within a material is:

$$K_a = k_a A / a$$

with A the contact surface with the adjoining volume and a the distance along the mesh between the centre of the neighbour and the control volume considered (Figure 2.8).

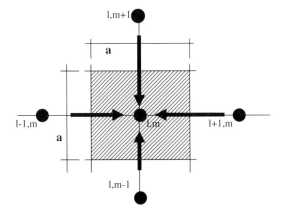

Fig. 2.8 Conservation of mass: sum of airflows from the adjoining to the central control volume zero

For p control volumes, the result is a system of p equations with p unknowns:

$$[K'_a]_{p,p}[P_a]_p = [K'_{a,i,j,k}P_{a,i,j,k}]_p \tag{2.26}$$

where $[K'_a]_{p,p}$ is a p rows, p columns permeance matrix, $[P_a]_p$ a column matrix of p unknown air pressures, and $[K'_{a,i,j,k}P_{a,i,j,k}]_p$ a column matrix of p known air pressures or air flows. Once the air pressures are known, then the flows follow from:

$$G_{a,i,j,k} = K'_{a,i,j,k}(P_{a,i+j} - P_{a,i}) \tag{2.27}$$

For anisotropic materials the same algorithm applies on condition that the lines linking the centres of the control volumes coincide with the main directions of the permeability tensor, allowing use of the related $(K'_a)_{x,y,z}$ values. Under non-isothermal conditions, thermal stack links the air with the heat balances. Most of the time, the solution requires iteration between both.

2.2.5 Airflow across assemblies with air-open layers, leaky joints, leaks and cavities

The flow equations were introduced above. For operable windows and doors, the exponent b is typically set at 2/3:

$$G_a = a\, L\, \Delta P_a^{2/3}$$

with L the length of the casement in m and a the air permeance per metre run of casement, with units kg/(m.s.Pa$^{2/3}$). Most assemblies combine air-open layers, open porous materials, joints and cavities. In such cases, writing the conservation law as a partial differential equation fails. One approach consists of transforming the assembly into an equivalent hydraulic circuit, with a combination of well-chosen points connected by air permeances (Figure 2.9).

Per point, the sum of the airflows from the neighbouring points must be zero. As each flow equals $K_a^x \Delta P_a$, insertion in the conservation law gives:

$$\sum_{\substack{i=l,m,n \\ j=\pm 1}} \left(K^x_{a,i+j} P_{a,i+j}\right) - P_{a,l,m,n} \sum_{\substack{i=l,m,n \\ j=\pm 1}} K^x_{a,i+j} = 0 \tag{2.28}$$

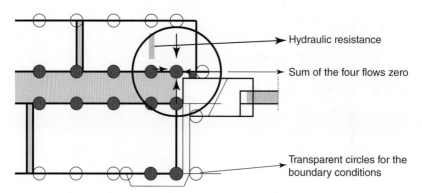

Fig. 2.9 Equivalent hydraulic circuit

2.2 Air

With p points, a system of p non-linear equations ensues. With known boundary conditions, the p unknown air pressures follow from solving that system. Insertion into the flow equations gives the airflows between points. Non-linearity makes iteration necessary. Assume values for the p air pressures, calculate the related air permeances and solve the system. Recalculate the permeances with the p air pressures found, solve the system again and continue until the standard deviation between preceding and new air pressures equals a preset value. In the exceptional case that the flow in a flat assembly is nonetheless one-dimensional, a series circuit of linear and non-linear air permeances emerges, with the same air flux g_a migrating across all:

$$\text{Layer 1} \quad g_a = K_{a_1} \Delta P_{a_1} = a_1 \Delta P_{a_1}^{b_1} \quad \text{or} \quad \left(\frac{g_a}{a_1}\right)^{1/b_1} = \Delta P_{a_1}$$

$$\text{Layer 2} \quad g_a = K_{a_2} \Delta P_{a_2} = a_2 \Delta P_{a_2}^{b_2} \quad \text{or} \quad \left(\frac{g_a}{a_2}\right)^{1/b_2} = \Delta P_{a_2}$$

$$\text{Layer } n \quad g_a = K_{a_n} \Delta P_{a_n} = a_n \Delta P_{a_n}^{b_n} \quad \text{or} \quad \left(\frac{g_a}{a_n}\right)^{1/b_n} = \Delta P_{a_n}$$

$$\text{Sum:} \quad g_a \left[\sum \frac{g_a^{1/b_i - 1}}{a_i^{1/b_i}}\right] = \Delta P_a$$

(2.29)

ΔP_a is the air pressure differential. The solution again requires iteration. Once the 'correct' air flux is known, the air pressure distribution in the assembly follows from the layer equations.

2.2.6 Air transfer at the building level

2.2.6.1 Definitions

Air in buildings may flow inter- and intra-zone. 'Inter' stands for air exchanges among spaces and the outdoors. Modelling starts from a grid of zone points linked by flow paths. 'Intra' relates to air movement within spaces and relates to questions such as 'What about looping? How does the zone and ventilation air mix? Which corners lack air washing?'. Solving requires computational fluid dynamics (CFD). The discussion here deals with interzone airflow only.

2.2.6.2 Thermal stack

Thermal stack acts vertically. With the outdoors as a reference, the equation is:

$$P_a - g\rho_a z = P_a - z\frac{gP_a}{R_a}\left(\frac{1}{T_e} - \frac{1}{T_i}\right) \approx P_a - 3460z\left(\frac{1}{T_e} - \frac{1}{T_i}\right) \quad (2.30)$$

with T_i the temperature in the space, T_e the temperature outdoors, both in K, and z the height. Between equally warm spaces at different heights coupled by a flow path, the pressure difference with the outdoors is:

$$\Delta_{12}(P_a - \rho g z) = P_{a1} - P_{a2} - 3460(z_1 - z_2)\left(\frac{1}{T_e} - \frac{1}{T_i}\right)$$

where z_1 and z_2 are the vertical distances from the zone points to a lower horizontal reference plane. If the spaces are at different temperatures ($T_{i,1}$, $T_{i,2}$), the equation changes to:

$$\Delta_{12}(P_a - \rho g z) = P_{a1} - P_{a2} - 3460\left[(z_o - z_1)\left(\frac{1}{T_e} - \frac{1}{T_{i,1}}\right) + (z_2 - z_o)\left(\frac{1}{T_e} - \frac{1}{T_{i,2}}\right)\right]$$

with z_o the height of the horizontal plane where the temperature changes.

2.2.6.3 Large openings

Open doors and windows act as large openings, of which the air permeance (K_a) changes with the driving force. While wind and fans give uniform flows, thermal stack activates a double flow, from warm to cold above and from cold to warm below the central horizontal, with their sum being zero (Figure 2.10). In other words, the same amount of air moves in both directions.

The question is how to calculate the air permeance? For wind and fans, the airflow is:

$$G_a = C_f B H \sqrt{2\rho_a \Delta P_a} \quad \text{(kg/s)} \tag{2.31}$$

giving as air permeance:

$$K_a = \left(C_f B H \sqrt{2\rho_a}\right) \Delta P_a^{-0.5} \tag{2.32}$$

with C_f a flow factor with value 0.33 to 0.7, B the width and H the height of the opening.

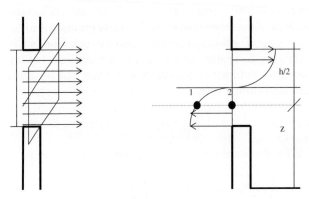

Fig. 2.10 Large opening, uniform flow by wind and fans (left), and stack-induced flow (right)

2.2 Air

With thermal stack the flow in both directions equals:

$$G_{a1} = -G_{a2} = \frac{C'_f B}{3} \sqrt{\frac{\rho_a g P_a H^3}{R_a} \left(\frac{1}{T_1} - \frac{1}{T_2}\right)} \quad \text{(kg/s)} \tag{2.33}$$

The air permance thus is:

$$\Delta P_{T,\max} = g P_a H / R_a \left(T_e^{-1} - T_2^{-1} - T_e^{-1} + T_1^{-1}\right) \approx 3450 \, H \left(T_1^{-1} - T_2^{-1}\right) :$$

$$K_{a1} = -K_{a2} = \frac{C'_f BH \sqrt{\rho_a}}{3} \left[3450 H \left(\frac{1}{T_1} - \frac{1}{T_2}\right)\right]^{-0.5} \tag{2.34}$$

with R_a the gas constant of air and C'_f the flow factor for this two-way move. The equation gives a parabolic air velocity along the opening's height, a consequence of Bernouilli's law, which for a flow path at height z above the floor states:

$$\rho_{a1} g z = \frac{\rho_{a2} v_{a,z}^2}{2} + \rho_{a2} g z$$

giving:

$$v_{a,z} = \sqrt{2 g z \left(\frac{\rho_{a1} - \rho_{a2}}{\rho_{a2}}\right)}$$

The air permeance equation follows from integrating the air velocity over half the height, then using the ideal gas law to convert density into temperature and multiplying the result by the flow factor.

2.2.6.4 The conservation law applied

Per node in a building (see Figure 2.11), the algebraic sum of the n inflows and the m outflows must be zero:

$$\sum_{i=1}^{m+n} G_{a,i} = 0 \tag{2.35}$$

The living room, zone 2 in the figure, has two windows, while doors separate it from the kitchen, zone 1, and the hall, zone 3. All partitions and opaque outer walls are assumed air-tight, which may differ from reality. In a first step the building has no fan-driven ventilation and the indoor and outdoor temperatures are equal. So, only wind matters. Air comes in and goes out of the living room across the operable window sashes ($G_{a,2\text{-e1}}$, $G_{a,2\text{-e2}}$) and the door bays with the kitchen ($G_{a,2\text{-1}}$) and hall ($G_{a,2\text{-3}}$). Wind pressures outdoors are supposed to be known and all four flows are assumed entering the living room:

$$G_{a,2\text{-e1}} = K_{a,2\text{-e1}} \left(P_{a,e1} - P_{a,2}\right) \qquad G_{a,2\text{-e2}} = K_{a,2\text{-e2}} \left(P_{a,e2} - P_{a,2}\right)$$

$$G_{a,2\text{-1}} = K_{a,2\text{-1}} \left(P_{a,1} - P_{a,2}\right) \qquad G_{a,2\text{-3}} = K_{a,2\text{-3}} \left(P_{a,3} - P_{a,2}\right)$$

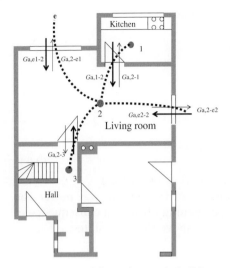

Fig. 2.11 Ground floor of example building

Summing and reshuffling gives:

$$-(K_{a,2\text{-}e1} + K_{a,2\text{-}e2} + K_{a,2\text{-}1} + K_{a,2\text{-}3})P_{a,2} + K_{a,2\text{-}e1}P_{a,e1} + K_{a,2\text{-}e2}P_{a,e2} + K_{a,2\text{-}1}P_{a,2}$$
$$+ K_{a,2\text{-}3}P_{a,3} = 0$$

or:

$$-P_{a,2}\sum\left(K_{a,i-j/e}\right) + \sum\left(K_{a,i-j}P_{a,j}\right) = -\sum\left(K_{a,i-e}P_{a,e}\right)$$

which is an equation with three unknowns: air pressure in the living room (P_{a2}), air pressure in the kitchen (P_{a1}) and air pressure in the hall (P_{a3}). Besides the living room, kitchen and hall, the ground floor also includes a garage and a toilet, while the staircase forms a link with the first floor and its three bedrooms and one bathroom. For each, all flows entering and leaving have sum zero. The result is as many equations as the building has spaces. In matrix notation:

$$[K_a]_{n,n}[P_a]_n = [K_{a,e}P_{a,e}]_n \qquad (2.36)$$

Solving that system of non-linear equations demands iteration. If the building is warmer or colder than outdoors, then all leaks must be located first.

All spaces get a node at each leak height, after which thermal stack with outdoors against a reference height complements the pressure differences between the neighbouring nodes at another height. Among nodes above each other in the same space, very large air permeances are assumed. With the indoor temperatures in all spaces and the outdoor temperature known, stack changes into a known term in the balance equations. Otherwise, in the unheated spaces, the temperature will depend on the transmission losses, the heat gains and the temperature of the entering air. This requires consideration of conservation of energy. The correct temperatures then follow from assuming temperatures first, followed by alternately solving the

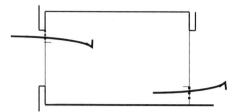

Fig. 2.12 Two openings in series

heat and air balances using the temperatures and airflows from the preceding iteration, until the differences between two successive solutions drop below a preset value. In the case that ventilation is fan-driven, the fans figure as sources with known pressure/flow characteristics.

2.2.6.5 Applications

Airflow across one narrow horizontal opening in a vertical wall belonging to an otherwise air-tight enclosed space, such as a trickle vent above a window, is barely possible. Large vertical openings such as operable windows instead activate air exchange. Not only does thermal stack cause twinned flow, but fluctuating wind pressures also induce pulsating air inflow and outflow when the window is set ajar.

Two openings in opposition instead act as a serial circuit. Let one have an air permeance $K_{a,1}$, and the other an air permeance of $K_{a,2}$. $P_{a,1}$ is the air pressure in front of one, $P_{a,2}$ the air pressure past the other. In isothermal conditions the air balance thus is (Figure 2.12):

$$-(K_{a,1} + K_{a,2})P_{a,x} + K_{a,1}P_{a,1} + K_{a,2}P_{a,2} = 0$$

with $P_{a,x}$ the unknown air pressure in the space.

For the air to flow, both pressures $P_{a,1}$ and $P_{a,2}$ must differ, which is most likely when the openings sit in outside walls with different orientations. Because the wind pressure increases with height, two openings in the same outside wall at different heights also generate airflow – how great depends on the air permeances. If constant, the mass balance gives:

$$G_{a,1} = -G_{a,2} = G_a = K_{a,1}(P_{a,1} - P_{a,x}) = \frac{1}{\frac{1}{K_{a1}} + \frac{1}{K_{a2}}}(P_{a,1} - P_{a,2})$$

Depending on the pressure difference but with the same exponent, the airflow becomes:

$$G_a = \frac{1}{\left(\frac{1}{a_1^{1/b}} + \frac{1}{a_2^{1/b}}\right)^b}(P_{a,1} - P_{a,2})^b$$

If the exponent differs, the airflow requires iteration starting from:

$$G_a = \frac{1}{\left[\left(\dfrac{G_a}{a_1}\right)^{\frac{1-b_1}{b_1}} + \left(\dfrac{G_a}{a_2}\right)^{\frac{1-b_2}{b_2}}\right]} (P_{a,1} - P_{a,2})$$

Solving this begins with guessing the airflow G_a, for example by setting the exponents of both air permeances to 1. Then the left-hand equation using this airflow gives a new airflow, which in turn is implemented. Iterating in this way continues until the difference between previous and actual airflow obeys a preset value. The equations anyhow underline that, when one of the two air permeances is really small, the airflow nears zero.

If there are several spaces coupled in series with an inlet in the first, an outlet in the last, and flow through openings in all partitions, the resulting airflow across becomes:

All air permeances constant:

$$G_a = \frac{1}{\sum_{j=1}^{n} \dfrac{1}{K_{aj}}} (P_{a,1} - P_{a,n+1})$$

All air permeance exponents b identical:

$$G_a = \frac{1}{\left(\sum_{j=1}^{n} \dfrac{1}{a_j^{1/b}}\right)^b} (P_{a,1} - P_{a,n+1})^b$$

Air permeance exponents different:

$$G_a = \frac{1}{\sum_{j=1}^{n} \left[\left(\dfrac{G_a}{a_j}\right)^{\frac{1-b_j}{b_j}}\right]} (P_{a,1} - P_{a,n+1})$$

with $P_{a,1}$ and $P_{a,n+1}$ the air pressures in front of the inlet and past the outlet. The lowest air permeance again defines the magnitude of the flow.

With openings in the same assembly at different heights and at different temperatures, thermal stack intervenes. Assume windless weather. The outside wall considered has two openings at a height ΔH one above the other. Both air permeances are constant. On the horizontals across, the air pressures are $P_{a,1}$ and $P_{a,2}$ respectively, while a very large but fictitious permeance $K_{a,z}$ couples related calculation points 1 and 2 (Figure 2.13). The mass balances are:

Node 1: $-(K_{a,1} + K_{a,z})P_{a,1} + K_{a,z}P_{a,2} = -K_{a,1}P_{a,e} + K_{a,z}\left[3460\Delta H\left(\dfrac{1}{T_e} - \dfrac{1}{T_i}\right)\right]$

Node 2: $-(K_{a,2} + K_{a,z})P_{a,2} + K_{a,z}P_{a,1} = -K_{a,2}P_{a,e} - K_{a,z}\left[3460\Delta H\left(\dfrac{1}{T_e} - \dfrac{1}{T_i}\right)\right]$

2.2 Air

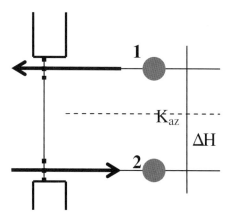

Fig. 2.13 Openings at different height in a same outer wall, thermal stack intervening

With zero air pressure outdoors, $P_{a,e}$ and $K_{a,z}$ are infinite, so the airflow becomes:

$$G_a = \frac{1}{\frac{1}{K_{a1}} + \frac{1}{K_{a2}}} \left[3460 \Delta H \left(\frac{1}{T_e} - \frac{1}{T_i} \right) \right] \approx \frac{1}{\frac{1}{K_{a1}} + \frac{1}{K_{a2}}} [0.043 \Delta H (\theta_i - \theta_e)]$$

Or, the two act as series with the lowest fixing the airflow.

Several openings in a partition between two equally warm spaces act as a circuit in parallel. If the air pressures at both sides differ, the resulting airflow is:

All air permeances constant:

$$G_a = \sum_{j=1}^{n} K_{aj}(P_{a,1} - P_{a,2})$$

All air permeance exponents b identical:

$$G_a = \sum_{j=1}^{n} a_j (P_{a,1} - P_{a,2})^b$$

Air permeance exponents different:

$$G_a = \sum_{j=1}^{n} \left(a_j \Delta P_a^{b_j - 1} \right) (P_{a,1} - P_{a,2})$$

Clearly, its value increases as more openings or leaks perforate the partition wall.

2.2.7 Combined heat and air flow in open-porous materials

2.2.7.1 Heat balance equation

Air displacement across an open-porous material adds an enthalpy flow to transmission. The air movement anyhow will be so slow that the material and the air passing through will have the same temperature in every spot. Energy conservation for an elementary volume dV so becomes:

$$\text{div}(\mathbf{q} + c_a \mathbf{g}_a \theta) = -\frac{\partial(\rho c + \rho_a c_a S_a \Psi_o)\theta}{\partial t} \pm \Phi'$$

with $\rho_a c_a$ the volumetric specific heat capacity of air, S_a the saturation degree for air and Ψ_o the open porosity of the material. In most cases the product $\rho_a c_a S_a \Psi_o$ is too small to be considered, which simplifies the balance to:

$$\text{div}(\mathbf{q} + c_a \mathbf{g}_a \theta) = -\frac{\partial \rho c \theta}{\partial t} \pm \Phi'$$

Introducing Fourier's law, assuming the thermal conductivity (λ) and the volumetric heat capacity (ρc) to be constant and expanding the operator term, wherein div $\mathbf{g}_a$ is zero, gives:

$$-\lambda \nabla^2 \theta + c_a \mathbf{g}_a \nabla \theta = -\rho c \frac{\partial \theta}{\partial t} \pm \Phi' \tag{2.37}$$

2.2.7.2 Steady state: flat assemblies

In steady state, the boundary conditions do not depend on time, heat sources and sinks never change, and material properties remain constant. The equation then simplifies to:

$$\lambda \nabla^2 \theta - c_a \mathbf{g}_a \nabla \theta = \pm \Phi' \tag{2.38}$$

If the air and heat fluxes develop perpendicular to the assembly's surface, and if neither a heat source nor a sink intervenes, it reduces to:

$$\frac{d^2 \theta}{dx^2} - \frac{c_a g_a}{\lambda} \frac{d\theta}{dx} = 0 \tag{2.39}$$

This second-order differential equation has, as a solution:

$$\theta = C_1 + C_2 \exp\left(\frac{c_a g_a x}{\lambda}\right) \tag{2.40}$$

with C_1 and C_2 the integration constants ensuing from the boundary conditions. The conduction-related, enthalpy-related and total heat flux then are:

Conduction	$q = -\lambda \dfrac{d\theta}{dx} = -c_a g_a C_2 \exp\left(\dfrac{c_a g_a x}{\lambda}\right)$
Enthalpy (convection)	$H = q_c = c_a g_a \left[C_1 + C_2 \exp\left(\dfrac{c_a g_a x}{\lambda}\right)\right]$
Total	$q_T = q + H = q + q_c = c_a g_a C_1$

Whereas conduction and convection change exponentially over the assembly, the total flux remains constant. In other words, the part functions as a heat exchanger between the conduction and enthalpy flow, thus heating or cooling the migrating air.

2.2 Air

If d is the thickness of a single layer assembly, rewriting the solution gives:

$$\theta = C_1 + C_2 \exp\left(\frac{c_a g_a d}{\lambda} \frac{x}{d}\right) \quad (2.41)$$

The dimensionless ratio $c_a g_a d/\lambda$, called the Péclet number, symbol Pe, links the enthalpy to the conductive flux. The more air flows across $1\,\text{m}^2$ and the higher the thermal resistance of the assembly, the higher the Peclet number.

Type 1 boundary conditions now presume that the air flux across and the temperatures on the assembly's end faces are known: $x=0$: $\theta=\theta_{s1}$; $x=d$: $\theta=\theta_{s2}$. The integration constants then follow from: $\theta_{s1} = C_1 + C_2$ and $\theta_{s2} = C_1 + C_2 \exp(\text{Pe})$, or:

$$C_1 = \frac{\theta_{s2} - \theta_{s1} \exp(\text{Pe})}{1 - \exp(\text{Pe})} \qquad C_2 = \frac{\theta_{s1} - \theta_{s2}}{1 - \exp(\text{Pe})}$$

The temperatures in the assembly and the conductive, convective and total heat flux become:

$$\theta = \theta_{s1} + F_1(x)(\theta_{s2} - \theta_{s1})$$

$$q = F_2(x)\lambda \frac{\theta_{s2} - \theta_{s1}}{d} \qquad q_c = \text{Pe}\frac{\lambda \theta}{d} \qquad q_T = \lambda \frac{\theta_{s2} F_2(0) - \theta_{s1} F_2(d)}{d} \quad (2.42)$$

with:

$$F_1(x) = \frac{1 - \exp\left(\text{Pe}\frac{x}{d}\right)}{1 - \exp(\text{Pe})} \qquad F_2(x) = \frac{\text{Pe}\exp\left(\text{Pe}\frac{x}{d}\right)}{1 - \exp(\text{Pe})}$$

The functions $F_1(x)$ and $F_2(x)$ depend on the material properties and the air flux. If moving from thickness d to 0, the flux has a negative sign, while in the opposite case it is positive. The airflow causes an exponential change in temperature, convex when flowing from warm to cold, and concave otherwise (Figure 2.14).

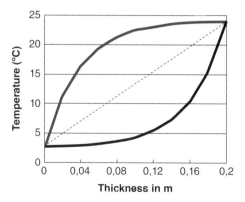

Fig. 2.14 Single-layer assembly, combined conductive and convective flow. The temperature curve is convex for warm air migrating to the cold side, and concave for cold air moving to the warm side

The higher the Péclet number, the more air passes through, the more concave or convex the exponential. Compared with pure conduction (q_o), the change in percentage of the total heat flux is (θ_{s1} for $x=0$ is 0 °C), θ_{s2} for $x=d$ is 1 °C):

$$100(q_T - q_0)/q_0 = 100[\text{abs}(F_2(0)) - 1]$$

For air flowing from warm to cold, abs($F_2(0)$) exceeds 1. In such case, the enthalpy flux increases the total heat flux. For air flowing from cold to warm, abs($F_2(0)$) drops below 1 and the enthalpy flux lowers the total. Conduction changes the opposite way: lower when air permeates from warm to cold, higher when moving from cold to warm.

Type 2 boundary conditions assume that the air flux across and the heat flux by convection and radiation at both end faces are known. Adding a fictitious surface layer with thermal resistance $1/h_1$ at the face $x=0$, and a fictitious surface layer with thermal resistance $1/h_2$ at the face $x=d$, converts type 2 to type 1. The reference temperatures θ_1 and θ_2 so become fictitious surface temperatures while the assembly changes from single-layer to composite. Rewriting the thermal balance with R as an independent variable gives:

$$\frac{d^2\theta}{dR^2} - c_a g_a \frac{d\theta}{dR} = 0 \qquad (2.43)$$

with as solution:

$$\theta = C_1 + C_2 \exp(c_a g_a R)$$

The boundary conditions are:

$$R = 0: \quad \theta = \theta_1 \qquad R = R_T = \frac{1}{h_1} + R + \frac{1}{h_2}: \quad \theta = \theta_2$$

The integration constants thus equal:

$$C_1 = \frac{\theta_2 - \theta_1 \exp(c_a g_a R_T)}{1 - \exp(c_a g_a R_T)} \qquad C_2 = \frac{\theta_1 - \theta_2}{1 - \exp(c_a g_a R_T)}$$

Temperatures, the conduction, convection and total heat fluxes become:

$$\theta = \theta_1 + (\theta_2 - \theta_1)F_1(R)$$

$$q = F_2(R)(\theta_2 - \theta_1) \qquad q_c = c_a g_a \theta \qquad q_T = \theta_2 F_2(0) - \theta_1 F_2(R_T) \qquad (2.44)$$

with:

$$F_1(R) = \frac{1 - \exp(c_a g_a R)}{1 - \exp(c_a g_a R_T)} \qquad F_2(R) = \frac{c_a g_a \exp(c_a g_a R)}{1 - \exp(c_a g_a R_T)} \qquad (2.45)$$

In these equations, R is the thermal resistance between the fictitious face $R=0$ and any parallel plane in the assembly. Again, the functions F_1 and F_2 depend on the material characteristics and the air flux.

The solution for type 2 boundary conditions also fits for a composite assembly. With R=0 outdoors or on the other side and R=R_T indoors, the boundary conditions become:

$$R = 0: \quad \theta = \theta_e \qquad R = R_T = \frac{1}{h_1} + \sum R + \frac{1}{h_2}: \quad \theta = \theta_i$$

For the temperatures in the interfaces, surface 1 being outdoors, interface 1 being closest to surface 1, and so on, and related heat fluxes θ_e ($R=0$) $< \theta_i$ ($R=R_T$), the result is:

Surface 1: $\theta_{s1} = \theta_e + (\theta_i - \theta_e)F_1(1/h_1)$
Interface 1: $\theta_1 = \theta_e + (\theta_i - \theta_e)F_1(1/h_1 + R_1)$
Interface 2: $\theta_2 = \theta_e + (\theta_i - \theta_e)F_1(1/h_1 + R_1 + R_2)$

...

$$q = F_2(R)(\theta_i \quad \theta_e) \qquad q_c = c_a g_a \theta \qquad q_T = \theta_i F_2(0) - \theta_e F_2(R_T)$$

with:

$$F_1(R) = \frac{1 - \exp(c_a g_a R)}{1 - \exp(c_a g_a R_T)} \qquad F_2(R) = \frac{c_a g_a \exp(c_a g_a R)}{1 - \exp(c_a g_a R_T)}$$

In the [θ, R] plane, the temperature curve resembles a single-layer assembly (Figure 2.15). If the airflow is to the colder outdoors, the curve turns convex, while vice versa it is concave. With inside insulation, outflow in winter will increase the wall temperature.

What consequences are there with a leaky envelope? In windy weather, infiltration occurs at the windward side, and exfiltration at the leeward side and the sides more or less parallel to the wind. A comparison of the sum of these combined heat flows with the total flow for conduction and infiltration separately, shows that the combined is lower. In other words, the passage of air through the envelope results in energy conservation. Indeed, part of the higher windward conduction losses warm up the

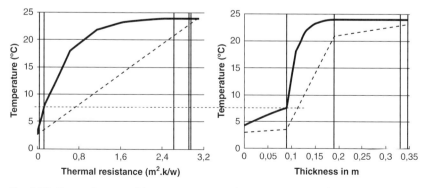

Fig. 2.15 Composite assembly, temperatures, indoor air moving to the cold outdoors

incoming air, while at the leeward side and the sides more or less parallel to the wind, the warm outgoing air lowers the conduction losses. Nonetheless, this advantage fades away compared with the many drawbacks that air passages create.

2.2.7.3 Steady state: two and three dimensions

Here, analytical solutions for the air balance equation are rare. When using CVM, first the airflows have to be calculated, then the temperatures and heat fluxes. If thermal stack intervenes, iteration is needed. The heat flows between every neighbouring point and each central point can be written using a purely numerical approach. Then the flow becomes:

$$\text{Airflow from 2 to 1}: \quad \Phi_{21} = P'_{21}(\theta_2 - \theta_1) + c_a G_a \theta_2 = \theta_2(P'_{21} + c_a G_a) - \theta_1 P'_{21}$$
$$\text{Airflow from 1 to 2}: \quad \Phi_{21} = P'_{21}(\theta_2 - \theta_1) - c_a G_a \theta_2 = \theta_2 P'_{21} - \theta_1(P'_{21} + c_a G_a)$$
(2.46)

with P'_{21} the thermal permeance. An alternative uses the analytical solution for flat assemblies, giving a series connection of thermal resistances:

$$\Phi_{21} = \theta_1 F_2(0) - \theta_2 F_2(R'_{21})$$

where R'_{21} is the surface-linked thermal resistance between two neighbouring points. In case of parallel serial paths, the flow per serial path is calculated first, after which the total flow is:

$$\Phi_{21} = \sum_{i=1}^{2} \Phi^i_{21}$$

2.2.7.4 Transient: flat assemblies

For a constant air flux and periodically varying temperatures, the complex calculation method for heat conduction is applicable, giving, per harmonic:

$$\theta(R) = \alpha(R) \exp \frac{2in\pi t}{T}$$

with $\alpha(R)$ the complex temperature. In the absence of a heat source or sink, this temperature ensues from:

$$\frac{d^2 \alpha}{dR^2} - c_a g_a \frac{d\alpha}{dR} - \frac{2in\pi \rho c \lambda}{T} \alpha = 0$$

with i the imaginary unit and T the period. The solution is:

$$\alpha(R) = C_1 \exp(r_1 R) + C_2 \exp(r_2 R) \tag{2.47}$$

where r_1 and r_2 are the roots of the quadratic equation $r^2 - c_a g_a r - 2in\pi \rho c \lambda / T = 0$:

$$r_1 = \frac{1}{2}\left[c_a g_a + \sqrt{(-c_a g_a)^2 + \frac{8in\pi \rho c \lambda}{T}}\right] \qquad r_2 = \frac{1}{2}\left[c_a g_a - \sqrt{(-c_a g_a)^2 + \frac{8in\pi \rho c \lambda}{T}}\right]$$

2.2 Air

or:

$$\alpha(R) = \exp\left(\frac{1}{2}c_a g_a R\right)\left[(C_1 - C_2)\sinh\left(\frac{1}{2}aR\right) + (C_1 + C_2)\cosh\left(\frac{1}{2}aR\right)\right]$$

with:

$$a = \sqrt{(-c_a g_a)^2 + \frac{8 i n \pi \rho c \lambda}{T}}$$

For air flux set to zero, the solution for conduction is obtained. The integration constants C_1 and C_2 follow from the boundary conditions, for example the complex temperature α_s and complex heat flux α'_s at the end face $R=0$. It is up to the reader to make these and all other calculations. The most important result is that the inflow decreases whereas the outflow increases the damping and the time shift of an assembly.

2.2.7.5 Transient: two and three dimensions

For two and three dimensional details, the energy balance per CVM control volume is:

$$\sum \Phi_m \approx \rho c \Delta V \Delta \theta / \Delta t \pm \Phi' \Delta V$$

with Φ' a heat source or sink and Φ_m the heat flow by conduction and convection during the time span Δt from a neighbour to any central point.

2.2.7.6 Air permeable layers, joints, leaks and cavities

With air permeable layers, leaks, joints and cavities, steady-state and transient problems are solved by combining the equivalent hydraulic network with CVM for the heat balances with the numeric steady-state equations for the heat flows given above.

2.2.7.7 Vented cavity

In a given assembly, a flat vented cavity, air velocity v m/s, has a width d_{cav} and a length L (Figure 2.16). The convective surface film coefficient at both cavity faces is h_c (W/(m².K)), R_1 is the thermal resistance between ambient 1 and cavity face 1, and R_2 is the thermal resistance between ambient 2 and cavity face 2 (m².K/W). The heat balance at both faces and along the z axis for the cavity equal:

$$\frac{\theta_1 - \theta_{s1}}{R_1} + h_c(\theta_{cav} - \theta_{s1}) + h_r(\theta_{s2} - \theta_{s1}) = 0$$

$$\frac{\theta_2 - \theta_{s2}}{R_2} + h_c(\theta_{cav} - \theta_{s2}) + h_r(\theta_{s1} - \theta_{s2}) = 0$$

$$[h_c(\theta_{s1} - \theta_{cav}) + h_c(\theta_{s2} - \theta_{cav})]\,dz = \rho_a c_a d_{sp} v\,d\theta_{cav}$$

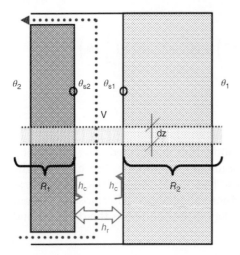

Fig. 2.16 Vented cavity in a flat assembly. Combines heat conduction through both leafs with enthalpy flow in the cavity

with h_r the surface film coefficient for radiation. The unknowns in these equations are the temperatures in the cavity and at both cavity faces (θ_{s1}, θ_{s2}, θ_{cav}).

To solve these, the temperatures θ_{s1} and θ_{s2} are rewritten as functions of the cavity temperature θ_{cav} using the surface balances, after which both are implemented in the cavity equation. To keep the equations looking simple, the constants D, A_1, A_2, B_1, B_2, C_1 and C_2 are defined as:

$$D = \left(h_c + h_r + \frac{1}{R_1}\right)\left(h_c + h_r + \frac{1}{R_2}\right) - h_r^2$$

$$A_1 = \frac{h_c + h_r + \dfrac{1}{R_2}}{D\,R_1} \qquad A_2 = \frac{h_r}{D\,R_1} \qquad B_1 = \frac{h_r}{D\,R_2} \qquad B_2 = \frac{h_c + h_r + \dfrac{1}{R_1}}{D\,R_2}$$

$$C_1 = \frac{h_c\left(h_c + h_r + \dfrac{1}{R_2}\right) + h_r h_c}{D} \qquad C_2 = \frac{h_c\left(h_c + h_r + \dfrac{1}{R_1}\right) + h_r h_c}{D}$$

The surface temperatures so become:

$$\theta_{s1} = A_1\theta_1 + B_1\theta_2 + C_1\theta_{cav} \qquad \theta_{s2} = A_2\theta_1 + B_2\theta_2 + C_2\theta_{cav} \qquad (2.48)$$

while the cavity balance turns into:

$$\left[\underbrace{\frac{(A_1+A_2)\theta_1 + (B_1+B_2)\theta_2}{2 - C_1 - C_2}}_{=a} - \theta_{cav}\right]dz = \underbrace{\frac{\rho_a c_a d_{cav} v}{h_c(2 - C_1 - C_2)}}_{=b}\,d\theta_{cav}$$

2.2 Air

with as solution:

$$\frac{z}{b} = -\ln\left(\frac{a - \theta_{cav}}{C}\right)$$

The integration constant C follows from the boundary conditions. For z nearing infinity, the ratio z/b_1 approaches infinity, turning the solution into:

$$\ln\left(\frac{a - \theta_{cav,\infty}}{C}\right) = -\infty$$

or $a_1 - \theta_{cav,\infty} = 0$, giving $a_1 = \theta_{cav,\infty}$. Consequently, a_1 equals the temperature asymptote in an infinite cavity, a value equal to the air temperature in the cavity if non-vented. Indeed, z/b_1 also becomes infinite for an air velocity of zero. For $z = 0$, θ_{cav} equals $\theta_{cav,0}$, giving:

$$\ln\left(\frac{\theta_{cav,\infty} - \theta_{cav,0}}{C}\right) = 0 \quad \text{or} \quad C = \theta_{cav,\infty} - \theta_{cav,0}$$

The cavity temperature thus equals:

$$\theta_{cav} = \theta_{cav,\infty} - (\theta_{cav,\infty} - \theta_{cav,0})\exp(-z/b)$$

The temperature in a ventilated cavity thus changes exponentially, from the inflow value to a value close to that if non-vented. The term b has a dimension of metres and is therefore called the ventilation length. The higher the air velocity in the cavity, the greater this length. The more heat crosses the cavity, the lower it becomes.

The temperatures at both cavity faces (θ_{s1}, θ_{s2}) are found by implementing the cavity temperature θ_{cav} in their temperature equations:

$$\theta_{s1} = A_1\theta_1 + B_1\theta_2 + C_1\left[\theta_{cav,\infty} - (\theta_{cav,\infty} - \theta_{cav,0})\exp\left(-\frac{z}{b}\right)\right]$$

$$\theta_{s2} = A_2\theta_1 + B_2\theta_2 + C_2\left[\theta_{cav,\infty} - (\theta_{cav,\infty} - \theta_{cav,0})\exp\left(-\frac{z}{b}\right)\right]$$

To find the heat loss, let '2' be the inside leaf. For the heat flow across, the following applies:

$$\Phi = \frac{1}{R_2}\int_0^L (\theta_2 - \theta_{s2})dz$$

$$= \frac{1}{R_2}\int_0^L \left\{-A_2\theta_1 + (1 - B_2)\theta_2 - C_2\left[\theta_{cav,\infty} - (\theta_{cav,\infty} - \theta_{cav,0})\exp\left(-\frac{z}{b}\right)\right]\right\}dz$$

$$= \frac{(\theta_2 - \theta_{s2\infty})L}{R_2} - \frac{C_2 b_1(\theta_{sp\infty} - \theta_{sp0})}{R_2}\left[\exp\left(-\frac{L}{b}\right) - 1\right]$$

The average thermal transmittance (U) thus becomes:

$$U = \frac{\Phi}{L(\theta_1 - \theta_2)} = U_o + \frac{C_2 b_1 (\theta_{sp\infty} - \theta_{spo})}{LR_2(\theta_1 - \theta_2)}\left[1 - \exp\left(-\frac{L}{b}\right)\right]$$

Since the ratio $(\theta_{sp\infty} - \theta_{spo})/(\theta_1 - \theta_2)$ is close to $R_1/R_a = R_1 U_o$, that formula simplifies to:

$$U = U_o\left\{1 + \frac{C_2 b_1 R_1}{LR_2}\left[1 - \exp\left(-\frac{L}{b}\right)\right]\right\} \qquad (2.49)$$

The effect of ventilation thus depends on the thermal resistances of the inside (R_2) and outside (R_1) leaves. If the inside one is well insulated, so R_2 is large, and the outside one conducts heat, so R_1 is small, meaning for example that in a cavity wall a partial insulation fill lines up with the inside leaf, then venting hardly has any impact on the thermal transmittance. On the other hand, if for any reason the partial fill lines up with the outside leaf, its impact becomes significant. At the air intake, the flux equals:

$$q_{x=0} = \frac{\theta_2 - \theta_{s2,0}}{R_2}$$

showing that a well-insulated inside leaf is of key importance there.

2.3 Water vapour

2.3.1 Water vapour in the air

2.3.1.1 Overview

Air contains water vapour, from now on simply called vapour. Evidence is provided by the damp condensing on single glass during winter. Those who wear glasses experience it when coming from the cold outdoors into a heated, crowded room. For reasons of convenience, moist air is seen as a mixture of two ideal gases, dry air and water vapour, although the term 'ideal' is less accurate for vapour than for dry air. In the airflow section, the gas law for dry air was analysed. For vapour as an ideal gas the equation of state is:

$$pV = m_v RT$$

with p the partial vapour pressure in Pa, V the volume in m³, T the vapour temperature in K, m_v the amount of vapour filling the volume V in kg, and R the gas constant for vapour, 461.52 J/(kg.K). More accurate is the following non-ideal gas equation:

$$\frac{pV}{n_v R_o T} = 1 + \frac{B_{vv}}{(V/n_v)} + \frac{C_{vvv}}{(V/n_v)^2}$$

with n_v the moles of vapour filling volume V, R_o is the general gas constant (8314.41 J/(mol.K)), and B_{vv} plus C_{vvv} the viral coefficients for vapour.

2.3 Water vapour

Normally[1] the gases in dry air do not react with vapour. As a result, Dalton's law applies for moist air: total air pressure in a volume V is equal to the sum of the partial dry air and vapour pressures:

$$P_a = p_1 + p \tag{2.50}$$

In the built environment, P_a is the atmospheric value, ≈ 101.3 kPa or 1 atm at sea-level. Thanks to the limited solubility in water, dry air does not impact upon the balance between liquid, vapour and ice (Raoult and Henry's law). Hence, in air, the diagram of state for water holds: triple point at 0 °C and so on. The kinetics, however, do change. In air, evaporation and condensation go much more slowly than in a vacuum.

2.3.1.2 Quantities

The following quantities describe the presence of vapour in the air:

Partial vapour pressure, symbol p, units Pa	Together with temperature and total air pressure, a fundamental variable of state.
Vapour concentration, symbol ρ_v, units kg/m³	Mass of vapour per unit volume of air. According to the ideal gas law, the quantity is given by: $$\rho_v = \frac{m_v}{V} = \frac{p}{RT} \tag{2.51}$$ showing that the vapour concentration is a function of the partial vapour pressure and temperature.
Vapour ratio, symbol x, units kg/kg	Mass of vapour per mass unit of dry air. Also the vapour ratio is a derived variable of state.

Following relationship exists between vapour ratio x and partial vapour pressure p:

$$x = \frac{\rho_v}{\rho_1} = \frac{R_a p}{R_v (P_a - p)} = \frac{0.62 p}{P_a - p} \tag{2.52}$$

In the remaining text, the notation <partial> is omitted.

2.3.1.3 Vapour saturation pressure

As previously stated, the presence of air hardly changes the equilibrium between liquid, vapour and ice. Therefore, looking to vapour pressures, for each temperature that equilibrium figures as the maximum attainable. The related vapour concentration and vapour ratio get the prefix 'saturation'. All increase with temperature. At 100 °C, the saturation pressure equals the standard atmospheric pressure at sea-level. Saturated values get the suffix 'sat': p_{sat}, $\rho_{v,sat}$, x_{sat}. Figure 2.17 and Table 2.4 link the

---

[1] When some compounds such as SO_2, NO_x and Cl contaminate air, there is interaction. SO_2 reacts with vapour to form sulfuric acid (H_2SO_3), although the SO_2 concentration is normally too low for any significant impact.

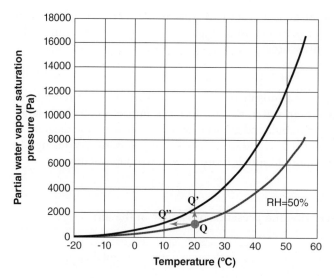

Fig. 2.17 Upper line: vapour saturation pressure as a function of temperature. Lower line: vapour pressure at 50% relative humidity

vapour saturation pressures to temperature, for the table in steps of 0.1 °C between $-30\,°C$ and $41\,°C$, and in steps of 5 °C for temperatures above 45 °C.

Temperature dependence follows a more or less exponential curve. The most accurate approximation between 0 and 50 °C is:

$$p_{sat} = p_{c,sat} \exp\left[2.3026\,\kappa\left(1 - \frac{T_c}{T}\right)\right] \qquad (2.53)$$

with T_c the temperature at the triple point above which water exists as vapour only, and $p_{c,sat}$ the related saturation pressure ($T_c = 647.4\,K$, $p_{c,sat} = 217.5 \times 10^5$ Pa). The parameter κ depends on the temperature in K:

$$\kappa = 4.39553 - 6.2442\left(\frac{T}{1000}\right) + 9.953\left(\frac{T}{1000}\right)^2 - 5.151\left(\frac{T}{1000}\right)^3$$

Less accurate are:

$-10 \leq \theta \leq 50\,°C$	$p_{sat} = \exp(65.8094 - 7066.27/T - 5.976 \ln T)$
$-30 \leq \theta \leq 0\,°C$	$p_{sat} = 611\exp(82.9 \times 10^{-3}\theta - 288.1 \times 10^{-6}\theta^2 + 4.403 \times 10^{-6}\theta^3)$
$0 \leq \theta \leq 40\,°C$	$p_{sat} = 611\exp(72.5 \times 10^{-3}\theta - 288.1 \times 10^{-6}\theta^2 + 0.79 \times 10^{-6}\theta^3)$
$0 \leq \theta \leq 80\,°C$	$p_{sat} = \exp(23.5771 - 4042.9(T - 37.58))$
$\theta \leq 0\,°C$ $p_{sat} = 611\exp\left(\dfrac{22.44\theta}{272.44 + \theta}\right)$	$\theta > 0\,°C$ $p_{sat} = 611\exp\left(\dfrac{17.08\theta}{234.18 + \theta}\right)$

The last two formulae are of interest because they allow the calculation of the related temperature for the known saturation pressure.

2.3 Water vapour

Table 2.4 Vapour saturation pressures in Pa, for temperatures between 0 °C and −30 °C and for temperatures between 0 °C and 40 °C

θ (°C)		Temperatures between 0 °C and −30 °C								
	Step −0.1	−0.0	−0.1	−0.2	−0.3	−0.4	−0.5	−0.6	−0.7	−0.8	−0.9
Step −1										
−0		611	606	601	596	591	586	581	576	572	567
−1		562	558	553	548	544	539	535	530	526	522
−2		517	513	509	504	500	496	492	488	484	479
−3		475	471	467	464	460	456	452	448	444	441
−4		437	433	430	426	422	419	415	412	408	405
−5		401	398	394	391	388	384	381	378	375	371
−6		368	365	362	359	356	353	350	347	344	341
−7		338	335	332	329	326	323	321	318	315	312
−8		310	307	304	302	299	296	294	291	289	286
−9		284	281	279	276	274	271	269	267	264	262
−10		260	257	255	253	251	248	246	244	242	240
−11		237	235	233	231	229	227	225	223	221	219
−12		217	215	213	211	209	207	206	204	202	200
−13		198	196	195	193	191	189	188	186	184	183
−14		181	179	178	176	174	173	171	170	168	167
−15		165	164	162	160	159	158	156	155	153	152
−16		150	149	148	146	145	143	142	141	139	138
−17		137	136	134	133	132	131	129	128	127	126
−18		125	123	122	121	120	119	118	116	115	114
−19		113	112	111	110	109	108	107	106	105	104
−20		103	102	101	100	99	98	97	96	95	94
−21		96	92	91	90	90	89	88	87	86	85
−22		84	84	83	82	81	80	80	79	78	77
−23		76	76	75	74	73	73	72	71	70	70
−24		69	68	68	67	66	66	65	64	64	63
−25		62	62	61	60	60	59	59	58	57	57
−26		56	56	55	55	54	53	53	52	52	51
−27		51	50	50	49	49	48	48	47	47	46

Table 2.4 *(Continued)*

θ (°C)		Temperatures between 0 °C and −30 °C								
	Step −0.1	−0.0	−0.1	−0.2	−0.3	−0.4	−0.5	−0.6	−0.7	−0.8	−0.9
Step −1											
−28		46	45	45	44	44	43	43	42	42	41
−29		41	41	40	40	39	39	38	38	38	37
−30		37	36	36	36	35	35	35	34	34	33
θ (°C)		Temperatures between 0 °C and 40 °C								
	Step +0.1	0.0	0.1	0.2	0.3	0.4	0.5	0.6	0.7	0.8	0.9
Step +1											
0		611	615	620	624	629	634	638	643	647	652
1		657	662	666	671	676	681	686	691	696	701
2		706	711	716	721	726	731	736	742	747	752
3		758	763	768	774	779	785	790	796	802	807
4		813	819	824	830	836	842	848	854	860	866
5		872	878	884	890	896	903	909	915	922	928
6		935	941	948	954	961	967	974	981	987	994
7		1001	1008	1015	1022	1029	1036	1043	1050	1057	1065
8		1072	1079	1087	1094	1101	1109	1117	1124	1132	1139
9		1147	1155	1163	1171	1178	1186	1194	1203	1211	1219
10		1227	1235	1243	1252	1260	1269	1277	1286	1294	1303
11		1312	1320	1329	1338	1347	1356	1365	1374	1383	1392
12		1401	1411	1420	1429	1439	1448	1458	1467	1477	1487
13		1497	1506	1516	1526	1536	1546	1556	1566	1577	1587
14		1597	1608	1618	1629	1639	1650	1661	1671	1682	1693
15		1704	1715	1726	1737	1748	1760	1771	1782	1794	1805
16		1817	1829	1840	1852	1864	1876	1888	1900	1912	1924
17		1936	1949	1961	1973	1986	1999	2011	2024	2037	2050
18		2063	2076	2089	2102	2115	2128	2142	2155	2169	2182
19		2196	2210	2224	2237	2251	2265	2280	2294	2308	2322
20		2337	2351	2366	2381	2395	2410	2425	2440	2455	2470

2.3 Water vapour

Table 2.4 (Continued)

θ (°C)		Temperatures between 0 °C and 40 °C									
	Step +0.1	0.0	0.1	0.2	0.3	0.4	0.5	0.6	0.7	0.8	0.9
Step +1											
21		2486	2501	2516	2532	2547	2563	2579	2595	2611	2627
22		2643	2659	2675	2691	2708	2724	2741	2758	2774	2791
23		2808	2825	2842	2859	2877	2894	2912	2929	2947	2965
24		2983	3001	3019	3037	3055	3073	3092	3110	3129	3148
25		3166	3185	3204	3224	3243	3262	3281	3301	3321	3340
26		3360	3380	3400	3420	3440	3461	3481	3502	3522	3543
27		3564	3585	3606	3627	3649	3670	3692	3713	3735	3757
28		3779	3801	3823	3845	3868	3890	3913	3935	3958	3981
29		4004	4028	4051	4074	4098	4122	4145	4169	4193	4218
30		4242	4266	4291	4315	4340	4365	4390	4415	4440	4466
31		4491	4517	4543	4569	4595	4621	4647	4673	4700	4727
32		4753	4780	4807	4835	4862	4889	4917	4945	4973	5001
33		5029	5057	5085	5114	5143	5171	5200	5229	5259	5288
34		5318	5347	5377	5407	5437	5467	5498	5528	5559	5590
35		5621	5652	5683	5715	5746	5778	5810	5842	5874	5907
36		5939	5972	6004	6037	6071	6104	6137	6171	6205	6239
37		6273	6307	6341	6376	6410	6445	6480	6516	6551	6587
38		6622	6658	6694	6730	6767	6803	6840	6877	6914	6951
39		6989	7026	7064	7102	7140	7178	7217	7255	7294	7333
40		7372	7412	7451	7491	7531	7571	7611	7652	7692	7733

	Temperatures between 45 °C and 95 °C							
θ (°C)	p' (Pa)	θ (°C)	p' (Pa)	θ (°C)	p' (Pa)	θ (°C)	p' (Pa)
45	9582	60	19 917	75	38 550	90	70 108
50	12 335	65	25 007	80	47 356	95	84 524
55	15 741	70	31 156	85	57 800		

2.3.1.4 Relative humidity

Normally, air contains less vapour than at saturation. The ratio between the actual (ρ_v) and saturated concentration ($\rho_{v,sat}$) at a given temperature is called the relative humidity, symbol ϕ, a dimensionless number with a value between 0 and 1 or expressed as a percentage, 0 to 100%:

$$\phi = 100 \rho_v / \rho_{v,sat} \quad (\%) \tag{2.54}$$

Moving to vapour pressure gives:

$$\phi = 100 \rho_v / \rho_{v,sat} = 100 p / p_{sat} \tag{2.55}$$

Or, the relative humidity is given by the ratio between the vapour pressure and vapour saturation pressure for the given temperature. For example, the relative humidity at point Q ($\theta = 20\,°C$, $p = 1169\,Pa$) in Figure 2.17 is 50%. All points with that relative humidity lie on a curve passing through Q, similar in shape to the saturation curve. By definition, relative humidity cannot exceed 100%. Indeed, for each temperature, related saturation pressure and concentration are the highest attainable. Any attempt to go beyond will see the pixel representing the vapour presence stop at saturation and give condensation while releasing the heat of evaporation, $2.5 \times 10^6\,J/kg$ at $0\,°C$. To evaporate the condensate again, the same amount of heat is needed.

A simple relationship exists between relative humidity, air pressure and air temperature. Consider an air mass with relative humidity ϕ_1, air pressure P_{a1} and temperature θ_1. Assume the air pressure changes from P_{a1} to P_{a2} and the air temperature from θ_1 to θ_2, while the quantity of vapour in the mass remains constant. Relative humidity ϕ_2 in the new situation then is:

$$\phi_2 = \phi_1 \frac{p_{sat1} P_{a2}}{p_{sat2} P_{a1}} \tag{2.56}$$

The proof goes as follows. Because the mass of dry air and vapour does not change, the ratio $m_v / (m_l + m_v)$ does neither. For $[P_{a1}, \theta_1]$, the following ratio holds:

$$\frac{m_v}{m_l + m_v} = \frac{\frac{p_1 V_1}{R T_1}}{\frac{P_{a1} V_1}{R_a T_1}} = \frac{p_1 R_a}{P_{a1} R} = \underbrace{\frac{\phi_1 p_{sat1} R_a}{P_{a1} R}}_{(4)}$$

For $[P_{a2}, \theta_2]$ that ratio becomes:

$$\frac{m_v}{m_l + m_v} = \frac{\frac{p_2 V_2}{R T_2}}{\frac{P_{a2} V_2}{R_a T_2}} = \frac{p_2 R_a}{P_{a2} R} = \underbrace{\frac{\phi_2 p_{sat2} R_a}{P_{a2} R}}_{(4)}$$

In both formulae, the moist air gas constant (R_a) equals $(m_v R + m_l R_l)/m_a$. Equating term (4) in both proves the relationship in Eq. (2.56). Under atmospheric conditions these changes of state become isobaric ($P_{a1} = P_{a2}$) and the relationship converts to:

$$\phi_2 = \phi_1 p_{sat1} / p_{sat2} \tag{2.57}$$

This simple equation shows what happens with relative humidity when the air temperature changes. Cooling the ventilation air in a building, for example, will give an increase, warming it a decrease in relative humidity.

2.3 Water vapour

For the correct quantification of a mixture of two gases, three variables of state must be known now. For moist air, obvious choices are the air pressure, the vapour ratio and the temperature. In most building physics applications, air pressure barely deviates from atmospheric, meaning that two variables will suffice. Put otherwise, relative humidity does not fully describe moist air. The only thing known is that the representing point in the diagram of state lies on a curve of equal relative humidity. To fully describe the dry air/vapour mixture, either temperature, vapour pressure, vapour saturation pressure or any other variable of state also has to be known.

2.3.1.5 Changes of state in humid air

Returning to the vapour saturation pressure versus temperature curve of Figure 2.17, to reach saturation starting from point Q, numerous possibilities exist but two are specific. If, without changing the air temperature, vapour is added, a move parallel to the vapour pressure axis will end in Q' on the saturation curve. Injecting more vapour will not result in further movement from Q' – the vapour pressure will remain the saturation pressure $p_{sat,Q'}$ and the extra vapour will condense. Such a change of state is called isothermal. Instead, lowering the air temperature will move the point parallel to the temperature axis up to Q" on the saturation curve. There, the temperature touches a value called the dew point (θ_d). Any further drop in temperature will again give condensation, with Q" descending along the saturation curve. The difference ($p_{sat,Q} - p_{sat,Q''}$) drives the condensation deposit but does not tell us anything about the kinetics. That change of state is called isobaric. All points on a line through Q parallel to the temperature axis have a same dew point. Thus, as for relative humidity, the dew point does not fully describe moist air. A second parameter is needed.

Mathematically, finding the dew point goes as follows. The relative humidity in each point x on the line Q–Q" is:

$$\phi_x = 100 \frac{p_Q}{p_{sat,x}}$$

At the dew point Q", the relative humidity (ϕ_x) reaches 100%, meaning that:

$$p_Q = \frac{100 \, p_{sat,Q''}}{100} = p_{sat,Q''}$$

In other words, the dew point is the temperature at which the actual vapour pressure becomes the saturation value. When temperatures drop below this point, condensation results.

Real changes of state combine isothermal with isobaric, so include both humidification or dehumidification and temperature change. A graph widely used to analyse such changes is the Mollier diagram, of which several versions exist (ASHRAE, VDI).

2.3.1.6 Enthalpy of humid air

The enthalpy of air with x kg vapour per kg of dry air is given by:

$$h = c_a \theta + x(c_v \theta + l_{bo}) \tag{2.58}$$

where c_a and c_v are the isobaric specific heats of dry air (≈ 1008 J/(kg.K)) and water vapour (≈ 1860 J/(kg.K)), and l_{bo} is the heat of evaporation at 0 °C.

2.3.1.7 Measuring air humidity

Direct logging uses the mirror dew point metre or the psychrometer. The first logs the temperature of a mirror that is cooled to the point where condensate forms – the dew point. The second measures the dry (θ) and wet bulb air temperatures (θ_w). Air at dry bulb temperature and given relative humidity reaches the wet bulb when cooled by adiabatic saturation. With both temperatures known, all other variables of state, included vapour pressure, can be deduced:

$$p = p_{sat}(\theta_w) - 66.71(\theta - \theta_w) \tag{2.59}$$

Indirect devices are the hair hygrometer and and the capacitance metre. The first determines the relative humidity from the hygric expansion or contraction of a hair bundle, while the second does it based on the change in dielectric constant of a hygroscopic polymer.

2.3.1.8 Vapour balance indoors

In non-conditioned spaces, the heat, air and vapour equilibrium determines the vapour pressure and relative humidity. Consider a room at constant temperature without sorption-active surfaces, where the ventilation air comes from outdoors or from neighbouring rooms and mixes ideally with the air present. In such a case, the air temperature, vapour pressure and relative humidity in the middle of the room, 1.8 m above floor level, will be representative for the indoor conditions.

Conservation of mass states that the vapour entering the room with the inflowing air, plus the vapour released or removed (G_{vP}) and the vapour gained from surface drying or withdrawn by surface condensation ($G_{vc/d}$) in the room, equals the vapour taken away by the air leaving leaving the room and by diffusion across the fabric, added the vapour stored in or removed from the room air. In the case that the inflowing air comes from outdoors only and diffusion across the fabric can be neglected, this balance is (Figure 2.18):

$$x_{ve} G_a + G_{vP} + G_{vc/d} = x_{vi} G_a + \frac{d(x_{vi} M_l)}{dt} \tag{2.60}$$

The air mass M_l in the room equals $\rho_a V$, with V being the room's volume in m³. The production G_{vP} includes vapour released by people, animals, plants, activities such as

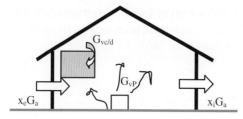

Fig. 2.18 Vapour balance in a room. Storage in the air is not shown

2.3 Water vapour

cooking, washing, and cleaning, evaporation of exposed water surfaces, and so on. Sometimes vapour release by wetted assemblies adds to the total. The ideal gas law allows us to write:

$$x_{ve} \dot{G}_a = \frac{\rho_{ve}}{\rho_{ae}} \rho_{ae}(nV) = \frac{p_e}{RT_e}(nV) \qquad x_{vi} \dot{G}_a = \frac{\rho_{vi}}{\rho_{ai}} \rho_{ai}(nV) = \frac{p_i}{RT_i}(nV)$$

while the storage term can be rewritten as:

$$\frac{d(x_{vi}\rho_{ai}V)}{dt} = \frac{d(\rho_{vi}\rho_{ai}V/\rho_{ai})}{dt} = \frac{V}{RT_i}\frac{dp_i}{dt}$$

In these equations, n is the ventilation rate in ach, the ratio between the air inflow in m^3 per time unit and the volume of the room, p_e and p_i are the vapour pressures outdoors and indoors, T_i is the temperature indoors in K, and R is the gas constant of vapour. The inflowing air experiences two changes of state: warmed from T_e to T_i, meaning:

$$x_{ve} \dot{G}_a = \frac{p_e}{RT_i}(nV)$$

and humidified or dehumidified from x_{ve} to x_{vi}. Both transformations convert the balance into a first-order differential equation with the vapour pressure indoors as a dependent variable. When neither surface condensation nor drying occurs, that equation is:

$$p_e + \frac{RT_i}{nV}G_{vP} = p_i + \frac{1}{n}\frac{dp_i}{dt}$$

In steady state, which represents the average over a long enough time span, vapour pressure indoors equals:

$$p_i = p_e + \frac{RT_i}{nV}G_{vP} \qquad (2.61)$$

Vapour release thus keeps its value above the one outdoors. The gap increases proportional to the average vapour release and inversely proportional to the average ventilation rate, a benefit that dies quickly (Figure 2.19).

Related relative humidity indoors is:

$$\phi_i = 100\frac{p_i}{p_{sat,i}} = \frac{p_e}{p_{sat,i}} + \frac{RT_i}{p_{sat,i}nV}G_{vP} \qquad (2.62)$$

Transient changes can occur in several situations. The vapour release indoors can increase suddenly while the vapour pressure outdoors and the ventilation rate remain constant. The initial conditions then are: $t < 0$: $G_{vP} = G_{vP1}$; $t \geq 0$: $G_{vP} = G_{vP2}$, giving as a solution:

$$p_i = p_{i\infty} + (p_{i0} - p_{i\infty})\exp(-nt) \qquad (2.63)$$

with p_{i0} the vapour pressure indoors at time zero and $p_{i\infty}$ its steady-state value for a vapour release G_{vP2}. Figure 2.20 shows a lag now, especially at high ventilation rates due to the time constant $1/n$ of the exponential.

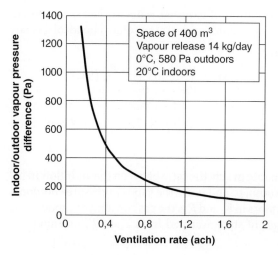

Fig. 2.19 Effect of better ventilation on the indoor/outdoor vapour pressure difference

The vapour pressure outdoors may change suddenly without variation in ventilation rate and vapour release ($n = C^t$, $G_{vP} = C^t$). The initial conditions then are: $t < 0$, $p_e = p_{e1}$, $t \geq 0$: $p_e = p_{e2}$. The solution resembles the sudden increase in vapour release, with p_{i0} as the vapour pressure indoors at time zero and $p_{i\infty}$ the steady-state value for a vapour pressure p_{e2} outdoors. The lag seen before remains.

Both the vapour release indoors and the vapour pressure outdoors can change periodically while the ventilation rate remains constant. The solution then

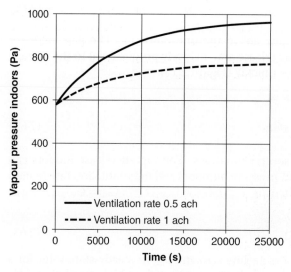

Fig. 2.20 Same case as Figure 2.19: sudden increase in vapour release. Air buffering retards the indoor vapour pressure

2.3 Water vapour

becomes:

$$\overline{p}_i = \overline{p}_i + \frac{1}{Z}\left(\mathbf{p_e} + \frac{RT_i\mathbf{G_{vP}}}{nV}\right)\mathrm{Re}\left(\exp\left(\frac{2i\pi t}{T}\right)\right) \quad (2.64)$$

with $\overline{p}_i$ the vapour pressure indoors for the mean vapour release and mean vapour pressure outdoors, $\mathbf{p_e}$ the complex vapour pressure outdoors, $\mathbf{G_{vP}}$ the complex vapour release indoors and Z the complex damping of the vapour pressure indoors, the three given by:

$$\mathbf{p_e} = \hat{p}_e \exp\left(i\phi_{p_e}\right) \quad \mathbf{G_{vP}} = \hat{G}_{vP} \exp\left(i\phi_{G_{vP}}\right) \quad Z = \left[1 + \left(\frac{2\pi}{nT}\right)^2\right]^{1/2} \quad \arg(Z) = bgtg\left(\frac{2\pi}{nT}\right)$$

with Z, $\hat{p}_e$ and $\hat{G}_{vP}$ the amplitudes of the damping factor, the vapour pressure outdoors and the vapour release indoors, $\phi_{p,e}$ the time lag of the vapour pressure outdoors and ϕ_{vP} the time lag of the vapour release indoors, both related to a cosine function. A falling ventilation rate combined with periodic changes in vapour release indoors and vapour pressure outdoors so induce a more time-shifted and dampened periodic change in vapour pressure indoors compared with outdoors. Of course, all parameters can vary randomly. Only finite difference calculus then gives the related indoor vapour pressure.

2.3.1.9 Relative humidity at a surface

An inappropriate relative humidity at a surface may have annoying consequences. When exceeding $52 + 1.2(\theta_s - 15)$ % for a long enough period of time, θ_s being the surface temperature, the dust mite *dermatophagoides farinae* multiplies explosively. Surface condensation occurs each time the relative humidity touches 100%, while the mould risk nears 1 if above 80% over a four-week period, which means that, assuming ideal air mixing, mould requires a four-week vapour pressure indoors (p_i) passing 0.8 times the four-week saturation pressure ($p_{sat,s}$) at a surface. For surface condensation, its surface temperature must pass the dew point indoors (Figure 2.21).

Fig. 2.21 Surface condensation on the ceiling of a water closet

Fig. 2.22 Edge between a concrete low-slope roof and two non-insulated outer walls. This acts as a thermal bridge with mould inside as a consequence

The vapour saturation pressure at an internal surface is fixed by its temperature, which anywhere on the envelope equals:

$$\theta_{i,s} = \theta_e + f_{h_i}(\theta_i - \theta_e) \tag{2.65}$$

with f_{h_i} increasing with a higher surface film coefficient (h_i). Mould and surface condensation risk goes up during colder weather (θ_e lower) in less heated spaces (θ_i lower) at locations where the temperature factor is low the case at thermal bridges in the envelope, which typically act as cold spots (Figure 2.22). Positive anyhow is a smaller ratio between the envelope and total wall surface in a room, because then the radiant part in the inside surface film coefficient turns larger, which is positive. Instead, behind cupboards against the envelope, the surface film coefficient drops drastically.

Turning to the average vapour pressure indoors, the higher it is, the more likely that mould and surface condensation will result. A higher value outdoors, less ventilation and more vapour released indoors are all important factors. In temperate and cold climates the highest values indoors are noted in summer, although the indoor temperatures are also highest then. The number of effective inhabitants, their living habits and the building's function determines vapour release, while ventilation is often uncontrolled.

What is responsible for all this? Mainly an energy-wasting envelope and the absence of a purpose-designed ventilation system. The first results in lower average temperatures in unheated bedrooms and bathrooms, low-temperature factors due to poor insulation without attention paid to thermal bridges, air looping and wind washing; the second produces varying infiltration, sometimes resulting in poor airing habits.

2.3.2 Vapour in open-porous materials

2.3.2.1 Different compared with air?

Under normal atmospheric conditions, humid air fills the open pores in dry porous materials. It therefore seems logical to apply what holds for air to the volume $\Psi_0 V$, with Ψ_0 the open porosity and V the material volume in m³. The amount of water vapour in equilibrium with the relative humidity in the pore air, representative for the

2.3 Water vapour

moisture content, should thus be:

$$G_m = \rho_v \Psi_0 V = \frac{p\Psi_0 V}{462\,T} \qquad w_H = \frac{G_m}{V} = \frac{p\Psi_0}{462\,T}$$

In other words, the content should vary linearly with relative humidity, whereas temperature dependence should reflect the saturation curve, thus steepening when warmer. Concrete, for example, has 15% open pores. At 20 °C and 65% relative humidity, 1 m³ should have as moisture content:

$$w_H = \frac{G_m}{V} = \frac{p\,\Psi_0}{462T} = \frac{0.15 \times 2340}{462 \times 293.16} = 0.0017 \text{ kg/m}^3$$

At 50 °C and 65% relative humidity, that value must be:

$$w_H = \frac{G_m}{V} = \frac{p\,\Psi_0}{462\,T} = \frac{0.15 \times 12335}{462 \times 293.16} = 0.0081 \text{ kg/m}^3$$

The reality, however, is completely different. At 20 °C and 65% relative humidity, the moisture content in 1 m³ of concrete equals 40–50 kg, while it varies in an S-shaped relationship with relative humidity and drops only a little at higher temperatures.

2.3.2.2 Sorption/desorption isotherm and specific moisture ratio

Generalities

A sorption/desorption isotherm is the name used for the S-shaped curve that the vapour response of open-porous materials to relative humidity generates with the moisture content at 98% called the hygroscopic maximum. Remarkably, desorption exceeds sorption at each relative humidity, inducing hysteresis with all points on and between representing equilibriums, depending on the moisture history (Figure 2.23). Open-porous material lacking sorption except at high relative humidity, such as bricks, synthetics and most insulation materials, are called non-hygroscopic. Those truly sorption-active, such as cement- and timber-based materials, are called hygroscopic.

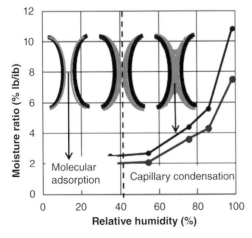

Fig. 2.23 Sorption/desorption isotherm

The literature advances several mathematical expressions for the sorption isotherm. Although none is universal, the following applies for a relative humidity below 95%, provided the roots of the denominator lie outside the interval [0, 1]:

$$w_H = \frac{\phi}{a_H \phi^2 + b_H \phi + c_H} \qquad (2.66)$$

The coefficients a_H, b_H and c_H are material-specific and differ between sorption and desorption. Another expression, applicable above 20% relative humidity, is:

$$w_H = w_c \left[1 - \ln(\phi)/b\right]^{-\frac{1}{c}} \qquad (2.67)$$

where w_c is the capillary moisture content and b and c again are material-specific constants. If the sorption and desorption isotherm is known, then the specific moisture ratio, which gives the amount of moisture a material absorbs or desorbs per unit change in relative humidity, units kg/(kg.RH), can be calculated, RH being the relative humidity on a scale from 0 to 1 now:

$$\xi_\phi = dX_H/d\phi \quad (\phi = \text{RH})$$

with X_H the moisture ratio. Compare this property, which measures the moisture storage per kg of dry material at any relative humidity, to the specific heat capacity. Multiplication by the density gives the equivalent of the volumetric specific heat capacity, called the specific moisture content, units kg/(m³.RH):

$$\rho \xi_\phi = dw_H/d\phi \qquad (2.68)$$

with w_H the hygroscopic moisture content. Both vary with relative humidity, as differentiation of the (de)sorption expressions above confirm:

$$\rho \xi_\phi = \frac{dw_H}{d\phi} = w_H^2 \left(\frac{c_H}{\phi^2} - a_H\right) \qquad \rho \xi_\phi = \frac{dw_H}{d\phi} = \frac{w_H}{cb\,\phi}\left(1 - \frac{\ln \phi}{b}\right)^{-1}$$

The fact that the momentary hygroscopic equilibrium could lie on or between the sorption and desorption curves has an impact on the specific moisture content.

Physics involved

The water molecules that diffuse into the pores and stick to the pore walls are responsible for building up hygroscopic moisture. At a relative humidity below 40%, molecular adsorption is the intervening phenomenon, which explains why the curve is convex there. At high relative humidity, between 40% and 100%, capillary condensation on the menisci in the pores, formed by adsorbed water, takes over. That changes the curve from convex into concave.

Molecular adsorption first gives a monolayer of water molecules, which the Langmuir equation describes:

$$w_H = \frac{M_w A_P}{A_w N} \frac{C\phi}{1 + C\phi} = 2.62 \times 10^{-7} A_P \frac{C\phi}{1 + C\phi} \quad (\phi \text{ on a scale from 0 to 1}) \qquad (2.69)$$

where M_w is the mass of 1 mol of water (0.018016 kg), A_w the area taken by a water molecule (11.4×10^{-20} m²), N the Avogadro number (6.023×10^{23} molecules/mol),

2.3 Water vapour

A_P the material's specific pore surface (m²/m³) and C the heat exchanged by adsorption between molecule and pore wall (J/kg):

$$C = k \exp[(l_a - l_b)/RT]$$

with k the adsorption constant, l_a the heat of adsorption (J/kg), l_b the heat of evaporation (J/kg) and R the gas constant of vapour. Around 20% relative humidity, multi-layer adsorption quantified by what is termed the Brunauer-Emmett-Teller (BET) equation starts:

$$w_H = 2.62 \times 10^{-7} A_P \frac{C\phi}{1-\phi} \left[\frac{1-(n+1)\phi^n + n\phi^{n+1}}{1+(C-1)\phi - C\phi^{n+1}} \right] \quad (2.70)$$

where n is the number of layers formed. Both the Langmuir equation and the BET equation underline that at low relative humidity the hygroscopic moisture content increases with the heat exchanged and the specific pore surface, albeit a higher temperature tempers that rise somewhat.

At constant open porosity, very small pores give a higher specific pore surface. Consider sand-lime stone. The total open porosity hardly differs from bricks (≈33%). However, the average pore diameter is only 0.1 μm, while for bricks it is 8 μm. Sand-lime stone thus has a 6000 times larger specific pore surface, and is therefore far more hygroscopic than bricks at low relative humidity.

As soon as the adsorbed layers touch each other in the smallest pores, capillary condensation begins. Surface tension in fact rearranges these layers to form a stable water-fill with a meniscus at both ends. The vapour saturation pressure above a water meniscus depends on its form. For a molecule, to escape from a concave meniscus is harder than escaping a flat one. For a convex one the inverse holds true. Therefore, the vapour phase above a concave meniscus will contain fewer water molecules than when flat, and above a convex meniscus it will contain more. Conversely, condensation onto a concave meniscus requires fewer vapour molecules above than condensation on a flat meniscus. Water will thus deposit on a concave meniscus at a relative humidity below 100%, while a convex meniscus will need a value above 100%, which Thompson's law states:

$$p'_{sat} = p_{sat} \exp\left[-\frac{\sigma_w \cos\vartheta}{\rho_w RT} \left(\frac{1}{r_1} + \frac{1}{r_2} \right) \right] \quad (2.71)$$

Here, p'_{sat} and p_{sat} are the saturation pressures above a curved and a flat meniscus respectively, ρ_w is the density and σ_w the surface tension of water. ϑ stands for the contact angle between the pore wall and the meniscus, while r_1 and r_2 are the curvature radii. Under the assumption that pores are circular with an equivalent diameter d_{eq}, the equation simplifies to:

$$p'_{sat} = p_{sat} \exp\left(-\frac{4\sigma_w \cos\vartheta}{\rho_w RT d_{eq}} \right) \quad (2.72)$$

As the ratio p'_{sat}/p_{sat} stands for the relative humidity (ϕ), it can also be written as:

$$\ln(\phi) = -\frac{4\sigma_w \cos\vartheta}{\rho_w RT d_{eq}} \quad (0 \le \phi \le 1) \quad (2.73)$$

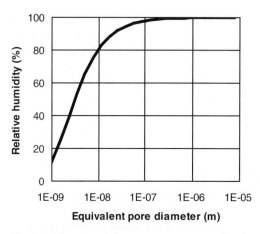

Fig. 2.24 Thompson's law at a temperature of 20 °C

The larger the equivalent pore diameter, the higher the relative humidity needed to achieve capillary condensation. Under 20%, the diameter drops below 10^{-9} m, the sphere of influence of a water molecule. Molecules then no longer behave statistically, which makes the capillary condensation theory meaningless. At 100% relative humidity, the equivalent diameter nears infinity. All open pores should then be filled with water. However, the air bubbles left will limit the storage to a value called the capillary moisture content, which prohibits the air from escaping. However, when this 100% continues for decennia, dissolution of the air bubbles will slowly shift the moisture content to saturation. Temperature, of course, also intervenes. The colder it is, the lower the relative humidity at which capillary condensation starts (see Figure 2.24).

In general, capillary condensation on a concave meniscus in a pore with equivalent diameter d_{eq} will start when the relative humidity in the air equals:

$$\phi = 100 p'_{sat}(d_{eq})/p_{sat}$$

In pores with smaller equivalent diameter than d_{eq}, condensate will already deposit. In those greater than two times that diameter $(2d_{eq})$, multi-layer adsorption will continue, while in intermediate-sized pores, the adsorbed layers will thicken due to condensation. Anyway, the dominant effect of the wide pores in the total pore volume explains the strong increase in hygroscopic moisture content beyond 90%.

The hysteresis between sorption and desorption has many causes. Water uptake and release progress so slowly that, when testing stops too early, the result gives too a low sorption and too high desorption moisture content. Testing in a vacuum gives a much smaller hysteresis, proving that between sorption and desorption the interaction between liquid and air in the pores changes. Also, menisci formed by sorption differ from those retreating during desorption, mainly because the contact angle ϑ changes.

Impact of salts

Salts in the pores increase sorption. This is a direct consequence of the vapour saturation depression that saturated salt solutions cause:

Saturated solution	Equilibrium relative humidity (%)
$MgCl_2$	33
NaCl	75
KaCl	86

With NaCl in the pores, a sudden increase in hygroscopic moisture content will show up at 75% relative humidity. As a consequence, masonry wetted by seawater hardly dries in a climate where the average relative humidity remains above 75%. A mixture of different salts even lifts the whole hygroscopic curve, which is often used as proof of their presence.

Consequences

(De)sorption plays an important role in the hygric response. In open-porous materials, relative humidity is the factual moisture potential, which the sorption isotherm links to moisture content. Sorption induces hygric inertia, with the specific moisture content in the same role as the volumetric specific heat capacity in thermal processes. Its inhibitory effect on so-called interstitial condensation, which is often nothing more than an increase in hygroscopic moisture content, can be favourable. Of course, there are also undesired effects. Due to sorption, materials see their dimensions change with relative humidity. A decrease causes shrinkage, an increase expansion. Thin layers of wood-based materials are very sensitive to this effect, but cement-bonded materials and bricks also respond this way. Above 75–80% relative humidity, the amount of water adsorbed may start favouring biological activity.

2.3.3 Vapour transfer in the air

Returning to moist air, suppose the dry air and vapour concentrations are ρ_{da} ($= m_1/V$) and ρ_v ($= m_v/V$) respectively. The moist air concentration then becomes:

$$\rho_a = (m_{da} + m_v)/V = \rho_{da} + \rho_v$$

A major cause of moist air displacement is convection. Let $\mathbf{v}_a$ be the moist air velocity. Remarkably, dry air can have a velocity $\mathbf{v}_{da}$ and vapour a velocity $\mathbf{v}_v$, both different from the moist air, with a relationship:

$$\mathbf{v}_a = \frac{\rho_{da}\mathbf{v}_{da} + \rho_v\mathbf{v}_v}{\rho_{da} + \rho_v} = \frac{\rho_{da}\mathbf{v}_{da} + \rho_v\mathbf{v}_v}{\rho_a}$$

Related dry air, vapour and moist air fluxes are:

$$\mathbf{g}_{da} = \rho_{da}\mathbf{v}_{da} \qquad \mathbf{g}_v = \rho_v\mathbf{v}_v \qquad \mathbf{g}_a = \mathbf{g}_{da} + \mathbf{g}_v = \rho_a\mathbf{v}_a$$

Why do the velocities differ? Because besides convection, diffusion intervenes. Brownian motion in fact tends to equalize the concentrations in liquid and gas

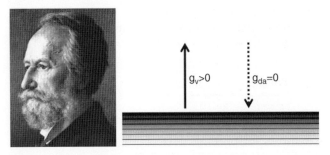

Fig. 2.25 Fick (1829–1901), German physicist; his diffusion law applied to evaporation from a water surface

mixtures. In moist air, this happens when overall differences in the vapour to dry air ratio exist. Diffusive flows also steer many processes such as carbonization of lime (CO_2/moist air), lime to gypsum transformation (SO_2/moist air), chlorine corrosion in concrete (Cl_2/moist air), and others.

In a right-angled coordinate system that moves at the velocity of the moist air, Fick's empirical diffusion law applies (Figure 2.25):

$$Vapour: \mathbf{g}_{v2} = -\rho_a D_{v,da} \, \mathbf{grad}\left(\frac{\rho_v}{\rho_a}\right) \quad Dry \, air: \mathbf{g}_{da2} = -\rho_a D_{v,da} \, \mathbf{grad}\left(\frac{\rho_{da}}{\rho_a}\right) \quad (2.74)$$

The total vapour flux ($\mathbf{g}_v$) in moist air thus becomes:

$$\mathbf{g}_v = \mathbf{g}_{v1} + \mathbf{g}_{v2} = \rho_v \mathbf{v}_a - \rho_a D_{v,da} \, \mathbf{grad}\frac{\rho_v}{\rho_a}$$

In the three equations, $D_{v,da}$ (units m^2/s) is the binary diffusion coefficient between vapour and dry air, given by Schirmer's relation:

$$D_{v,da} = \frac{2.26173}{P_a}\left(\frac{T}{273.15}\right)^{1.81} \quad (2.75)$$

For moist air at 1 atmosphere (1 bar, 101,300 Pa) this relationship becomes:

$$D_{v,da} = 8.69 \times 10^{-10} T^{1.81}$$

Convective flows usually dominate, but a diffusive vapour flow in one direction generates an equally large diffusive dry air flow in the opposite direction. If diffusion is the only mode, then no moist air movement, nor a change in its concentration, will be noted.

Under atmospheric conditions and at temperatures below 50 °C, the atmospheric pressure P_a largely exceeds the vapour pressure p, which makes the sum $p/R + p_{da}/R_{da}$ close to constant. The general expression for the total vapour flux in atmospheric moist air then becomes:

$$\mathbf{g}_v = \frac{p}{\rho_a RT}\mathbf{g}_a - \frac{D_{v,da}}{T}\left(\frac{p}{R} + \frac{p_{da}}{R_{da}}\right) \mathbf{grad} \frac{\frac{p}{R}}{\frac{p}{R} + \frac{p_{da}}{R_{da}}} \approx \frac{\mathbf{g}_a}{\rho_a RT} p - \frac{D_{v,da}}{RT} \mathbf{grad} \, p \quad (2.76)$$

2.3 Water vapour

an equation that includes vapour pressure as well as its gradient (see the analogy with the heat flux by conduction and enthalpy displacement). A first specific case, then, is a vapour flux in one direction without an opposing dry air flux, take evaporation from a water surface (Figure 2.25). For $R_a \approx R_{da}$ the total vapour flux then converts into:

$$\mathbf{g}_v = -\frac{1}{1-\frac{\rho_v}{\rho_a}} \frac{D_{v,da}}{RT} \mathbf{grad}\, p \approx -\frac{D_{v,da}}{RT} \frac{P_a}{P_a - p} \mathbf{grad}\, p$$

The second specific case is diffusion in stagnant moist air. The vapour flux then becomes:

$$Vapour: \mathbf{g}_v \approx -\frac{D_{v,da}}{RT} \mathbf{grad}\, p \qquad (2.77)$$

To maintain the analogy with heat transfer, the ratio $D_{v,da}/(RT)$ is called the vapour permeability of air, symbol δ_a, units s. The fluxes for the three cases can then be written as:

$$\mathbf{g}_v = \frac{\mathbf{g}_a}{\rho_a RT} p - \delta_a\, \mathbf{grad}\, p \qquad \mathbf{g}_v = -\delta_a \frac{P_a}{P_a - p} \mathbf{grad}\, p \qquad \mathbf{g}_v = -\delta_a\, \mathbf{grad}\, p$$

This last equation resembles heat conduction. It suffices to replace the vapour permeability by the thermal conductivity and the vapour pressure by the temperature. But, should the thermal conductivity be assumed to be constant, vapour permeability remains temperature dependent.

2.3.4 Vapour flow by diffusion in open-porous materials and assemblies

2.3.4.1 Flow equation

At first sight, vapour fluxes in open-porous materials should not differ from those in air. However, movement can only develop in pores that are accessible for water molecules, while convection will only take place in materials with really large pores. As in air, the convective vapour flux is written as:

$$\mathbf{g}_{v1} = \frac{\mathbf{g}_a}{\rho_a RT} p \qquad (2.78)$$

while below 50 °C and for air pressures much larger than the vapour pressures present, diffusion obeys:

$$\mathbf{g}_{v2} = -\delta\, \mathbf{grad}\, p \qquad (2.79)$$

where δ, units s, is the vapour permeability of the material. As the area taken by the open pores in a unit area of material is much smaller than that unit area, related vapour flux and vapour permeability must stay lower than for a unit area in air. Krischer therefore introduced the vapour resistance factor μ as a material characteristic, a number telling how much larger the vapour permeability of stagnant air is at the same temperature and total pressure:

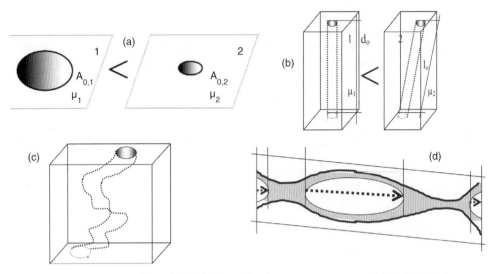

Fig. 2.26 Vapour resistance factor (μ): (a) higher with a lower open pore area; (b) higher with longer path length l_o; (c) higher with more deviousness; (d) lower thanks to capillary condensation

$$\mu = \delta_a/\delta \quad (0 \leq \mu \leq \infty)$$

The value depends on the pore system, with the open-pore area A_o per unit area of material as the first significant parameter (Figure 2.26a):

$$\mu \propto 1/A_o$$

For a zero open-pore area, the factor becomes infinite. In such a material, neither diffusion nor convection can develop.

A second parameter intervening is the path length l_o, which stands for the average distance the molecules must travel across the pores compared with the thickness of the material (d_o). The higher the ratio, the greater the diffusion resistance factor (Figure 2.26b):

$$\mu \propto l_o/d_o$$

A third factor is the 'deviousness' of the pore system (Figure 2.26c). In general, the vapour resistance factor increases with a higher ratio between the deviousness (Ψ_1) and the total open porosity (Ψ_o):

$$\mu \propto \Psi_1/\Psi_o$$

In hygroscopic materials, other transport mechanisms have an effect. In pores with diameter close to the free path length of a water molecule, friction diffusion occurs. At high relative humidity, the adsorbed water layers and the pores filled with capillary condensate add some water transfer, while diffusion in the remaining pore air allows the vapour to move between the condensed water islets. This shortens the path length considerably (Figure 2.26d). The result is an additional impact of temperature

2.3 Water vapour

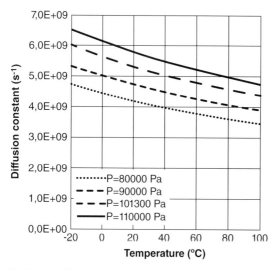

Fig. 2.27 Diffusion constant N as a function of temperature and air pressure

but above all a decreasing vapour resistance factor at higher relative humidity, or, to be more exact, at higher moisture content. Therefore the term 'equivalent' is added to the vapour resistance factor. But because vapour movement still dominates, Fick's law continues to be used. As the vapour permeability of air is so small that diffusion looks unimportant ($\approx 1.87 \times 10^{-10}$ s at 20 °C), the inverse number, called the diffusion constant N, units s^{-1}, takes over (see Figure 2.27). This and the 'equivalent' vapour resistance factor redefines the already simplified Fick's law as:

$$g_{v2} = -\frac{1}{\mu N} \text{grad } p \qquad (2.80)$$

2.3.4.2 Mass conservation

Mass conservation, with the vapour content converted into vapour pressure, gives:

$$\text{div } \mathbf{g}_v \pm G'_c = -\frac{\partial}{\partial t}\left(\frac{\Psi_o p}{RT}\right) \qquad (2.81)$$

Whenever vapour condenses, evaporates, or sublimates, the source term G'_c differs from zero. In the buffering term, the open porosity Ψ_o decreases when water vapour condenses and increases when liquid evaporates. In non-hygroscopic materials, the source term G'_c remains zero until 100% relative humidity. In hygroscopic materials, changes in sorption moisture replace both the source and the buffering term.

Under isothermal conditions and for a relative humidity below a pivot value ϕ_M close to 100%, which marks the end of a combined diffusion and surface flow, the mass equilibrium is:

$$\text{div } \mathbf{g}_v = -\frac{\partial w_H}{\partial t} = -\rho \xi_\phi \frac{\partial \phi}{\partial t} \tag{2.82}$$

At a relative humidity beyond the pivot, often set at 95–98% even though measurements may suggest lower values, liquid flow starts to complement vapour movement.

2.3.4.3 Applicability of the <equivalent> diffusion concept

The <equivalent> diffusion concept will hold as long as no convective flows in or across assemblies develop. The presumption thus is that air layers, cracks, leaks and other air paths are absent, which restricts the applicability to compact assemblies wherein at least one layer is air-tight (i.e. consists of a material with pores that are not accessible to air). Such materials are often also vapour-tight. Otherwise, air pressure differences should remain absent, a hypothetical assumption as different temperatures at both sides of vertical and inclined assemblies automatically already activate stack. Thus there is no way that <equivalent> diffusion is the rule. The following paragraphs therefore have a theoretical value, especially for when the assemblies start dry, though in reality construction moisture short-circuits that assumption.

2.3.4.4 Steady state: flat assemblies

In steady state the derivative $\partial w_H/\partial t$ is zero, so mass conservation converts to:

$$\text{div}(\mathbf{g}_v) = 0$$

Introducing the flux equation gives:

$$\text{div}\left(\frac{1}{\mu N} \mathbf{grad}\, p\right) = 0$$

In flat assemblies, this relationship simplifies to:

$$\frac{d}{dx}\left(\frac{1}{\mu N}\frac{dp}{dx}\right) = 0$$

Even if the vapour resistance factor has a fixed value, the diffusion constant remains dependent on temperature and air pressure. Under atmospheric conditions, a distinction must therefore be made between isothermal and non-isothermal.

Isothermal

Consider a single-layer assembly. When isothermal, the diffusion constant N becomes a fixed number, which simplifies the mass balance to

$$\frac{d^2 p}{dx^2} = 0$$

with as a solution:

$$p = C_1 x + C_2$$

2.3 Water vapour

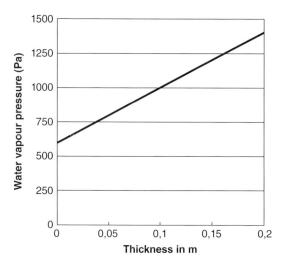

Fig. 2.28 Vapour pressure across single-layer assembly, isothermal regime

Assume the vapour pressures at both end faces (p_{s1}, p_{s2}), the thickness (d) and the vapour resistance factor (μ) of the material used are known. The boundary conditions ($x = 0 \rightarrow p = p_{s1}$, $x = d \rightarrow p = p_{s2}$, $p_{s1} < p_{s2}$) then give: $C_2 = p_{s1}$, $C_1 = (p_{s2} - p_{s1})/d$. Or, vapour pressure becomes (Figure 2.28):

$$p = \frac{p_{s2} - p_{s1}}{d} x + p_{s1} \tag{2.83}$$

The vapour flux equals:

$$g_v = -\frac{1}{\mu N}\frac{dp}{dx} = -\frac{p_{s1} - p_{s2}}{\mu N d} \tag{2.84}$$

The isothermal steady state vapour pressure thus changes linearly, while the vapour flux is proportional to the difference in vapour pressure and inversely proportional to the product of the thickness with the vapour resistance factor. The value $\mu N d$ with the diffusion constant N, equal to $\approx 5.4 \times 10^9$ s^{-1}, is called the diffusion resistance, symbol Z, units m/s. The larger that resistance, the smaller the vapour flux. A high diffusion resistance means a thick assembly or the use of thin materials with a high vapour resistance factor. The diffusion constant is a number so large that only very small fluxes are generated, even across vapour-permeable assemblies. Displacement of significant quantities so requires long time spans.

Turning to composite assemblies in isothermal regime, assume the vapour pressure at both end faces (p_{s1}, p_{s2}), all layer thicknesses (d_i) and their vapour resistance factors (μ_i) are known. In contrast to heat conduction, the likelihood of having notable contact resistances Z_{ci} is now high. Their size depends on how the layers touch each other; for example, being glued together over the whole interface gives large contact resistances. Each interface thus generates two vapour pressures, which is why, per layer, the following holds:

Layer 1	$g_v = \dfrac{p_{11} - p_{s1}}{\mu_1 N d_1}$	Contact resistance 1	$g_v = \dfrac{p_{12} - p_{11}}{Z_{c1}}$
Layer 2:	$g_v = \dfrac{p_{21} - p_{12}}{\mu_2 N d_2}$	Contact resistance 2	$g_v = \dfrac{p_{22} - p_{21}}{Z_{c2}}$
Layer n:	$g_v = \dfrac{p_{s2} - p_{n-1,2}}{\mu_n N d_n}$	Contact resistance $n-1$:	$g_v = \dfrac{p_{n-1,2} - p_{n-1,1}}{Z_{c,n-1}}$

with $p_{11}, p_{12}, \ldots, p_{n-1,1}, p_{n-1,2}$ the unknown interface vapour pressures, and g_v the unknown vapour flux. Rearranging and adding gives:

$$g_v = \frac{p_{s2} - p_{s1}}{N \sum (\mu_i d_i) + \sum Z_{ci}} \tag{2.85}$$

The denominator is the total diffusion resistance of the assembly, symbol Z_T, units m/s, while $\mu_i N d_i$ and $\mu_i d_i$ are the diffusion resistance and diffusion thickness respectively, units m, of layer i. The higher the total diffusion resistance, the lower the vapour flux. Inserting a layer with major diffusion thickness gives a large total diffusion resistance. If that layer is a thin foil, the terms vapour retarder and vapour barrier are used.

Vapour pressure follows from the same vapour flux traversing each layer and each contact resistance. The related algorithm is:

$$p_x = p_{s1} + g_v Z_{s1x} \qquad p_x = p_{s2} - g_v Z_{s2x} \tag{2.86}$$

In the $[Z, p]$ plane, these equations form a straight line between the vapour pressures at both end faces $(0, p_{s1}; Z_T, p_{s2})$, with the flux as slope. The curve in the $[x, p]$ plane is then obtained by transposing the intersections with all interfaces and contact resistances to the $[x, p]$ plane in the correct order, and connecting these with line segments (Figure 2.29).

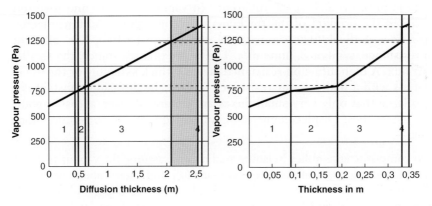

Fig. 2.29 Vapour pressures in a composite assembly, isothermal regime. Note the high contact resistance in grey between layers 3 and 4

2.3 Water vapour

The greatest difference in vapour pressure occurs over the most vapour-tight layer or that with the highest contact resistance, which then acts as a vapour retarder. In the [Z, p] plane, a composite assembly resembles one with a single layer.

As the contact resistances are usually unknown, in practice they are ignored, which presumes fictitiously free contact between layers.

Non-isothermal

What changes in non-isothermal regime? Reconsider a single-layer assembly. In steady state, the temperature changes linearly with thickness x. The temperature-dependent diffusion constant will thus also change with x. Dividing the assembly into infinitesimally thin layers, each with thickness dx and at temperature $\theta(x) = \theta_1 + (\theta_2 - \theta_1)x/d$, gives as constant vapour flux (Figure 2.30):

$$g_v = -\frac{1}{\mu N(\theta(x))}\frac{dp}{dx}$$

Integration over the thickness d gives:

$$g_v \mu \int_0^d N\left[\theta_1 + (\theta_2 - \theta_1)\frac{x}{d}\right]dx = -\int_{p_{s1}}^{p_{s2}} dp \qquad (2.87)$$

The diffusion constant N now equals:

$$N = \frac{RT}{D_{vl}} = (5.25 \times 10^6 P_a) T^{-0.81}$$

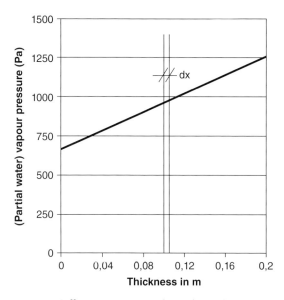

Fig. 2.30 Diffusion in a non-isothermal, steady-state regime for a single-layered assembly

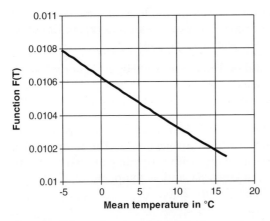

Fig. 2.31 F(T) as a function of the mean temperature, a straight line

Implementing in the integral equation and solving both integrals gives:

$$g_v = -\frac{p_{s2} - p_{s1}}{\mu d\, 5.25 \times 10^6\, P_a\, F(T)} \tag{2.88}$$

with:

$$F(T) = (T_{s2}^{0.19} - T_{s1}^{0.19})/[0.19(T_{s2} - T_{s1})]$$

Numerically, this expression hardly differs from the integral of a straight line between the end face temperatures θ_{s1} and θ_{s2}, divided by the difference between the two (Figure 2.31).

With the temperature a linear function of x, the outcome for $5.25 \times 10^6\, P_a\, F(T)$ is N linked to the mean (θ_m) of the two surface temperatures. The vapour flux thus becomes:

$$g_v = \frac{p_{s2} - p_{s1}}{\mu d N(\theta_m)} \tag{2.89}$$

This equation resembles the isothermal equation, but with the diffusion constant calculated for the mean temperature in the layer. Vapour pressures ensue from:

$$p = p_{s1} + \mu g_v \int_0^x N(x)dx$$

with:

$$\int_0^x N(x)dx = \left\{(5.25 \times 10^6\, P_a) d\left\{\left[T_{s1} + (T_{s2} - T_{s1})\frac{x}{d}\right]^{0.19} - T_{s1}^{0.19}\right\}\right\}\bigg/[0.9(T_{s2} - T_{s1})]$$

an expression that closely matches the product $N(\theta_m)x$, or:

$$p = \frac{p_{s2} - p_{s1}}{d}x + p_{s1} \tag{2.90}$$

2.3 Water vapour

As in isothermal regime, the vapour pressure evolves almost linearly across the assembly. It is not quite linear because the simple product remains an approximation.

For a composite assembly with the vapour pressures and temperatures at the end faces, the thicknesses (d_i), vapour resistance factors (μ_i) and thermal conductivities (λ_i) of all layers known, per layer the diffusion constant equals the value at that layer's mean temperature and at the contact temperature per contact resistance. The vapour flux thus becomes:

$$g_v = \frac{p_{s2} - p_{s1}}{\sum [\mu_i d_i N(\theta_{mi})] + \sum Z_{ci}(\theta_{i,i+1})} \tag{2.91}$$

Again, the denominator represents the total diffusion resistance Z_T, with the first term summing up the diffusion resistances Z_i of all layers. Vapour pressure, in turn, hardly differs from isothermal, provided the diffusion constant is calculated as mentioned. The algorithm is:

$$p_x = p_{s1} + g_v Z_{s1x} \qquad p_x = p_{s2} - g_v Z_{s2x} \tag{2.92}$$

In the $[Z, p]$ plane, the vapour pressures still form a straight line between the values at both end faces (0, p_{s1}; Z_T, p_{s2}) with the flux as slope. The curve in the $[x, p]$ plane is again found by transposing the intersections with the interfaces to the $[x, p]$ plane in the correct order and connecting them with line segments. In practice, the contact resistances Z_{ci} are mostly unknown and therefore not considered.

Yet, the named vapour pressures are only correct as long they do not intersect the saturation pressure in the assembly. Otherwise, a phenomenon called 'interstitial condensation' occurs. To control this, the temperature line traced in the $[Z, p]$ plane is transposed into saturation pressures. If there is intersection of the vapour pressure line, then interstitial condensate will form, and the correct solution requires a return to mass conservation. If the assembly is dry, the vapour balance in the zones where the vapour pressure passes saturation becomes:

$$\frac{d^2 p_{sat}}{dZ^2} = \pm \frac{G'_c}{\mu N} \tag{2.93}$$

with G'_c the deposit in kg/(m.s) per square metre, acting as source term, and p_{sat} the saturation pressure in the zone considered.

Outside these zones, the balance remains:

$$\frac{d^2 p}{dZ^2} = 0 \tag{2.94}$$

There, the vapour pressure keeps its linear relationship. Where it condenses, the liquid in kg/(m^3.s) is deposited proportionately to the second derivative of the saturation curve. Suppose now that the correct vapour pressure combines these straight lines outside with the saturation curves inside the condensation zones. Where both meet, the vapour flux, given by their slope, must be equal whether approached from the vapour pressure or the saturation curve. Keeping the isothermal vapour pressure line would also require vapour sources in these meeting points, which conflicts with

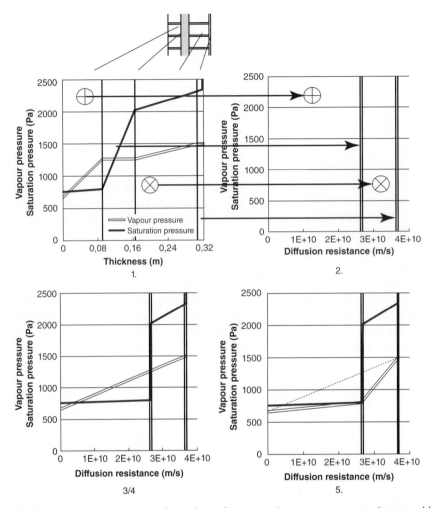

Fig. 2.32 Diffusion in a non-isothermal steady-state regime across a composite assembly, tangent method: (1) vapour pressure passes saturation pressure, giving interstitial condensation; (2) transposing the assembly into the [Z, p] plane; (3/4) the vapour (p) and saturation pressure (p_{sat}) in the assembly intersect, which is physically impossible; (5) correct solution: the vapour pressure line is replaced by the tangents

the dryness assumption. Thus, the lines to, from and between saturated zones must change direction. Or, as the derivatives of both the vapour pressure outside and the saturation curve inside the condensation zones represent vapour fluxes, the following must hold:

$$\left(\mathrm{d}p_{sat}/\mathrm{d}Z\right)_{c\to} = \left(\mathrm{d}p/\mathrm{d}Z\right)_{c\leftarrow} \tag{2.95}$$

meaning that in each contact, the slope of the saturation curve must equal the slope of the vapour pressure line arriving or setting off. Effectively, these lines must become tangents to the saturation curve (Figure 2.32).

2.3 Water vapour

To find the point of tangency of the ingoing vapour pressure line from end face 2, we use the fact that the vapour resistance Z_T, and the vapour saturation pressure must be equal there, or:

$$p_{sat,c2} = p_{s2} - g_{v2}(Z_T - Z_{c2})$$

with g_{v2} the unknown inward vapour flux and Z_{c2} the diffusion resistance at the point of tangency. Now, as the vapour pressure line in the no-condensation zone and the saturation curve in the condensation zone must have the same slope, the following holds:

$$g_{v2} = \left[\frac{dp_{sat}}{dZ}\right]_{c2} = \left[\frac{dp_{sat}}{d\theta}\frac{q}{\lambda \mu N}\right]_{c2}$$

where q is the heat flux across the assembly, λ is the thermal conductivity and μ the vapour resistance factor of the layer containing the point of tangency. For the outgoing tangent the same reasoning applies. In many cases, the two points of tangency coincide at an interface between layers, the condensation interface. Moreover, often only one such interface is found.

The tangent method, named after H. Glaser, who published it in 1959, provides an answer to several questions. How does the vapour pressure curve look in cases of interstitial condensation? Outside the condensation zones, it coincides with the tangent lines. Inside it coincides with the vapour saturation curve. Where does condensate deposition take place? Where the vapour and saturation pressures coincide, thus between each related in- and outgoing point of tangency. When coinciding on the same interface, condensate deposits there. How much vapour condenses? The amount follows from the difference in slope between each couple of inward and outward tangents. If the diffusion resistance between end face 1 ($Z=0$), where the vapour pressure and temperature is lowest, and the point of tangency of the last outward tangent is Z_{c1}, if the diffusion resistance between end face 2 ($Z=Z_T$), where the vapour pressure and temperature are highest, and the point of tangency of the inward tangent is Z_{c2}, and if the saturation pressures in the points of tangency are $p_{sat,c1}$ and $p_{sat,c2}$, the condensing flux equals:

$$g_c = \frac{p_2 - p_{sat,c2}}{Z_T - Z_{c2}} - \frac{p_{sat,c1} - p_1}{Z_{c1}} \tag{2.96}$$

In the case of several condensation interfaces, calculating the distribution of condensate follows the same course, starting at the end face where the vapour pressure and temperature is highest, jumping from condensation interface to condensation interface and calculating the difference between the arriving and departing vapour flux, For one of these interfaces:

$$g_c = \frac{p_{sat,i-1} - p_{sat,i}}{Z_{i-1} - Z_i} - \frac{p_{sat,i} - p_{i+1}}{Z_i - Z_{i+1}}$$

If the condensation deposit is spread over a zone, then its distribution in the $[x, p]$ plane over that zone is calculated by:

$$\frac{dg_c}{dx} = G'_c = \mu N \frac{d^2 p_{sat}}{dZ^2} = \frac{1}{\mu N}\frac{d^2 p_{sat}}{dx^2} = \frac{1}{\mu N}\frac{d^2 p'}{d\theta^2}\left(\frac{d\theta}{dx}\right)^2 \tag{2.97}$$

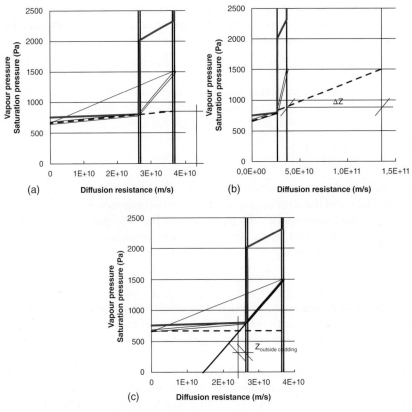

Fig. 2.33 Tangent method: (a) highest vapour pressure and warmest indoors. Two measures to avoid interstitial condensation shown graphically: (b) diffusion resistance to be added at the warm side; (c) diffusion resistance to be added as the outside finish

The same condensation approach applies for single-layer assemblies with variable vapour resistance factors. They in fact behave as if composed of n different layers of infinitesimal thickness dx.

How do we avoid interstitial condensation? The $[Z, p]$ graph suggests, as a first possibility, lowering the vapour pressure on the warm side until no intersection is left. Better ventilation of the building or lower vapour release indoors is the action needed (Figure 2.33a). A second solution is increasing the vapour resistance on the warm side until no intersection is left. This demands adding a resistance ΔZ (= $Z_{\text{vapour retarder}}$) to the existing $Z_T - Z_{c2}$ value somewhere between the condensation interface closest to the warm side and that warm side. The value can be read on the Z axis between the diffusion resistance of the assembly and the intersection of the outward tangent with the horizontal through the vapour pressure at the warm side (p_2, Figure 2.33b).

$$\Delta Z = Z_{\text{vapour retarder}} \geq \frac{p_2 - p_1}{p_{\text{sat,c1}} - p_1} Z_{c1} - Z_T \qquad (2.98)$$

2.3 Water vapour

In practice, a vapour retarder at the warm side of the thermal insulation takes cares of it – in cold and temperate climates on the end face inside, and in climates that require cooling and dehumidification, the end face outside.

A third possibility consists of lowering the diffusion resistance on the cold side (Z_{c1}) until no intersection is left. This lower value (Z_{c1}^i) appears on the Z axis, now between the point of tangency of the outward tangent and the intersection of the inward tangent with the horizontal through the vapour pressure at the cold end face (p_1) (Figure 2.33c):

$$Z'_{c1} = Z_{\text{outside cladding}} \leq \frac{p_{\text{sat,c1}} - p_1}{p_2 - p_{\text{sat,c1}}}(Z_T - Z_{c1}) \tag{2.99}$$

In cold and moderate climates, the implication is a vapour-permeable outside finish, and for climates that require cooling and dehumidification, a vapour-permeable inside finish.

Evaluating diffusion as a cause of interstitial condensation does not necessarily require calculations (Figure 2.34). Of course, the warm side with the highest vapour pressure changes between cold or moderate climates, where it is the indoors, and warm and humid climates, where in air-conditioned buildings it is the outdoors.

No interstitial condensation means no intersection between vapour and saturation pressure. The conditions needed for that, between warm and cold sides, are a convex saturation curve and concave vapour pressure curve. As the saturation pressure curves and temperature curves are quite identical in shape, a convex temperature line assures convexity. Thermal conductivity and vapour resistance factor must thus decrease from warm to cold. In other words, locate the best insulating but more vapour-permeable layers at the cold side, and the less insulating, most vapour-retarding layers at the warm side. Interstitial condensation highly likely translates

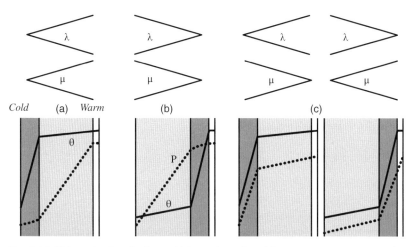

Fig. 2.34 Risk evaluation for interstitial condensation: (a) temperature convex and vapour pressure concave, risk zero; (b) temperature concave and vapour pressure convex, high risk; (c) temperature and vapour pressure convex or concave, risk unclear

into an intersection between vapour and saturation pressures. The conditions required for that are a concave temperature curve and convex vapour pressure curve from warm to cold. The rule of thumb again is simple. Condensation can be a problem if the thermal conductivity and the vapour resistance factor increase from warm to cold – in other words, if the best insulating but more vapour-permeable layers sit at the warm side and the least insulating but vapour-retarding layers at the cold side. Of course, whether interstitial condensation will occur still depends on the boundary conditions. Perhaps interstitial condensation finally demands either concave or convex vapour and temperature lines from warm to cold. Whether interstitial condensation will occur then depends on the boundary conditions and the sequence of thermal conductivities and vapour resistance factors.

These simple rules of thumb, however, do not guarantee a moisture-tolerant assembly. Many vapour-permeable layers are unsuitable as outside finish because they are rain permeable, too capillary or not strong and rigid enough. Warm and cold sides often switch with the seasons. In moderate climates, the outdoors forms the cold side in winter but in the summer, indoors may be colder – a problem in assemblies where the outside finish acts as a rain buffer. Some assemblies are quite air permeable, meaning that air flow in and across may overthrow the diffusion-based measures.

2.3.4.5 Steady state: two and three dimensions

In an isothermal regime, conservation of mass leads to equations of the form:

$$\sum_{\substack{i=l,m,n \\ j=l\pm1,m\pm1,n\pm1}} G_{v,i,j} = 0$$

The vapour flows $G_{v,i,j}$ are equal to the vapour pressure difference between each neighbour and the central control volume, multiplied by the vapour permeance (P'_d) which connects both (see heat conduction). In a non-isothermal regime, the temperatures are calculated first and and converted into saturation pressures. Then, using the same mesh, vapour pressures are quantified. If nowhere passing saturation, interstitial condensation is excluded. Otherwise, to find the amounts deposited and the correct vapour pressures, equality between vapour and saturation pressure in all volumes where saturation is exceeded is assumed, and related condensation flows are introduced as new variables ($G'_c \Delta V$). This changes the mass equilibrium in these volumes to:

$$\sum_{\substack{i=l,m,n \\ j=l\pm1,m\pm1,n\pm1}} G_{v,i,j} = G'_c \Delta V \qquad (2.100)$$

which gives a new system of n equations with $n-y$ unknown vapour pressures and y unknown condensation flows (p and G'_c). In the volumes where solving gives negative deposits, vapour pressure is reintroduced as an unknown. Recalculation then changes both distributions. Iterating goes on until only volumes with positive deposits or vapour pressures below saturation are left.

2.3.4.6 Transient regime

Conservation of mass gives:

Hygroscopic materials	Non-hygroscopic materials
$\operatorname{div} \mathbf{g}_v = -\dfrac{\partial w_H}{\partial t} = -\rho \xi_\phi \dfrac{\partial \phi}{\partial t}$	$\operatorname{div} \mathbf{g}_v \pm G'_c = \dfrac{\partial}{\partial t}\left(\dfrac{\Psi_0 p}{RT}\right)$

The vapour flux remains:

$$\mathbf{g}_v = -\frac{1}{\mu N}\operatorname{grad} p$$

Hygroscopic materials
Below the pivot relative humidity ϕ_M, the vapour flux can be rearranged to:

$$\mathbf{g}_v = -\frac{1}{\mu N}\operatorname{grad}(p_{sat}\phi) = -\frac{1}{\mu N}\left(p_{sat}\operatorname{grad}\phi + \phi\frac{dp_{sat}}{d\theta}\operatorname{grad}\theta\right) \quad (2.101)$$

with temperature and relative humidity as driving forces. Provided there are no contact resistances, relative humidity in all interfaces is continuous, which makes it a real potential. Stepping from relative humidity and temperature to moisture content goes via the sorption isotherm. Mass conservation changes to:

$$\operatorname{div}(D_\phi \operatorname{grad}\phi + D_\theta \operatorname{grad}\theta) = \frac{\partial(\rho\xi_\phi\phi)}{\partial t} \quad (2.102)$$

with:

$$D_\phi = p_{sat}/(\mu N) \qquad D_\theta = \frac{\phi(dp_{sat}/d\theta)}{\mu N}$$

Both coefficients and the specific moisture content $\rho\xi_\phi$ are functions of relative humidity and temperature. So, this mass balance must be solved together with the heat balance, using FEM or CVM. In an isothermal regime, it reduces to:

$$\operatorname{div}(D_\phi \operatorname{grad}\phi) = \frac{\partial(\rho\xi_\phi\phi)}{\partial t}$$

Isothermal conditions, however, are only possible in a material with infinite thermal conductivity. Otherwise, the conversion of latent to sensible heat irrevocably results in temperature differences.

In case the transport coefficient D_ϕ and specific moisture content $\rho\xi_\phi$ should be constants, the balance further simplifies to:

$$\nabla^2 \phi = \frac{\rho\xi_\phi}{D_\phi}\frac{\partial \phi}{\partial t} \quad (2.103)$$

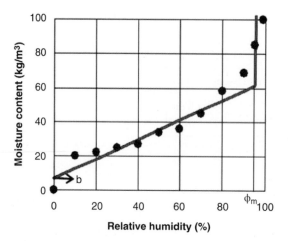

Fig. 2.35 Simplified sorption curve. At 0% relative humidity, moisture content jumps from 0 to b, then increases linearly to the pivot ϕ_M, rising linearly from there to the capillary moisture content at 100% relative humidity. The dots represent the moisture content values as measured

A constant specific moisture content requires the sorption isotherm to be a straight line between relative humidity 0 and the pivot ϕ_M. An acceptable value results from the least square line in the interval 30–86% (see Figure 2.35).

Only non-hygroscopic materials have constant transport coefficients. Indeed, the vapour resistance factor of hygroscopic materials drops and their transfer coefficient D_ϕ increases with relative humidity. That limits the applicability of the equation above to small differences in relative humidity. Even then, the transport coefficient D_ϕ still depends on temperature via the saturation pressure and the diffusion constant N.

The simplest balance equation, if usable, resembles Fourier's second law, and the solution follows the same path. For a periodic change in relative humidity, the characteristics per harmonic and temperature are the assembly matrix for diffusion W_{nd}, moisture damping D^n_ϕ with phase shift ϕ^n_ϕ, dynamic moisture resistance D^n_g with phase shift ϕ^n_g and the hygric admittance Ad^n_v with phase shift $\phi^n_{Ad_v}$. For single-layer assemblies the last three are given by:

$$D^n_\phi = \cosh\left(\frac{\omega_{vn} d}{D_\phi}\right), \quad \phi^n_\phi = \arg\left(\cosh\frac{\omega_{vn} d}{D_\phi}\right)$$

$$D^n_g = \frac{\sinh\left(\frac{\omega_{vn} d}{D_\phi}\right)}{\omega_n}, \quad \phi^n_g = \arg\left(\frac{\sinh\frac{\omega_{vn} d}{D_\phi}}{\omega_n}\right)$$

$$Ad^n_v = \frac{D^n_\phi}{D^n_g}, \quad \phi^n_{Ad_v} = \phi^n_\phi - \phi^n_g$$

with ω_{vn} the complex hygroscopic pulsation:

$$\omega_{vn} = \sqrt{\frac{2D_\phi \rho \xi_\phi i n \pi}{T}}$$

2.3 Water vapour

For composite assemblies, the same multiplications as for heat conduction govern the layer and assembly matrices.

The relative humidity inside a semi-infinite solid after a sudden change in relative humidity at the surface ($\Delta\phi_s$) can be written as:

$$\phi = \phi_{s0} + \Delta\phi_s \left[1 - \operatorname{erf} \frac{x}{\sqrt{4a_H t}} \right]$$

with a_H the moisture diffusivity in m²/s, equal to $D_\phi/\rho\xi_\phi$. The vapour flux at the surface is:

$$g_v = -D_\phi(\operatorname{grad} \phi)_{x=0} = \Delta\phi \sqrt{\frac{\rho\xi_\phi D_\phi}{\pi t}}$$

with $\sqrt{\rho\xi_\phi D_\phi}$ the water vapour sorption coefficient or vapour effusivity, units kg/(m².s$^{1/2}$). For two- and three-dimensional problems, the analysis requires FEM or CVM.

Non-hygroscopic materials
For non-hygroscopic materials, inserting the flux equation into the vapour balance gives:

$$\operatorname{div}\left(\frac{1}{\mu N}\operatorname{grad} p\right) \pm G'_c = \frac{\partial}{\partial t}\left(\frac{\Psi_0 p_{\text{sat}}}{RT}\right)$$

This differential equation must be solved together with the heat balance and the equation of state linking saturation pressure to temperature. When the vapour pressure equals saturation, the condensation/evaporation term (G'_c) differs from zero, or:

$$\pm G'_c = -\operatorname{div}\left(\frac{1}{\mu N}\operatorname{grad} p_{\text{sat}}\right) + \frac{\partial}{\partial t}\left(\frac{\Psi_0 p_{\text{sat}}}{RT}\right)$$

In an isothermal, dry assembly, the mass balance equation simplifies to:

$$\nabla^2 p = -\left(\frac{\Psi_0 \mu N}{RT}\right)\frac{\partial p}{\partial t} \tag{2.104}$$

a relationship identical to the simple isothermal vapour balance for hygroscopic materials. Analogous characteristics are defined: the vapour diffusivity (a_v) and effusivity (b_v):

$$a_v = RT/(\Psi_0 \mu N) \qquad b_v = \sqrt{\Psi_0/(RT\mu N)}$$

In a periodic regime, the complex properties that define the response of any flat assembly are vapour pressure damping, the dynamic diffusion resistance and the vapour admittance.

2.3.5 Vapour flow by diffusion and convection in open-porous materials and assemblies

As previously stated, <equivalent> diffusion is the exception. Combined convection and diffusion often reflects reality much better. In fact, many assemblies are more or less air-permeable. Wind, stack, and fans then may suffice to activate airflow, which changes the vapour balance into:

$$\text{div}(\mathbf{g}_{v1} + \mathbf{g}_{v2}) \pm G'_c = -\frac{\partial}{\partial t}\left(\frac{\Psi_0 p}{RT}\right) \tag{2.105}$$

with $\mathbf{g}_{v1}$ the convective and $\mathbf{g}_{v2}$ the diffusive flux. Introducing related flux equations gives:

$$\text{div}\left(\frac{1}{\mu N}\text{grad }p - \frac{\mathbf{g}_a}{\rho_a RT}p\right) \pm G'_c = \frac{\partial}{\partial t}\left(\frac{\Psi_0 p}{RT}\right) \tag{2.106}$$

Solving this relationship is not straightforward. Indeed, CVM is only possible in combination with the air and heat balances. Further discussion is therefore limited to inflow or outflow across flat assemblies normal to their surface, and to assemblies with a ventilated cavity, both steady state. All contact resistances are presumed to be zero. The vapour balance for inflow or outflow is then:

$$\frac{d}{dx}\left(\frac{1}{\mu N}\frac{dp}{dx} - \frac{g_a}{\rho_a RT}p\right) = 0 \tag{2.107}$$

The vapour resistance factor is considered constant, meaning hygroscopicity is overlooked. The diffusion constant (N) of course remains a function of temperature, but for reasons of simplicity, it is kept constant at $\approx 5.4 \times 10^9 \, s^{-1}$.

First, the isothermal regime is considered. Transposing the equation above to the $[Z, p]$ plane gives:

$$\frac{d^2 p}{dZ^2} - \frac{g_a}{\rho_a RT}\frac{dp}{dZ} = \frac{d^2 p}{dZ^2} - a_P \frac{dp}{dZ} = 0$$

with:

$$a_P = \frac{g_a}{\rho_a RT} \approx \frac{0.62 g_a}{P_a}$$

The result is a second-order differential equation that holds for single-layer as well as composite assemblies. The solution is:

$$p = C_1 + C_2 \exp(a_P Z) \tag{2.108}$$

The diffusive, convective and total vapour fluxes equal:

Diffusion	$g_v = -\frac{dp}{dZ} = -a_P C_2 \exp(a_P Z)$
Convection	$g_{vc} = a_P p = a_P[C_1 + C_2 \exp(a_P Z)]$
Total	$g_{vT} = g_v + g_{vc} = a_P C_1$

Whereas the diffusive and convective fluxes vary across the assembly, their sum remains constant. The two integration constants C_1 and C_2 ensue from the boundary

2.3 Water vapour

conditions. With known vapour pressures at both end faces ($Z=0$, $p=p_{s1}$ and $Z=Z_T$, $p=p_{s2}$), the system $p_{s1}=C_1+C_2$ and $p_{s2}=C_1+C_2\exp(a_P Z_T)$ gives:

$$C_1 = \frac{p_{s2}-p_{s1}\exp(a_P Z_T)}{1-\exp(a_P Z_T)} \qquad C_2 = \frac{p_{s2}-p_{s1}}{1-\exp(a_P Z_T)}$$

The vapour pressures and vapour fluxes across the assembly thus become:

$$p = p_{s1} + (p_{s2}-p_{s1})F_{v1}(Z)$$

$$g_v = F_{v2}(Z)(p_{s2}-p_{s1}) \qquad g_{vc} = a_P p \qquad g_{vT} = p_{s2}F_{v2}(0) - p_{s1}F_{v2}(Z_T)$$

with:

$$F_{v1}(Z) = \frac{1-\exp(a_P Z)}{1-\exp(a_P Z_T)} \qquad F_{v2}(Z) = \frac{a_P \exp(a_P Z)}{1-\exp(a_P Z_T)} \qquad (2.109)$$

In the [Z, p] plane the vapour pressure changes from linear to a convex exponential between [0, p_{s1}] and [Z_T, p_{s2}] when the air goes from high to low, and to a concave exponential in the opposite case (Figure 2.36).

The higher the coefficient a_P, meaning more air moving across, the more concave or convex these exponentials. Vapour pressures in the assembly in the [x, p] plane follow from transferring the intersections with the successive interfaces in the right order to the [x, p] plane and linking them with exponential segments.

At first sight, the results resemble combined heat conduction and enthalpy flow. However, there is a difference. Consider a 10 cm thick mineral fibre layer, with

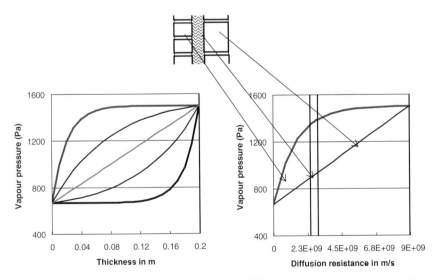

Fig. 2.36 Steady-state combined convection and diffusion in an isothermal regime. Vapour pressure across (a) a single-layer assembly in the [x, p] plane, and (b) across a composite assembly in the [Z, p] plane

thermal conductivity 0.036 W/(m.K), air permeability 8×10^{-5} s, and vapour resistance factor 1.2. For a 1 Pa air pressure difference across, the air flux touches 0.0008 kg/(m².s) or 2.4 m³/(m².h). For $x = d$ the exponent becomes 2.2 for the temperature and 3.2 for the vapour pressure, or, even for an insulating, extremely vapour-permeable layer, airflow makes the vapour pressure curve most convex or concave. For less insulating, air-permeable assemblies, this is even more pronounced.

The increase in total vapour flux compared with diffusion (g_{vo}) is:

$$100(g_{vT} - g_{vo})/g_{vo} = 100[Z_T \, \text{abs}(F_{v2}(Z_T)) - 1] \tag{2.110}$$

When the air migrates from high to low vapour pressure, $\text{abs}(F_{v2}(Z_T))$ exceeds 1 and the total vapour flux increases. In the opposite case, even a limited inflow can force the total flux to move from low to high vapour pressure.

In a non-isothermal regime, the combination of the saturation (p_{sat}) course, deduced from the temperatures, and the vapour pressure course defines what will happen. Without intersection, no condensate will deposit and the isothermal vapour pressure exponential remains. When intersecting, condensation will occur. As the amounts are usually too small to consider the heat of evaporation released, only the vapour pressure changes. Indeed, where condensation starts and stops, the vapour and saturation curves must have a common value and the vapour flux there for both must be the same, or:

$$p = p_{sat} \qquad \frac{dp_{sat}}{dZ} - a_P p_{sat} = \frac{dp}{dZ} - a_P p$$

The unknown in both equations are the saturation pressure and the diffusion resistance. The combination gives:

$$\left[\frac{dp}{dZ}\right] = \left[\frac{dp_{sat}}{dZ}\right] \tag{2.111}$$

or, the second condition presumes a common tangent in the $[Z, p]$ plane. In practice, first the saturation and vapour pressure curves are calculated. Then all interfaces where vapour pressure exceeds saturation are tested for condensation, starting where the difference is highest. Let this be interface j. The inward vapour pressure exponential from the face s1 where the temperature and vapour pressure is highest then becomes:

$$p = p_{s1} + (p_{sat,j} - p_{s1}) F_{v1}(Z_{s1}) \quad \text{with} \quad F_{v1}(Z_{s1}) = \frac{1 - \exp[a_P Z_{s1}^z]}{1 - \exp[a_P(Z_{s1}^j)]}$$

where Z_{s1}^j is the diffusion resistance between s1 and j, and Z_{s1}^z the diffusion resistance from s1 to any interface z closer by. If that exponential does not intersect saturation, interface j is where condensate will deposit. Moving outwards, an analogous reasoning applies, except that the saturation pressure $p_{sat,j}$ in interface j and the vapour pressure at the outside face s2 now are the bounding values:

$$p = p_{sat,j} + (p_{s2} - p_{sat,j}) F_{v1}(Z)$$

2.3 Water vapour

with:

$$F_{v1}(Z) = \frac{1 - \exp(a_P Z_j^z)}{1 - \exp(a_P Z_j^{s2})}$$

where Z_j^{s2} is the diffusion resistance between j and s2 and Z_j^z is the diffusion resistance from j to an interface z closer by. Without intersection with saturation, the j to s2 curve is the outgoing exponential. In case the inward exponential intersects saturation before j, testing each interface in between s1 and j will give the solution. If the zone with higher vapour pressure than saturation contains no interfaces, then the point of tangency (L) on the saturation curve is where:

$$\left(p_{\text{sat,L}} - p_{s1}\right) \frac{a_P \exp\left[a_P Z_{s1}^L\right]}{1 - \exp\left[a_P Z_{s1}^L\right]} = \left[\frac{dp_{\text{sat}}}{dZ}\right]_L$$

There, condensation starts. The same applies between L and s2, additional condensation interfaces, or the end point of tangency k of the condensation zone:

$$\left(p_{s2} - p_{\text{sat,k}}\right) \frac{a_P \exp\left[a_P Z_k^{s2}\right]}{1 - \exp\left[a_P Z_k^{s2}\right]} = \left[\frac{dp_{\text{sat}}}{dZ}\right]_k$$

In case k and L coincide or delimit one zone, only that one interface or zone sees condensate. Additional vapour pressure tangents between k and L mean more interfaces or zones that suffer.

The air moving from low to high temperature and vapour pressure excludes interstitial condensation because a more concave vapour pressure exponential can never intersect a less concave saturation curve. Instead, when air goes from high to low temperature and high to low vapour pressure, a high enough vapour pressure at the warm side will unavoidably give condensation, as Figure 2.37 illustrates. Airflow also changes the accumulation speed, from very slow for diffusion only to almost instantaneously.

Combined diffusion and convection allows us to answer several questions. The way in which interstitial condensation changes the vapour pressure line is explained. Air ingress can increase the number of condensation zones compared with diffusion. When only a single interface is wetted, then, with s2 the outside and s1 the inside face, the condensation flux equals:

$$g_c = \left[p_{\text{sat},x} F_{v2}(0) - p_{s2} F_{v2}\left(Z_{s2}^{\text{sat},x}\right)\right] - \left[p_{s1} F_{v2}\left(Z_{s2}^{\text{sat},x} - Z_{s2}^{\text{sat},x}\right) - p_{\text{sat},x} F_{v2}\left(Z_{s1}^{s2} - Z_{s2}^{\text{sat},x}\right)\right]$$

with $p_{\text{sat},x}$ the saturation pressure in the condensation interface and:

$$F_{v2}(0) = \frac{a_P}{1 - \exp\left(a_P Z_{s2}^{\text{sat},x}\right)} \qquad F_{v2}\left(Z_{s2}^{\text{sat},x}\right) = \frac{a_P \exp\left(a_P Z_{s2}^{\text{sat},x}\right)}{1 - \exp\left(a_P Z_{s2}^{\text{sat},x}\right)}$$

$$F_{v2}\left(Z_{s2}^{\text{sat},x} - Z_{s2}^{\text{sat},x}\right) = \frac{a_P}{1 - \exp\left[a_P\left(Z_{s1}^{s2} - Z_{s2}^{\text{sat},x}\right)\right]} \qquad F_{v2}\left(Z_{s1}^{s2} - Z_{s2}^{\text{sat},x}\right) = \frac{a_P \exp\left[a_P\left(Z_{s1}^{s2} - Z_{s2}^{\text{sat},x}\right)\right]}{1 - \exp\left[a_P\left(Z_{s1}^{s2} - Z_{s2}^{\text{sat},x}\right)\right]}$$

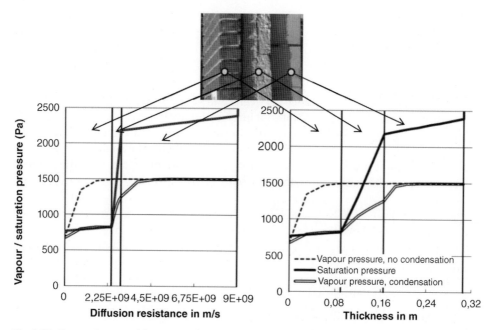

Fig. 2.37 Composite assembly, combined diffusion and convection under non-isothermal, steady-state conditions. The vapour pressure (thin line) intersects saturation, resulting in interstitial condensation. The correct vapour pressure line gives condensate at the backside of the veneer wall (see the assembly drawn in the [Z, p] and [d, p] coordinate systems)

Compared with diffusion, the deposit increases significantly. Simultaneously, more air outflow warms the condensation interface, which slows the boom in deposit until a maximum is reached, after which the amount drops to become zero once the temperature in the condensation interface touches the dew point at the warm side (Figure 2.38).

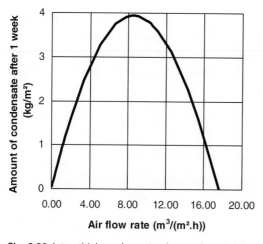

Fig. 2.38 Interstitial condensation by combined diffusion and convection (advection) in a composite assembly: deposit in kg/week as a function of the outflow rate

2.3 Water vapour

A radical measure to avoid interstitial condensation, and optimal from an energy efficiency point of view, is caring for air-tightness and ordering the composing layers per assembly correctly in terms of thermal conductivity and vapour resistance factor. The wrong approach, as it negates energy efficiency and thermal comfort, is to apply such leaky assemblies that air egress heats all likely condensation interfaces to above the dew point at the warm side. Of course, in cold and temperate climates, combining good ventilation with low vapour release may keep vapour pressure indoors below saturation at all likely condensation interfaces. Depressurizing the indoors to ensure air inflow is an alternative in winter. In warm, humid climates, moderate pressurization helps.

In diffusion mode, the tangent method allowed us to fix an upper pivot for the vapour pressure at the warm side that excluded interstitial condensation. Also, we could deduce which vapour retarder is needed at the warm side, or how vapour-permeable the outside finish should be. Now, the additional measures required are either limiting vapour pressure at the warm side to a value that excludes interstitial condensation, or designing for an air-tight envelope. Vapour retarders and vapour-permeable outside finishes in fact lose their effect when air-tightness fails.

2.3.6 Surface film coefficients for diffusion

Up to now, the vapour pressures at the end faces had to be known. But measuring vapour pressure at a surface is difficult, whereas in the air it is easy – one only has to log temperature and relative humidity. Therefore we prefer to use the air-related values as boundary conditions, so describing vapour transfer ambient-to-ambient. The question, of course, is whether vapour, and in general any gas, experiences resistance when migrating from the air to a surface or vice versa. The answer is yes. Against each surface there is a laminar boundary layer, wherein diffusion dominates, while in the air around convection dominates (Figure 2.39).

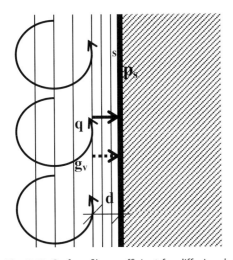

Fig. 2.39 Surface film coefficient for diffusion, boundary layer

Vapour migration across that boundary layer is written as:

$$g_v = \beta(p - p_s) \tag{2.112}$$

with p the vapour pressure outside the layer, with as reference indoors the value in the centre of the space, 1.7 m above floor level, and outdoors the value measured in the nearest weather station; p_s the vapour pressure at the surface; and β the surface film coefficient for diffusion, units s/m. The inverse $1/\beta$ is the surface resistance for diffusion, symbol Z, units m/s. The suffix $_i$ indicates indoors (β_i, Z_i), the suffix $_e$ outdoors (β_e, Z_e).

The surface film coefficient for diffusion is calculated in the same way as the thermal one for convection. The basics include the mass balance for air, the momentum equilibrium according to the Navier-Stokes equations, the turbulence equations and the mass balance for vapour. Solving that system of seven partial differential equations is done numerically or by similarity. The last generates dimensionless numbers that quantify the surface film coefficient:

Forced convection	Mixed convection	Free convection
Reynolds number: $\text{Re} = \dfrac{vL}{\nu}$	$\text{Re}/\text{Gr}^{1/2}$	Grasshof number: $\text{Gr} = \dfrac{\Delta \rho_a g L^3}{\nu^2}$
$\longleftarrow$ The Sherwood number, $\text{Sh} = \dfrac{\beta L}{\delta_a}$, replacing the Nusselt number $\longrightarrow$		
$\longleftarrow$ The Schmidt number, $\text{Sc} = \dfrac{\nu}{RT\delta_a}$, replacing the Prandtl number $\longrightarrow$		

The Sherwood number relates the total vapour transfer to diffusion, while the Schmidt number underlines that the field of flow velocities is similar to that of vapour pressures.

Between these numbers, identical relationships as for convective heat transfer exist. Thus, forced laminar flow along a horizontal flat assembly results in a surface film coefficient for diffusion, given by:

$$\text{Sh}_L = 0.664\, \text{Re}_L^{1/2}\, \text{Sc}^{1/3}$$

That similarity led to the following relationship, called Lewis' equation:

$$\beta = \frac{h_c}{\rho_a R_a T\, c_p} \left(\frac{R_a T\, c_p \delta}{\lambda_a} \right)^{0.67} \tag{2.113}$$

with h_c the thermal convective surface film coefficient, ρ_a the density, R_a the gas constant, c_p the specific heat at constant pressure, and λ_a thermal conductivity of air. Under atmospheric conditions, Lewis' equation simplifies to:

$$\beta = h_c \frac{1}{\lambda_a N} \tag{2.114}$$

2.3 Water vapour

Table 2.5 Indoor and outdoor surface film coefficients for diffusion

Indoors, β_i ($p_a = 1$ atm, $0 \leq \theta_i \leq 20\,°C$)		Outdoors, β_e ($p_a = 1$ atm, $-20 \leq \theta_i \leq 30\,°C$)	
$\theta_i - \theta_{si}$, K	$\beta_i, \times 10^{-9}$ s/m	v_a, m/s	$\beta_e, \times 10^{-9}$ s/m
2	28.6	<1	≤110
4	30.0	5	212
6	31.4	5–10	280
8	32.8	25	849
10	34.2		
12	36.0		
$\beta_i = [27 + 0.73(\theta_i - \theta_{si})] \times 10^{-9}$, $r^2 = 1$		$\beta_e = 49.9 \times 10^{-9} v_a^{0.875}$, $r^2 = 0.998$	

an expression that assumes that for heat and vapour the air layers involved are equally thick, a condition met when the Schmidt and Prandtl numbers have the same value. Let d be that common thickness. Heat gives:

$$q = \lambda_a \frac{\theta - \theta_s}{d} = h_c(\theta - \theta_s) \quad \text{or} \quad d = \lambda_a/h_c$$

Vapour results in:

$$g_v = \frac{p - p_s}{Nd} = \beta(p - p_s) \quad \text{or} \quad d = \frac{1}{N\beta}$$

Equating both thicknesses gives the simple equation above. For air, with diffusion constant $N \approx 5.4 \times 10^9$ s^{-1} and thermal conductivity $\lambda_a \approx 0.025$ W/(m.K), the relation reduces to (see Table 2.5):

$$\beta \approx 7.7 \times 10^{-9} h_c \tag{2.115}$$

In practice, the following values are used:

	Indoors	Outdoors
β (s/m)	18.5×10^{-9}	140×10^{-9}
Z (m/s)	54×10^6	7.2×10^6

The diffusion resistance, environment to environment, across a flat assembly thus becomes:

Indoors to outdoors: $\quad Z_a = Z_i + Z_T + Z_e \quad$ *Indoors to indoors:* $\quad Z_a = 2Z_i + Z_T$

giving as steady-state vapour flux:

Indoors to outdoors: $\quad g_v = \frac{p_i - p_e}{Z_i + Z_T + Z_e} \quad$ Indoors to indoors: $\quad g_v = \frac{p_{i1} - p_{i2}}{2Z_i + Z_T}$

Vapour pressures on the end faces of an envelope assembly thus are:

Indoors: $\quad p_{si} = p_i - g_v Z_i \quad$ Outdoors: $\quad p_{se} = p_e + g_v Z_e$

When compared with the surface film resistances for heat (R_i and R_e), differences appear. Consider a 0.3 m thick air-dry aerated concrete wall ($\rho = 480$ kg/m^3, $\lambda = 0.15$ W/(m.K), $\mu = 5$). The thermal and diffusion resistances are $R = 0.3/0.15 = 2$ m^2.K/W, $Z = 0.3 \times 5 \times 5.4 \times 10^9 = 8.1 \times 10^9$ m/s. Between both environments, they are $R_a = 2 + 0.17 = 2.17$ m^2.K/W, $Z_a = 8.1 \times 10^9 + 61 \times 10^6 = 8.17 \times 10^9$ m/s. Or, the surface resistances take 7.7% of the total resistance thermally and 0.7% diffusively. The share in the total is much smaller for vapour than for heat, and the error for negating them when analysing diffusion and advection across assemblies is so small that vapour pressures in the air may be transposed to surfaces without affecting the result, except when fluxes at surfaces are looked for. Then, the surface film coefficients play their role.

2.3.7 The surface film coefficient for diffusion applied

2.3.7.1 Diffusion resistance of an unvented cavity

When the air everywhere in a cavity has the same temperature and pressure, only diffusion will move vapour, giving, as diffusion resistance ($\mu_a = 1$):

$$Z_c = Nd \qquad (2.116)$$

If, instead, both vary, convection will act as a short-circuit, reducing the diffusion resistance to the sum of the surface resistances at the bounding faces:

$$Z_c = Z_1 + Z_2 = \frac{\beta_1 + \beta_2}{\beta_1 \beta_2} \qquad (2.117)$$

This value is so small compared with the diffusion resistances of the other layers in any assembly that it may be discounted. Of course, in cases where the vapour fluxes at the bounding faces are the quantities that matter, it cannot.

2.3.7.2 Do vented cavities enhance drying?

Consider a partially filled cavity wall with open head joints in the veneer at the bottom and the top. The effect of cavity venting on drying is approximated by combining vapour diffusion across both leafs with the vapour picked up by the venting air (Figure 2.40).

2.3 Water vapour

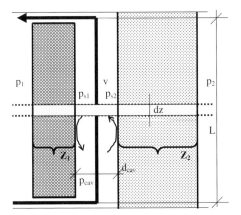

Fig. 2.40 Vented cavity with diffusion across both leafs, airflow along the cavity

Assume the cavity is d_{cav} metres wide and L metres high. The ordinate of the height is z. The air velocity in the cavity is v (m/s), while the surface film coefficient for diffusion at each cavity face is β (s/m). The diffusion resistance from ambient 1 (vapour pressure p_1) to face 1 is Z_1 (m/s), and the diffusion resistance from ambient 2 (vapour pressure p_2) to face 2 is Z_2 (m/s).

The vapour balances resemble the heat balances, with vapour moving from the ambient to the bounding faces. This gives:

Face 1: $\frac{p_1 - p_{s1}}{Z_1} + \beta(p_{cav} - p_{s1}) = 0$ Face 2: $\frac{p_2 - p_{s2}}{Z_2} + \beta(p_{cav} - p_{s2}) = 0$

with as a solution:

$$p_{s1} = \left(\frac{p_1}{Z_1} + \beta p_c\right) / \left(\frac{1}{Z_1} + \beta\right) \qquad p_{s2} = \left(\frac{p_2}{Z_2} + \beta p_c\right) / \left(\frac{1}{Z_2} + \beta\right) \qquad (2.118)$$

where p_c is the vapour pressure in the cavity. The first term in each describes the vapour transfer across the leafs, and the second term the flow between the air in the cavity and its faces. The balance in the cavity is:

$$[\beta(p_{s1} - p_{cav}) + \beta(p_{s2} - p_{cav})]dz = \frac{d_{cav} v}{RT_{cav}} dp_{cav}$$

The fluxes from the cavity faces change the vapour content in the venting air. If the cavity temperature T_{cav} is assumed equal to the asymptote $T_{cav,\infty}$, the vapour pressures p_{s1} and p_{s2} (see the two flux equations above) turn the balance into:

$$\left[\frac{\frac{p_1}{Z_1}\left(\frac{1}{Z_2}+\beta\right) + \frac{p_2}{Z_2}\left(\frac{1}{Z_1}+\beta\right)}{\frac{1}{Z_1}\left(\frac{1}{Z_2}+\beta\right) + \frac{1}{Z_2}\left(\frac{1}{Z_1}+\beta\right)} - p_{cav}\right] dz = \frac{d_{cav} v}{RT_{cav,\infty}\left(\frac{1}{Z_1+\beta^{-1}} + \frac{1}{Z_2+\beta^{-1}}\right)} dp_{cav}$$

If the vapour pressure on bounding face 1 is known, which is the case for a wet veneer, this balance simplifies to:

$$\underbrace{\frac{p_1\left(\frac{1}{Z_2}+\beta\right)+\frac{p_2}{Z_2}}{\frac{2}{Z_2}+\beta}}_{a} - p_{cav}\Bigg] dz = \underbrace{\frac{d_{cav}v}{RT_{cav,\infty}\left(\beta+\frac{1}{Z_2+\beta^{-1}}\right)}}_{b} dp_{cav}$$

with as a solution:

$$z/b = -\ln\left[(a - p_{cav})/C\right]$$

where C is the integration constant. An infinite cavity height or length gives $z/b = \infty$, or:

$$\ln\left[(a - p_{cav,\infty})/C\right] = -\infty$$

which transposes to $a - p_{cav,\infty} = 0$ or $a_1 = p_{cav,\infty}$. So, the parameter a represents the vapour pressure asymptote in an infinitely high cavity, equal to the value in the non-vented cavity. Indeed, for $v = 0$, z/b also becomes infinite. For $z = 0$, p_{cav} equals $p_{cav,0}$, giving:

$$\ln\left[(p_{cav,\infty} - p_{cav,0})/C\right] = 0 \quad \text{or} \quad C = p_{cav,\infty} - p_{cav,0}$$

Hence, the solution rewrites as:

$$p_{cav} = p_{cav,\infty} - \left(p_{cav,\infty} - p_{cav,0}\right)\exp(-z/b) \qquad (2.119)$$

In a vented cavity, vapour pressure changes exponentially from the value at the inflow to the value without venting. The term b is called the venting length. The slower the air moves along and the higher the vapour transfer to the cavity, the shorter that length.

A question with envelope assemblies is: how much vapour does a vented cavity carry away? The approximate answer is:

$$G_v = b_{cav}v\frac{p_{cav,L} - p_{cav,0}}{RT_{cav,\infty}} = \frac{b_{cav}v}{RT_{cav,\infty}}\left(p_{cav,\infty} - p_{cav,0}\right)\left[1 - \exp\left(-\frac{L}{b_1}\right)\right] \qquad (2.120)$$

Drying is minimal in the case that vapour pressure in the incoming air is little different from the value in the non-vented cavity.

2.3.7.3 Surface condensation and the vapour balance indoors

Without hygroscopic buffering, the vapour balance in a ventilated room where surface condensation or drying occurs is:

$$x_{ve}G_a + G_{vP} + G_{vc/d} = x_{vi}G_a + \frac{d(x_{vi}M_1)}{dt}$$

2.4 Moisture

If restricted to one surface, this balance can be written as:

$$p_e + \frac{RT_i}{nV}\left[G_{vP} - \beta_i A\left(p_i - p_{sat,A}\right)\right] = p_i + \frac{1}{n}\frac{dp_i}{dt} \tag{2.121}$$

In steady state ($G_{vP} = C^t$, $n = C^t$, $p_e = C^t$, $dp_i/dt = 0$), the solution is:

$$p_i = \frac{p_e + \frac{RT_i}{nV} G_{vP} + \frac{RT_i}{nV}\beta_i A p_{sat,A}}{\underbrace{\left(1 + \frac{RT_i}{nV}\beta_i A\right)}_{c}} = p_e + \frac{RT_i\left[G_{vP} + \beta_i A\left(p_{sat,A} - p_e\right)\right]}{nVc}$$

or:

$$p_i = p_{io} - \frac{RT_i}{nVc}\left[G_{vP}(c-1) + \beta_i A\left(p_{sat,A} - p_e\right)\right] \tag{2.122}$$

where p_{io} is the vapour pressure indoors without surface condensation or drying. For the same vapour release and ventilation rate, surface condensation brings the vapour pressure indoors down compared with no condensation. The difference increases with an increase in the size of the condensing surface and a reduction in its saturation pressure. Some therefore believe that replacing well-insulating glass by single glass may cure mould problems elsewhere indoors. This is false. Surface condensation per square metre of area gives only a limited reduction in vapour pressure, thus requiring installation of really large single glass surfaces – a bad idea in terms of end energy use and comfort, and because at the same time, surface drying will increases the indoor vapour pressure.

Transient solutions equal those without surface condensation or drying, provided the steady-state equation is used for the asymptote $p_{i\infty}$ and the product of ventilation rate, and c forms the exponential. The two go hand in hand with latent heat release and uptake (l_b in J/kg), which per square metre equals:

$$q = \beta\left(p_i - p_{sat,A}\right) l_b$$

2.4 Moisture

2.4.1 Overview

Up to now, moisture content in materials and assemblies was assumed low enough to quote any moisture movement as <equivalent> vapour diffusion and convection, which in capillary-porous materials is below a threshold ϕ_M in relative humidity, often set equal to 95%. Yet, interstitial condensation, rain absorption, construction moisture and rising damp all give moisture contents touching 100% relative humidity. Unsaturated water flow then develops with capillary suction, gravity, external pressure heads and internal pressure differences as driving forces. The term moisture transfer covers both vapour and unsaturated water flow, the last including the displacement of dissolved substances.

Air and vapour transfer have been analysed in a phenomenological way, introducing macroscopic properties such as the vapour resistance factor, the air permeability and others, which were meaningful for a representative material volume, a term referring to the smallest volume with the same properties as the material as a whole. The size of such representative volumes differs between materials. If fine-porous and homogeneous, it can be microscopically small. For highly heterogeneous materials, such as concrete, hardly any volume is really representative.

Although a pore filled with moist air, water and dissolved substances is a true simplification, discussing moisture flow nonetheless starts at that scale, after which the results are generalized to all materials.

2.4.2 Water flow in a pore

2.4.2.1 Capillarity

Consider the triplet air/water/pore wall. In a pore filled with water and air, capillary action originates from the cohesion between the water molecules, the cohesion impairment through contact with air and the adhesion of the water and air molecules to the pore wall.

A water molecule in water is attracted by all neighbouring molecules. The attraction weakens for those further away – once beyond 1 nanometre (nm = 10^{-9} m), not much is left. In other words, a water molecule creates a kind of cohesive sphere with radius 1 nm (Figure 2.41).

When these spheres lie far below the contact plane with air, no resulting attraction develops. However, if close to the air contact plane, then, due to the attraction to air being smaller than the cohesion in water, a force directed towards the water attaches those spheres to the underlying liquid. To release, this demands work, meaning that the potential energy of the surface layer, called the meniscus, is greater than that of the liquid below, a difference representing the surface energy per unit of surface, symbol σ_w, units J/m². As 1 J/m² equals 1 N/m, the term 'surface tension' is also used. The surface tension of water contacting air decreases with temperature:

$$\sigma_w = (75.9 - 0.17\,\theta)10^{-3}$$

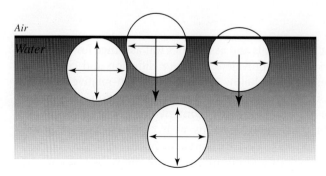

Fig. 2.41 Surface tension for water molecules in liquid water

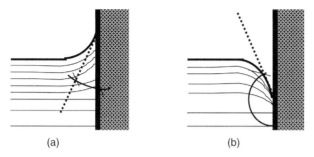

Fig. 2.42 (a) Water creeps up the pore wall, giving a contact angle of 0° to 90°. (b) Water is pushed away from the pore wall, giving a contact angle of 90° to 180°

When the meniscus touches a pore wall, provided the adhesion to the wall is greater than the attraction to air, it creeps up the wall at an angle ϑ, called the contact angle, which is between 0 and 90° (Figure 2.42a). Practically all stony and wood-based materials show such behaviour, and are therefore called hydrophilic. When wall adhesion is less than air attraction, the meniscus is pushed away. This produces a contact angle between 90° and 180° (Figure 2.42b). This is the case for most synthetics and all water-repellent surfaces, which are termed hydrophobic.

When circular pores in hydrophilic material are in contact with water, in sufficiently small pores the meniscus creeps up the pore wall to form a concave water surface, a phenomenon named capillarity. Related capillary traction by the meniscus per unit surface, called suction (symbols p_c or s) equals (Figure 2.43):

$$p_c = s = -4\sigma_w \cos\vartheta/d \quad (\text{Pa}) \tag{2.123}$$

with d being the pore diameter.

In a crack with width b, suction changes to (Figure 2.43)

$$p_c = s = -2\sigma_w \cos\vartheta/b \tag{2.124}$$

Suction is thus inversely proportional to the pore radius or crack width. If the surface tension is linked to water contacting air, the pore material intervenes

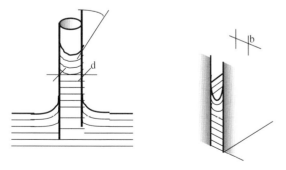

Fig. 2.43 Capillary suction in a circular pore and in a crack

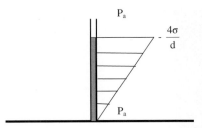

Fig. 2.44 Suction in an ideally hydrophilic pore (cos $\vartheta = 1$)

through the contact angle ϑ. Suction is an important driving force. In an ideal hydrophilic pore ($\vartheta = 0$) with diameter $1\,\mu m$, it touches 300 kPa, which is the pressure exerted by a 30 m high water head. To compare, a 175 km/h wind gives a pressure on a windward wall hardly passing 1 kPa, a value that nevertheless will press water across a 0.15 mm wide crack.

Figure 2.44 shows what suction looks like in an ideally hydrophilic pore with wet length L in contact with a water surface. It increases linearly from 0 Pa at the surface to $p_c = -4\sigma_w/d$ Pa at length L, with a value $p_c x/L$ Pa in between, x being the distance from the water surface. Because the air above the meniscus exerts the same pressure as at the water surface, the resulting suction reflects the difference between the water pressure right below and the air pressure just above the meniscus:

$$p_c = s = P_{w,L} - P_a$$

Capillary repulsion or negative suction is also inversely proportional to the pore diameter or crack width. Wind may so short-circuit a water-repellent surface once there are microcracks wider than 0.15 mm. Repulsion again equals the difference between the water pressure just below and the air pressure just above the meniscus.

2.4.2.2 Poiseuille's law

The pressure differences that suction induces in capillary-active pores allow the movement of water. The nature of the flow – laminar, transitional or turbulent – depends on the Reynolds number (Re). Water at 10 °C has a kinematic viscosity of 1.25×10^{-6} m/s², giving Re = 800 000 vd_H, with d_H the hydraulic diameter, being the diameter of a circular pore or twice the width of a crack. Flows are laminar as long as Re stays below 2000, which coincides with a water velocity of $0.0025/d_H$ in a pore. Therefore the flow in a $1\,\mu m$ wide circular pore should turn turbulent beyond a supersonic 2500 m/s! Water movement in pores thus clearly remains laminar, with the flux equal to:

$$g_w = -\frac{\rho_w d_H^2}{32 \eta_w} \operatorname{grad} P_w \qquad (2.125)$$

2.4 Moisture

an equation called Poiseuille's law, with η_w the dynamic viscosity of water, with units N.s/m^2:

Temperature, °C	η_w, N.s/m^2
0	0.001 820
20	0.001 025
100	0.000 288

For a circular pore the proof goes as follows. Laminar flow without surface slip gives a water velocity of zero at the pore wall. For each concentric water cylinder with length dy, the equilibrium between the pressure differential and viscous friction along the perimeter is (Figure 2.45):

$$\pi r^2 dP_w = 2\pi r \eta_w dy \frac{dv}{dr}$$

or:

$$dv = \frac{1}{2\eta_w} \frac{dP_w}{dy} (r\, dr)$$

Integration gives:

$$\int_v^0 dv = \frac{1}{2\eta_w} \frac{dP_w}{dy} \int_r^R r\, dr$$

or:

$$v(r) = -\frac{1}{4\eta_w} (R^2 - r^2) \frac{dP_w}{dy}$$

The water flow through the circular pore thus becomes:

$$G_w = \rho_w \int_0^R v(r) 2\pi r\, dr = -\rho_w \frac{\pi R^4}{8\eta_w} \frac{dP_w}{dy}$$

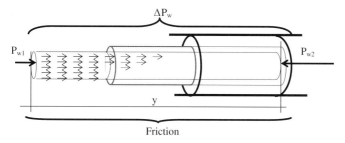

Fig. 2.45 Equilibrium between pressure differential and viscous friction

resulting in an average flux equal to:

$$g_w = -\frac{4G_w}{\pi d^2} = -\rho_w \frac{d^2}{32\eta_w}\frac{dP_w}{dy}$$

As, for a circular pore, the diameter d is also the hydraulic one (d_H), this closes the proof. The term $32\eta_w/d_H^2$ stands for the specific flow resistance of the pore, symbol W_m, units kg/(m^3.s). In case air flows through, it suffices to implement its density ρ_a and dynamic viscosity η_a, which gives a specific flow resistance 70 times smaller than for water ($\eta_a \approx 1.74 \times 10^{-5}$ N.s/m^2 versus $\eta_w \approx 1.025 \times 10^{-3}$ N.s/m^2).

The specific flow resistance becomes fluid-independent by rewriting the equation as:

$$g_w = -\frac{\rho\, d_H^2}{\eta\, 32}\frac{dP_w}{dy} = -\frac{\rho}{\eta}\left[\frac{1}{W'}\frac{dP_w}{dy}\right] \tag{2.126}$$

Rearranging Poiseuille's law gives for the pressure differential in water:

$$dP_w = -\frac{32\eta_w dy}{d^2}\left(\frac{g_w}{\rho_w}\right) = -\frac{8\eta_w dy}{r^2}v_m$$

where v_m is the average velocity in the pore (= dy/dt).

2.4.2.3 Isothermal water flow in a pore contacting water

Consider a circular pore, radius r, length L. The slope to the vertical is α. One side contacts water, and the other air. Applying Newton's law to the water undergoing suction gives (Figure 2.46):

$$\sum F_i = (\rho_w \pi r^2 l)\frac{d^2 l}{dt^2} \tag{2.127}$$

with $\rho_w \pi r^2 l$ the mass of the water column, l the ordinate of the meniscus in the pore, and d^2l/dt^2 being the acceleration of the moving meniscus.

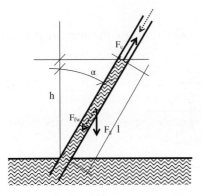

Fig. 2.46 Suction by a pore from a flat water surface

2.4 Moisture

The most important forces F_i that play a role are:

Gravity	$F_g = -\rho g \pi r^2 l \cos \alpha$
Capillary suction	$F_c = \pi r^2 \dfrac{2\sigma_w \cos \vartheta}{r} = 2\pi r\, \sigma_w \cos \vartheta$
Water friction:	$F_{fw} = -\dfrac{8\eta_w l}{r^2}\left(\dfrac{dl}{dt}\right)\pi r^2 = -8\pi \eta_w l \left(\dfrac{dl}{dt}\right)$
Air friction above the meniscus	$F_{fa} \approx -8\pi \eta_a (L-l)\left(\dfrac{dl}{dt}\right)$

The sign $\approx$ indicates that Poiseuille's law is an approximation for air. Indeed, air is compressible, so when suction begins, a peak in pressure will build up, impacting upon water flow for a while. In any case, introducing the four forces into the dynamic equilibrium gives, as a fair approximation of the kinetics:

$$\frac{d^2l}{dt^2} + \frac{8\eta_w}{\rho_w r^2}\left(1 + \frac{\eta_a(L-l)}{\eta_w l}\right)\frac{dl}{dt} - \left(\frac{2\sigma_w \cos \vartheta}{\rho_w r l} - g \cos \alpha\right) = 0 \quad (2.128)$$

a second-order differential equation with variable coefficients, which cannot be solved analytically. However, the second derivative has little effect because, due to the significant friction term ($\div 1/r^2$!), the meniscus moves so slowly that the acceleration is negligible. So, shortly after the start of the suction, the kinetics reduces to:

$$v_m = \frac{dl}{dt} = \frac{\rho_w r^2}{8}\underbrace{\left(\frac{1}{\eta_w + \eta_a \frac{(L-l)}{l}}\right)}_{\eta_r}\left[\frac{2\sigma_w \cos \vartheta}{\rho_w r l} - g \cos \alpha\right] \quad (2.129)$$

The term η_r acts as variable viscosity, infinite at the start ($l=0$) and approaching the viscosity of water as suction proceeds. The equation proves the existence of a maximum suction height h_{max}. Indeed, velocity becomes zero for:

$l_{max} = 2\sigma_w \cos \vartheta / (\rho_w g\, r \cos \alpha)$

giving (Figure 2.47):

$h_{max} = l_{max} \cos \alpha$

The general solution of the reduced equation links time to the wet length, as Figure 2.48 shows for a vertical pore:

$$t = \frac{8(\eta_w - \eta_a)}{\rho_w g\, r^2 \cos \alpha}\left[-l - \left(\frac{(\eta_w - \eta_a)l_{max} + \eta_a L}{\eta_w - \eta_a}\right)\ln\left(1 - \frac{l}{l_{max}}\right)\right] \quad (2.130)$$

When a pore with length L and radius r sucking water stretches horizontally, a case seen as representative for rain absorption by capillary finishes (Figure 2.49), the kinetics reduce to:

$$\frac{dl}{dt} = \frac{dx}{dt} = \frac{r\sigma_w \cos \vartheta}{4\left[(\eta_w - \eta_a)x + \eta_a L\right]}$$

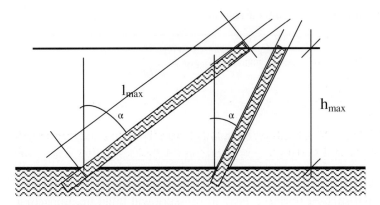

Fig. 2.47 Maximum suction height

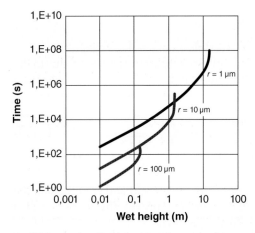

Fig. 2.48 Wet height versus time for a 15 m long vertical pore (both on a logarithmic scale): $r = 1\,\mu m$, $h_{max} = 14.8$ m; $r = 10\,\mu m$, $h_{max} = 1.48$ m; $r = 100\,\mu m$, $h_{max} = 0.148$ m

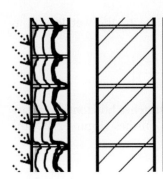 

Fig. 2.49 Rain absorption

2.4 Moisture

with as a solution:

$$2(\eta_w - \eta_a)x^2 + 4\eta_a L x - r\sigma_w \cos\vartheta\, t = 0$$

The positive root of this quadratic equation is:

$$x = -\frac{\eta_a L}{\eta_w - \eta_a} + \sqrt{\left(\frac{\eta_a L}{\eta_w - \eta_a}\right)^2 + \frac{r\sigma_w \cos\vartheta}{2(\eta_w - \eta_a)}t}$$

So the wet length (x) depends not only on time, but also on the radius and total length of the pore. The longer and narrower it is, the slower it sucks. In other words, the velocity at which porous facade materials will pick up rain will decrease with thickness and the fineness of the pore structure. If friction in the air column ahead is negligible, the solution simplifies to:

$$x = \sqrt{\underline{\frac{r\sigma_w \cos\vartheta}{2\eta_w}}}\sqrt{t} \qquad (2.131)$$

For short but wide pores, air friction is of little impact. For finer pores, the error from neglecting air friction increases, especially at the start. The underlined term in the equation represents the capillary water penetration coefficient, symbol B, units m/s$^{1/2}$. Multiplying both sides by water density gives the water absorbed by the pore in kg/m^2:

$$m_w = \rho_w B \sqrt{t} = A\sqrt{t}$$

with A the capillary water absorption coefficient, units kg/(m^2.s$^{1/2}$). The smaller its value, the slower a horizontal pore will suck and the longer it takes before water will reach the far side. Both the correct and simplified $\sqrt{t}$ relations also show that horizontal suction never stops. Whatever the length of a pore may be, the end will be reached. Once there, the water meniscus straightens, making suction zero and equilibrating the process. This means that an outflow of water by capillary suction is impossible. So, rain absorbed by porous materials will give wet stains on the inside surface but never water flowing down the surface. On the other hand, exterior pressures or gravity can cause seeping (see Figure 2.50).

Fig. 2.50 Seeping rain

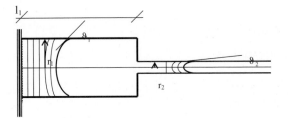

Fig. 2.51 Model of two serial pores, in the case of water contact (e.g. stucco on whatever substrate)

The model of a horizontal pore with invariable section is easily expanded to two pores in series with different sections and contact angles. Plaster on a substrate functions in this way. Assume the first pore, the plaster, has a radius r_1, a contact angle ϑ_1 and a length l_1, and the second pore, the substrate, has a radius r_2, a contact angle ϑ_2 and a length l_2. Air friction is neglected. When the first pore is widest (Figure 2.51), the horizontal flow equation applies as long as the meniscus stays in it. A smaller capillary water absorption coefficient (A_1), meaning a finer pore and/or a larger contact angle, will slow the water uptake. Once the meniscus reaches the second pore, that one starts sucking while the first becomes a hydraulic resistance. If instead the second pore is the widest, then, as soon as the meniscus in the first makes contact, it will straighten until fitting the second, which then takes over while the first again turns into a hydraulic resistance.

Once the water enters the second pore, total friction becomes:

$$F_{fw} = \left(\frac{8\pi \eta_w l_1 r_2^2}{r_1^2} + 8\pi \eta_w x \right) \frac{dx}{dt} \quad (2.132)$$

with x being the wet length in the second pore. Solving the force equilibrium $F_{fw} = F_c$ for x gives:

$$x^2 + \left(\frac{2l_1 r_2^2}{r_1^2} \right) x - \left(\frac{r_2 \sigma_w \cos \vartheta}{2\eta_w} \right) t = 0$$

This quadratic equation has as positive root:

$$x = -\underbrace{\frac{l_1 r_2^2}{r_1^2}}_{a} + \sqrt{\left(\frac{l_1 r_2^2}{r_1^2} \right)^2 + B_2^2 t}$$

Water absorption by the second pore (x) thus slows down when the first is longer and finer (see the ratio above brace a). Hence, a coarse plaster will barely increase the rain resistance of a wall made of fine porous material. Only if the layer is water repellent will wetting take longer than the duration of a long wind-driven rain event.

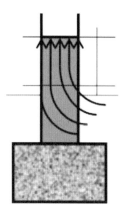

Fig. 2.52 Rising damp

With air friction, the solution of the quadratic equation for the wet length in the first pore becomes:

$$x = -\underbrace{\frac{\eta_a d_1}{\eta_w - \eta_a}}_{(1)} + \underbrace{\frac{\eta_a d_2 r_1}{(\eta_w - \eta_a) r_2^2}}_{(2)} \sqrt{\left[\frac{\eta_a d_1}{\eta_w - \eta_a} + \frac{\eta_a d_2 r_1}{(\eta_w - \eta_a) r_2^2}\right]^2 + \frac{r \sigma_w \cos \vartheta}{2(\eta_w - \eta_a)} t}$$

For the radius r_2 smaller than r_1 and the thickness d_2 greater than d_1, term (2) overwhelms term (1). So thanks to air friction, the meniscus will move a lot more slowly in the first wider pore. A coarse plaster on a fine porous substrate will thus absorb water more slowly than when lacking a substrate. But, take care. The pore model is one-dimensional, whereas pore systems in materials are three-dimensional. Surface wetting also never occurs uniformly. Even the outflow of air is three-dimensional.

Consider now a vertical pore, a case representative for rising damp (Figure 2.52).

The cosine of the slope α now is 1, converting the kinetics into:

$$v_m = \frac{dz}{dt} = \frac{\rho_w r^2}{8} \frac{1}{\eta_w + \eta_a(L - z)/z} \left(\frac{2\sigma_w \cos \vartheta}{\rho_w r z} - g \right)$$

The velocity becomes zero for:

$$2\sigma_w \cos \vartheta / (\rho_w r z) = g$$

which is when the maximum suction height is reached:

$$z = h_{\max} = \frac{2\sigma_w \cos \vartheta}{\rho_w g r} \qquad (2.133)$$

A smaller pore diameter and lower contact angle both increase that maximum. Thus, the finer the pores and the more hydrophilic a material, the higher the damp rises. Instead, for a material with a contact angle nearing 90°, with very wide pores or no open pores, that maximum becomes zero. So rising damp in walls can be prevented by either inserting a coarse porous layer or injecting a water-repellent substance that moves the contact angle beyond 90°. In both cases, suction should remain below $\rho_w g h$, with h the height above grade of the layer or zone. A safe alternative consists of inserting a waterproof layer just above ground level.

2.4.2.4 Isothermal water flow in a pore after water contact

Consider a pore with a circular section contacting water. After a time, the contact is broken, giving a water isle in the pore with a concave meniscus on both sides (Figure 2.53). If horizontal, both meniscuses pull equally, excluding water flow. If inclined, gravity makes the lower meniscus less concave and the upper more concave. Flow will start if the water isle is heavy enough to straighten the lower one, that is, when its length equals the maximum rise:

$$l_{max} = h_{max}/\cos \alpha$$

Hence, the wider a pore, the more easily gravity and external pressures will move water.

For a pore with variable section (Figure 2.53b) the meniscus with smallest diameter d_1 (the other is d_2) will pull the most, moving the water isle in the direction of greatest suction:

$$4\sigma_w \cos \vartheta / d_1 > 4\sigma_w \cos \vartheta / d_2$$

The related flux will increase when the isle shortens and the difference in suction increases, thus with higher suction gradient, or the flux writes as:

$$\mathbf{g}_w = -\left(\frac{\rho_w d^2}{32\eta_w}\right) \mathbf{grad}\, p_c = -k_w\, \mathbf{grad}\, s \qquad (2.134)$$

with k_w the water permeability of the pore (units s). The movement will last until the meniscuses equalize. The suction gradient then becomes zero, stopping the flow. A water isle partly located in a narrow pore, partly in a wider pore, reacts in the same way, with the narrow draining the wider.

If the contact angle is variable but the section constant, the water isle moves in the direction of the smaller contact angle, giving the greatest suction. When the section also varies, the isle goes on moving in the direction of the greatest suction, which

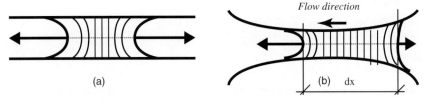

Fig. 2.53 Isothermal water transport in a pore after water contact is interrupted: constant contact angle. (a) Constant section; (b) variable section

2.4 Moisture

matches with the largest ratio between the cosine of the contact angle and the diameter (cos ϑ/d).

2.4.2.5 Non-isothermal water flow in a pore after water contact

Returning to the case with a constant contact angle and constant section, as surface tension of colder water increases and suction does with higher surface tension, a temperature difference will push the water isle in the colder direction. With suction written as:

$$s = s_o \sigma_w / \sigma_{wo} = s_o \frac{75.9 - 0.17\,\theta}{72.5} \qquad (2.135)$$

where s_o is the value at the reference temperature, for example 20 °C, then the gradient becomes:

$$\mathbf{grad}\,s = \frac{75.9 - 0.17\,\theta}{72.5}\,\mathbf{grad}\,s_o + 0.17 \times 10^{-3}\,\frac{s_o}{72.5 \times 10^{-3}}\,\mathbf{grad}\,\theta \qquad (2.136)$$

Since $\mathbf{grad}\,s_o$ is zero in a pore with constant section and contact angle, the flux equation reduces to:

$$\mathbf{g}_w = -\left[0.17\,\frac{s_o}{72.5}\left(\frac{\rho_w d^2}{32\eta_w}\right)\right]\mathbf{grad}\,\theta = -K_{\theta,w}\,\mathbf{grad}\,\theta \qquad (2.137)$$

with $K_{\theta,w}$ the thermal water permeability of the pore, units kg/(m.s.K). In pores with variable section and contact angle, both suction and temperature act as driving forces, giving as flux:

$$\mathbf{g}_w = -k_w\,\mathbf{grad}\,s - K_{\theta,w}\,\mathbf{grad}\,\theta \qquad (2.138)$$

2.4.2.6 Remark

Besides temperature and suction, two additional gradients may drive the water flux: gravity and external pressures. In narrow pores they can be neglected, but not so in wider pores. There, instead of suction, the driver in isothermal conditions is:

$$\mathbf{grad}\,s + \mathbf{grad}(\rho_w g\,z) + \mathbf{grad}\,P$$

Also remember that suction indicates the difference in water and air pressure over a meniscus. When a water isle starts moving, it expels air, so the air pressure in front will rise and behind will drop a little.

2.4.3 Vapour flow in a pore that contains water isles

Moving isles give water transfer. At the same time, vapour transfer may develop in the air inclusions in between, so the two develop in series. A succession of isles and air inclusions occurs when the contact with water is broken repeatedly. Linked to Thompson's law, two factors in vapour transfer have an impact: gradients in pore diameter and temperature.

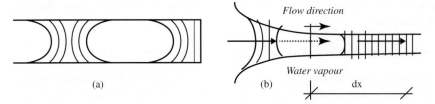

Fig. 2.54 Vapour transport in the air inclusions between water isles, with isothermal conditions and a constant contact angle: (a) constant section; (b) variable section

2.4.3.1 Isothermal

When the contact angle and section of a pore are constant, so is the vapour saturation pressure, or:

$$p'_{sat} = p_{sat}(\theta)\exp[s/(\rho_w RT)] = C^t$$

That excludes vapour transfer (Figure 2.54), although vapour exchange between the isles nearest to and the ambient will proceed each time the relative humidity in the ambient differs from the ratio between the vapour saturation pressures in the pore (p'_{sat}) and in the ambient.

In pores with a variable section but constant contact angle, suction differences exist between the menisci of the sequential water isles (Figure 2.54b), resulting in a difference in saturation pressure and vapour flow in the air inclusions in between. Related vapour flux will be:

$$\mathbf{g}_v = -\delta_a \, \mathbf{grad} \, p'_{sat} = -\delta_a \frac{\partial p'_{sat}}{\partial s} \, \mathbf{grad} \, s = -\delta_a \frac{\rho_v}{\rho_w} \, \mathbf{grad} \, s \quad (2.139)$$

That last equality is based on Thompson's law, stating that for a meniscus with constant contact angle in equilibrium, vapour pressure and suction are interchangeable. In principle, vapour transfer disturbs the water isle's equilibrium. However, for small fluxes, the effect is too small to have an impact.

If the contact angle varies but not the section, or vice versa, again a difference in suction between the menisci of sequential water isles exists, giving vapour transfer in the air inclusions in between. Related vapour flux obeys the equation above.

2.4.3.2 Non-isothermal

With the contact angle and the section constant, different temperatures give different saturation pressures (p'_{sat}) between the meniscuses of sequential water isles, resulting in a vapour flux given by:

$$\mathbf{g}_v = -\delta_a \, \mathbf{grad} \, p'_{sat} = -\delta_a \phi \left[\frac{dp_{sat}}{d\theta} + \underbrace{\frac{p_{sat}}{\rho_w RT} \left(\frac{\partial s}{\partial \theta} - \frac{s}{T} \right)}_{a} \right] \mathbf{grad} \, \theta$$

2.4 Moisture

For very small pore diameters the term a is of importance, otherwise the equation reduces to:

$$\mathbf{g_v} = -\delta_a \phi \frac{dp_{sat}}{d\theta} \mathbf{grad}\,\theta \qquad (2.140)$$

If both the contact angle and the section vary, suction and temperature join forces, turning the general equation for the vapour flux between the sequential water isles, the vapour flux between the first isle and the adjoining ambient, and the last isle and its adjoining ambient into:

$$\mathbf{g_v} = -\delta_a \left[\frac{\rho_v}{\rho_w} \mathbf{grad}\,s + \phi \frac{dp_{sat}}{d\theta} \mathbf{grad}\,\theta \right] = -\delta_a \left(p_{sat} \mathbf{grad}\,\phi + \phi \frac{dp_{sat}}{d\theta} \mathbf{grad}\,\theta \right) \qquad (2.141)$$

2.4.4 Moisture flow in a pore that contains water isles

Moisture flow in a pore implies a serial transfer of liquid in the water isles and vapour in between. As a consequence, the isles with the coldest and smallest meniscuses will grow, and those with the warmer and larger meniscuses contract. Transfer stops once all isles are rearranged in a way that achieves equilibrium.

2.4.5 Moisture flow in materials and assemblies

2.4.5.1 Transport equations

Three moisture intervals characterize capillary porous materials: dry to hygroscopic at relative humidity ϕ_M, from there up to capillary, and between capillary and saturation. Only the first and second are discussed here.

Materials consist of a matrix traversed by a labyrinth of pores that are neither circular nor straight, change section continuously, deviate, come together, split, and so on. Each representative volume hides the whole labyrinth with the same driving forces activating moisture transfer as in a pore. Any cut through a representative volume will pass across water isles in some pores and air inclusions in others. Or, due to suction and temperature gradients, each will see water and vapour flowing in parallel. Replacing the water permeability and the thermal water permeability in the flux equation at pore level by analogous properties for the material, and exchanging vapour permeability for air in the general vapour flux equation by the value for the material, gives for moisture transfer:

$$\begin{array}{c} \mathbf{g_w} = -k_w\,\mathbf{grad}\,s - K_{\theta,w}\,\mathbf{grad}\,\theta \\ + \quad \mathbf{g_v} = -\delta \dfrac{\rho_v}{\rho_w}\,\mathbf{grad}\,s - \delta\phi \dfrac{dp_{sat}}{dT}\,\mathbf{grad}\,\theta \\ \hline \mathbf{g_m} = -(k_m\,\mathbf{grad}\,s + K_\theta\,\mathbf{grad}\,\theta) \end{array} \qquad (2.142)$$

with:

$$k_m = \left(k_w + \delta \frac{\rho_v}{\rho_w} \right) \qquad K_\theta = K_{\theta,w} + \delta\phi \frac{dp_{sat}}{d\theta}$$

The quantity k_m is called the moisture permeability (units s), and the quantity K_θ is called the thermal moisture diffusion coefficient (units kg/(m.s.K)). Unsaturated water flow governs moisture permeability, while vapour transfer dominates the thermal moisture diffusion coefficient. For most capillary-porous materials, both properties are tensors. Like temperature, suction (s) is a real potential, unequivocally defined in each interface, at least in the case of ideal contact between the materials involved. Beyond capillary, the transport equation still governs drying, but wetting reduces to vapour diffusion between water islands and dissolution of the air inclusions in the pore water.

Besides materials that are hygroscopic, capillary (type 1), and non-hygroscopic, capillary (type 2), materials can be non-hygroscopic, non-capillary (type 3) and hygroscopic, non-capillary (type 4). Type 3 has such wide pores that capillarity does not play a role, while gravity and pressure only intervene when the material is soaked, limiting moisture flow to vapour diffusion and convection. Type 4 has such fine pores that the flow resistance becomes too great to see vapour move by convection or water by capillarity, gravity or external pressures. Vapour diffusion is possible in type 1 and 2 materials. In other words, below ϕ_M for type 1 and 2 materials, and in general for type 3 and 4 materials, capillary suction fails. Relative humidity does not, giving the equations:

$$\text{Diffusion only} \quad \mathbf{g_m} = -\left(k_{\varphi,m}\,\mathbf{grad}\,\phi + K_\theta\,\mathbf{grad}\,\theta\right) \tag{2.143}$$

$$\text{Diffusion and convection} \quad \mathbf{g_m} = -\left(k_{\phi,m}\,\mathbf{grad}\,\phi + K_\theta\,\mathbf{grad}\,\theta\right) + 6.21\times 10^{-6}\,\mathbf{g_a}\,\phi p_{sat} \tag{2.144}$$

For the transfer coefficients, the following apply:

Material	$k_{\phi,m}$	K_θ
Type 1, 2: capillary porous	$k_m \dfrac{ds}{d\phi}$	$\delta\phi \dfrac{dp_{sat}}{d\theta}$
Type 3: non-hygroscopic, non-capillary	δp_{sat}	$\delta\phi \dfrac{dp_{sat}}{d\theta}$
Type 4: hygroscopic, non-capillary	δp_{sat}	$\delta\phi \dfrac{dp_{sat}}{d\theta}$

External water heads (P_w), if large enough, may saturate all open-porous materials. The water flux then equals:

$$\mathbf{g_w} = -k_w\,\mathbf{grad}\,P_w$$

with k_w the water permeability at saturation (sometimes called the Darcy coefficient). Saturated fluxes lack inertia. Across a single-layer assembly with thickness d they become:

$$g_w = k_w \frac{\Delta P_w}{d} = \frac{\Delta P_w}{d/k_w} = \frac{\Delta P_w}{W_w}$$

2.4 Moisture

with W_w the water resistance (units m/s). A composite assembly gives as flux and water pressure distribution:

$$g_w = \frac{\Delta P_w}{\sum_{i=1}^{n} W_{w,i}} \qquad P_{w,x} = P_{w1} + g_w W_{w,1}^x (P_{w1} < P_{w2})$$

Making an assembly waterproof requires a layer with infinite water resistance. Alternatively, preventing the water from reaching a side that must stay dry could be done by activating one-side drying to such an extent that it equals the water flux at an interface.

2.4.5.2 Moisture permeability

For capillary materials, moisture permeability is a function of either suction or relative humidity. The higher the suction and the lower the relative humidity (i.e. the finer the pores and the smaller the contact angles), the more the permeability drops. For the water part in a single pore the expression was:

$$k_w = \frac{\rho_w d^2}{32\eta_w} = \frac{\rho_w \sigma_w^2}{2\eta_w} \left[\frac{\cos\vartheta}{s}\right]^2$$

For materials the property is determined experimentally as the product of the moisture diffusivity D_w (see hereafter) and the derivative of the relation between moisture content and either suction or relative humidity, called the suction characteristic $w(s)$ (see Figure 2.55). An alternative is using a pore model.

The water part in the thermal moisture diffusion coefficient is often set to zero, which requires to consider the temperature dependency of suction and moisture permeability.

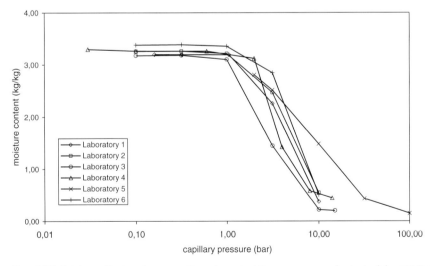

Fig. 2.55 Calcium-silicate stone, suction characteristic as measured at six laboratories. Capillary pressure on the abscissa equals suction (s)

2.4.5.3 Mass conservation

The mass balance per representative material volume (V_{rep}) gives:

$$(\text{div } \mathbf{g_m} \pm G') dV_{rep} = \left(-\frac{\partial w}{\partial t}\right) dV_{rep}$$

with w the average moisture content and G' a local moisture source or sink. In type 1, 2 and 4 materials, vapour convection can barely develop. For all infinitely small representative volumes, a combination with the general transfer equation gives:

$$\text{div}(k_{\phi,m} \,\textbf{grad}\, \phi + K_\theta \,\textbf{grad}\, \theta) \pm G' = \frac{\partial w}{\partial t} = \rho_w \xi_\phi \frac{\partial \phi}{\partial t} \qquad (2.145)$$

In type 3 materials, vapour convection is possible. Implementation there gives:

$$\text{div}\left[(k_{\phi,m} \,\textbf{grad}\, \phi + K_\theta \,\textbf{grad}\, \theta) + 6.21 \times 10^{-6} \, \mathbf{g_a} \phi p_{sat}\right] \pm G' = \rho_w \xi_\phi \frac{\partial \phi}{\partial t} \qquad (2.146)$$

In both formulae, $\rho_w \xi_\phi$ is the specific moisture content, with ξ_ϕ the derivative of the suction characteristic multiplied by the derivative of suction versus relative humidity ϕ:

$$\xi_\phi = \xi_s \partial s / \partial \phi$$

The result equals the derivative of the hygroscopic curve between 0 and 100%, except for type 3 materials where the specific moisture content is:

$$\rho_w \xi_\phi = \frac{\Psi_o}{\rho_w R} \frac{\partial(\phi p_{sat}/T)}{\partial \phi}$$

In type 1 and type 4 materials, hygroscopic moisture embodies the source and sinks term (G'). This way, local interstitial condensation and local evaporation disappear as phenomena.

2.4.5.4 Starting, boundary and contact conditions

In most cases, the starting condition is a wet assembly thanks to construction moisture, which gives wetness far beyond the hygroscopic equilibrium at a relative humidity ϕ_M (95–98%). A likely boundary condition is contact with moist air, giving as vapour flux at any end face:

$$g_v = \beta(p - p_s)$$

with p the vapour pressure in the air, p_s the vapour pressure at the end face and β the surface film coefficient for vapour diffusion there. With relative humidity as potential, vapour pressure at the end face is rewritten as:

$$p_s = p_{sat,s} \, \phi$$

with $p_{sat,s}$ the saturation pressure there. Contact with moist air has an effect when drying, sorption, desorption and interstitial condensation are considered. Other likely boundary conditions are a known moisture flux, a situation typical for

2.4 Moisture

impinging rain without water runoff, and contact with water, in the case of surface condensation and water films formed by precipitation.

A possible contact condition is ideal suction exchange in the interfaces, so ensuring continuity of the water and vapour flow. In reality this seldom happens. Most of the time suction contacts introduce a contact resistance, sush as the mortar-to-brick and brick-to-plaster interfaces. Small air gaps are also likely between layers, allowing diffusion contact. Then only vapour keeps continuity, as it does between non-capillary materials. Interstitial condensation and rain penetration may create contact with water between materials each time gravity and water pressure initiate runoff and capillarity immobilizes that water layer. In real contacts, the three often occur in parallel.

2.4.5.5 Remarks

The balance equations given are correct as long as capillary porous materials do not contact liquid water. If so, water not only moves from the wider to the narrower pores, but wider pores also suck water from the narrower, as shown when discussing isothermal water transfer in a serial two-pore system that contacts water at one side. That difference in moisture movement is accounted for by considering the moisture permeability and the specific moisture content between dry and capillary. For non-capillary materials, water contact means a 100% relative humidity in the contact plane, while for capillary ones it equates to zero suction, 100% relative humidity and the moisture content capillary.

2.4.6 Simplified moisture flow model

2.4.6.1 How it looks

All properties in the balance equations given are dependent on suction, relative humidity and temperature. This excludes analytical solutions. Applying CVM of course is possible, which requires software tools commercialized under various names. A disadvantage of these tools is their 'black box' nature. This opens the door for easy-to-understand, simple models with the moisture content as an 'improper' driving force – 'improper' because, contrary to a 'proper' driving force, its value changes at all interfaces. Water transfer then becomes:

$$\mathbf{g}_w = -D_w \, \mathbf{grad} \, w \tag{2.147}$$

with D_w the water diffusivity of the material (units m^2/s) linked to the water permeability:

$$D_w = k_w ds/dw = k_w/(\rho_w \xi_s) \tag{2.148}$$

For the relationship with moisture content, a preset function is mostly used:

$$D_w = a(A/w_c)^2 \exp(bw/w_c)$$

with A the capillary water absorption coefficient, w_c the capillary moisture content, and a and b as constants, which differ between materials.

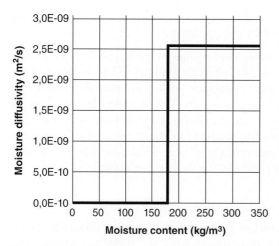

Fig. 2.56 Moisture diffusivity as a function of moisture content, simplified

The simplified model proposed reduces the water diffusivity to a step function: zero for a moisture content below critical (w_{cr}), which corresponds to the pivot relative humidity ϕ_M, and constant above w_{cr} (Figure 2.56).

Below the critical moisture content, only vapour is moving:

$$\mathbf{g_w} = -\delta\, \mathbf{grad}\, p$$

Above the critical moisture content, unsaturated water flow dominates, with vapour added when non-isothermal:

	$w < w_{cr}$	$w \geq w_{cr}$
Isothermal	$\mathbf{g_v} = -\delta\, \mathbf{grad}\, p$	$\mathbf{g_w} = -D_w\, \mathbf{grad}\, w$
Non-isothermal	$\mathbf{g_v} = -\delta\, \mathbf{grad}\, p$	$\mathbf{g_w} = -D_w\, \mathbf{grad}\, w - \delta \dfrac{dp_{sat}}{d\theta}\, \mathbf{grad}\, \theta$

Typical for the model is a sharp moisture front between wet and air-dry zones in materials that are partially above-critical moist. In fact, when assuming a one-dimensional, hygrically steady-state isothermal regime, each elementary material layer sees the same moisture flux passing through. Thus, between below and above-critical the fluxes must be equal. Since:

$$g_{m,below} = -\delta(\mathbf{grad}\, p)_{below} = -\delta dp/dw(\mathbf{grad}\, w)_{below} = -D_{w,below}(\mathbf{grad}\, w)_{below}$$

and:

$$g_{m,below} = g_{mm\ above} = -D_{w,above}(\mathbf{grad}\, w)_{above}$$

with $D_{w,below}$ (≈ 0) being very small compared with $D_{w,above}$, so $(\mathbf{grad}\, w)_{below}$ must be much larger than $(\mathbf{grad}\, w)_{above}$. Therefore, moisture content at the subcritical side

2.4 Moisture

Fig. 2.57 Moisture front between above-critical wet and air-dry zones in brickwork

will increase suddenly in the contact, stay constant again on the above-critical side, resulting in a moisture front between dry and wet – a fact confirmed by observation. An ink stain on blotting paper spreads out for a while and then stops, with the ink content in the stain critical and outside hardly visible. The front between dry and wet sand in the outflow zone at beaches is sharp. The same is seen on brick walls (see Figure 2.57).

Although the transfer equations look similar, the moisture front makes unsaturated water flow different from heat conduction, air transfer or vapour diffusion. Analytical solutions for example are only possible when assuming the shape moisture content takes known.

2.4.6.2 Applying the simplified model

Capillary suction
Capillary suction is most important as a wetting mechanism. If a sample of a hydrophilic, homogeneous material is weighed regularly while sucking water, just as for a pore, a linear relationship emerges between the quantity sucked per m^2 of contact surface and the square root of time (Figure 2.58):

$$m_c = A\sqrt{t} \qquad (2.149)$$

with A the water absorption coefficient of the material, units kg/(m^2·s$^{1/2}$).

Monitoring the moving moisture front gives the following relationship between location and time:

$$h = B\sqrt{t} \qquad (2.150)$$

with B the water penetration coefficient, units m/s$^{1/2}$. As soon as that front reaches the far side of the sample, suction switches to a second [$\sqrt{t}$, m] line with a much smaller slope, equal to the coefficient of secondary water absorption A'. Related water uptake then is the result of diffusion and solution of the air inclusions in the pore water. The intersection with the capillary line fixes the capillary moisture content

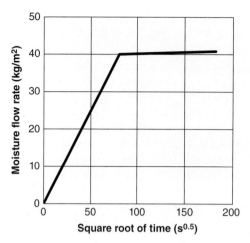

Fig. 2.58 Capillary suction, $mw = A\sqrt{t}$

(w_c). The following relationship exists between that moisture content, the water absorption coefficient and the water penetration coefficient:

$$A \approx Bw_c \qquad (2.151)$$

an equation that presumes that the capillary moisture content extends over the whole height of the sample. The water penetration coefficient and capillary moisture content are pure material characteristics – the higher the first, the more capillary the material; the higher the second, the more moisture it buffers. The capillary water absorption coefficient instead is a composite property, comparable with the contact coefficient in heat transfer. Its value instructs how easily a material absorbs water, just like the contact coefficient shows how easily it picks up heat.

The linear relationship $[\sqrt{t}, m_c]$ does not account for air transfer. As shown for a pore, air egress induces anomalies. This is seen when samples of fine-porous materials or samples sealed at the top are tested. In the first case the curve bends towards the time axis. In the second case, water sorption is retarded as time progresses, while at the water side, air bubbles leave the sample. In non-homogeneous materials the curve also deviates from linear. Capillary moisture uptake by hydrophobic materials, if any, advances very slowly without a linear link to the square root of time. Calculating a kind of water sorption coefficient is only possible from the slope of the tangent to the curve at time zero.

The question now is whether the simplified model may predict that linear relationship between water uptake and the square root of time. Assuming a one-dimensional capillary flow, mass conservation gives:

$$D_w \frac{\partial^2 w}{\partial x^2} = \frac{\partial w}{\partial t} \qquad (2.152)$$

an equation analogous to heat conduction in a flat layer without source or sink. Capillary water sorption is comparable now to a sudden temperature increase at the surface of a semi-infinite medium, although replaced by a moisture increase from

2.4 Moisture

hygroscopic to capillary. The critical moisture content however complicates the solution. Besides the starting and boundary conditions, the existence of a front imposes two extra conditions. During suction, moisture content at the front must remain critical and related water flux must supply what is needed to move that front over a distance dx_{fr} per time unit, or:

$$t \leq 0, 0 \quad = \quad x \leq \infty : w = w_H$$
$$t \geq 0, x \quad = \quad 0 : w = w_c$$
$$t > 0, x \quad = \quad x_{fr} : w = w_{cr}, \quad -D_w \left(\frac{\partial w}{\partial x}\right)_{x=x_{fr}} = w_{kr}\left(\frac{dx_{fr}}{dt}\right)_{x=x_{fr}}$$

Mathematically, the front at $x=x_{fr}$ is a singular point that excludes an analytical solution, although two extreme cases allow one. First, both the hygroscopic and critical moisture content nears zero. Once sucking (w_c), moisture content in an infinitely long sample at a distance x away from the contact face with water then becomes:

$$w = w_c \left(\underbrace{\frac{2}{\sqrt{\pi}} \int_{q=\frac{x}{2\sqrt{D_w t}}}^{\infty} \exp(-q^2) dq}_{} \right) \tag{2.153}$$

with the term above the brace equal to the inverse error function (see Figure 2.59).

In the wet zone, the lowest critical moisture content is found at the moisture front. At the water contact, the moisture content is capillary. For the quantity of water

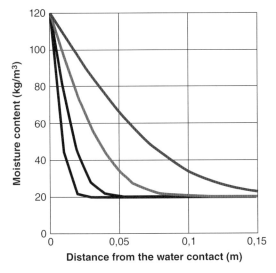

Fig. 2.59 Capillary suction, moisture content in the material for a critical moisture content close to zero ($w_{cr} \ll w_c$)

absorbed, the following holds:

$$m_c = 2w_c \sqrt{D_w/\pi} \sqrt{t}$$

Or, the relationship between capillary water uptake and the square root of time emerges as linear. The water sorption coefficient is:

$$A = 2w_c \sqrt{D_w/\pi}$$

which gives as moisture diffusivity:

$$D_w = \frac{\pi}{4} \left(\frac{A}{w_c}\right)^2 \tag{2.154}$$

The water penetration coefficient becomes:

$$B = 2\,\mathrm{erfc}^{-1}(w_{cr}/w_c)\sqrt{D_w}$$

Materials that have critical moisture content close to zero are rare.

For the second extreme case, the critical moisture content hardly differs from capillary. Presumed then is that the moisture profile between the sucking face and the moisture front remains linear. The mass balance so becomes:

$$\frac{w_{cr} + w_c}{2} dx = D_w \frac{w_c - w_{cr}}{x} dt$$

Solving for x produces the desired linear relationship between the moisture front and the square root of time (Figure 2.60):

$$x = 2\sqrt{D_w \frac{w_c - w_{cr}}{w_c + w_{cr}}} \sqrt{t}$$

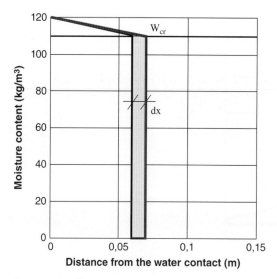

Fig. 2.60 Capillary suction, moisture content in the material when critical and capillary moisture content hardly differ ($w_{cr} \approx w_c$)

2.4 Moisture

The water penetration coefficient becomes:

$$B = 2\sqrt{D_w \frac{w_c - w_{cr}}{w_c + w_{cr}}} \qquad (2.155)$$

The amount of absorbed water between 0 and t reaches (m_c, kg/m^2):

$$m_c = x \frac{w_c + w_{cr}}{2} = \sqrt{D_w(w_c + w_{cr})(w_c - w_{cr})}\sqrt{t}$$

giving as the water sorption coefficient:

$$A = \sqrt{D_w(w_c^2 - w_{cr}^2)}$$

The moisture diffusivity thus equals:

$$D_w = \frac{A^2}{(w_c^2 - w_{cr}^2)} \qquad (2.156)$$

Of course, in reality, moisture diffusivity is a function of moisture content. Rewriting mass conservation gives:

$$\frac{\partial}{\partial x}\left(D_w \frac{\partial w}{\partial x}\right) = \frac{\partial w}{\partial t} \qquad (2.157)$$

If the moisture diffusivity is unequivocally moisture-content related, if the material volume is close to semi-infinite and if moisture content is w_o at time zero and remains constant in the contact face for $t \geq 0$, then a Boltzmann transformation allows the conversion of the partial differential equation into a second-order differential equation:

$$\lambda = x/\sqrt{t} \qquad -\frac{\lambda}{2}\frac{dw}{d\lambda} = \frac{d}{d\lambda}\left[D_w(w)\frac{dw}{d\lambda}\right] \qquad (2.158)$$

Moisture content here depends on the variable λ. In a [λ, w] coordinate system, all time-related curves come together into one curve, with as surface to the λ axis:

$$\int_0^{\lambda_m} w(\lambda)d\lambda = C^{te} \qquad (2.159)$$

The moisture absorbed since time zero is now:

$$m_w(t) = \left[\int_0^{x_m} w(x)\,dx\right]_t$$

with:

$$x_m = \lambda_m \sqrt{t}$$

As the integration for λ is independent of time, a combination with the surface equation gives as water sorption coefficient:

$$A = C^{te} = \int_0^{\lambda_m} w(\lambda) d\lambda \qquad (2.160)$$

Or, the relationship $m_c = A\sqrt{t}$ looks general, provided air egress has no influence and the function $D_w(w)$ is univocal. When assuming a profile $w(\lambda)$, the last equation allows calculation of the moisture diffusivity as a function of the water sorption coefficient and moisture content.

Wind-driven rain

Rain creates two successive boundary conditions. As long as the moisture content at the wetted surface stays below capillary, the sucked water flux (g_{ws}) equals the wind-driven rain intensity. Once that face is capillary wet, a water film forms and capillary suction begins (Figure 2.61). Assume a critical moisture content much lower than capillary. As long as the wetness at the surface (w_s) stays below capillary, the boundary conditions are: $t=0$, $0 \leq x \leq \infty$: $w=0$; $t \geq 0$, $x=0$: $-D_w(dw/dx)_{x=0} = g_{ws}$, with g_{ws} the constant wind-driven rain intensity in kg/(m².s). The solution of the mass conservation equation then gives:

$$w = \frac{2g_{ws}}{D_w}\left[\sqrt{\frac{D_w t}{\pi}}\exp\left(-\frac{x^2}{4D_w t}\right) - \frac{x}{2}\text{erfc}\left(\frac{x}{2\sqrt{D_w t}}\right)\right]$$

On the wet surface ($x=0$), the moisture content equals:

$$w_s = \frac{2g_{ws}}{D_w}\sqrt{\frac{D_w t}{\pi}} \qquad (2.161)$$

Once capillary wet, a water film forms and water uptake changes into horizontal suction. The moment of film formation is given by:

$$t_f = \frac{\pi D_w w_c^2}{4g_{ws}^2} = \frac{\pi^2 A^2}{16 g_{ws}^2} \qquad (2.162)$$

In other words, the smaller the water absorption coefficient and the higher the wind-driven rain intensity, the quicker the rain runs off to give an extra water load on lower

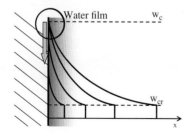

Fig. 2.61 Rain absorption before a water film forms ($w_{cr} \ll w_c$, $w_s \ll w_c$ until $w_s = w_c$)

2.4 Moisture

parts. Brick veneers and stuccoed walls reflect that difference. Most bricks have a high water absorption coefficient ($0.2 \leq A \leq 0.8 \, \text{kg}/(\text{m}^2.\text{s}^{1/2})$), making runoff exceptional. Water-repellent stucco instead has a very low water absorption coefficient ($A < 0.005 \, \text{kg}/(\text{m}^2.\text{s}^{1/2})$), giving an almost instant runoff, with water spreading over the lower parts.

Until runoff starts, the water absorbed is $m_w = g_{ws} t_f$. Just before film formation, the flux sucked still equals the constant wind-driven rain intensity:

$$g_{ws} = w_c \sqrt{\frac{\pi D_w}{4}} \frac{1}{\sqrt{t_f}} = 0.89 \, w_c \sqrt{\frac{D_w}{t_f}} = 0.79 \frac{A}{\sqrt{t_f}}$$

If the solution for capillary sorption applied to the water film after runoff started, the flux sucked must change into:

$$g_{ws2} = w_c \sqrt{\frac{D_w}{\pi}} \frac{1}{\sqrt{t_f}} = 0.56 \, w_c \sqrt{\frac{D_w}{t_f}} = \frac{A}{2\sqrt{t}} = g_{ws} \frac{2}{\pi} < g_{ws}$$

or, at film formation, the material suddenly picks up less, which is physically impossible. Therefore, a transition regime must exist between a constant water flux and sucking from the water-loaded surface. Mathematically, this transition results from the condition in the material at the moment the film forms: neither dry nor uniformly wet but a moisture profile across.

To avoid that anomaly, further discussion considers the critical and capillary moisture content to be nearly equal. In the material, moisture is assumed to change linearly from capillary at the wet face down to critical at depth x. As long as the wet face (w_s) is below capillary (w_c), the water flux in the material equals:

$$g_{ws} = D_w \frac{w_s - w_{cr}}{x}$$

From the mass balance, $g_{ws} dt = w_s dx$ follows as the position of the moisture front:

$$x = \frac{D_w w_{cr}}{g_{ws}} \left(\sqrt{1 + \frac{2 g_{ws}^2 t}{D_w w_{cr}^2}} - 1 \right)$$

while moisture content at the impinged face equals:

$$w_s = w_{cr} + g_{ws} x / D_w$$

Once capillary wet ($w_s = w_c$), a water film forms, while the moisture front in the material has then penetrated to a depth (x_{fr}):

$$x_{fr} = [D_w(w_c - w_{cr})]/g_{ws} = \frac{A^2}{(w_c + w_{cr})g_{ws}}$$

Film formation thus happens at:

$$t_f = A^2 / (2 g_{ws}^2)$$

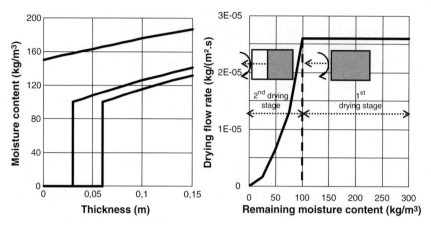

Fig. 2.62 The two stages of isothermal drying

Drying

Drying demands an ambient relative humidity below the one at the material's surface. To test drying experimentally, one of the small faces of a water-saturated beam-like sample is left unsealed. Then the sample is hung in a climate chamber kept at constant temperature and constant relative humidity and weighed regularly. When depicting the weight loss per unit of time and surface, giving the drying flux, as function of the average moisture content left in the sample, a curve such as Figure 2.62 is obtained. Two drying stages appear with a transition in between: the first at higher average moisture content and quasi-constant drying flux, and a second at lower average moisture content and rapidly decreasing drying flux.

The question now is whether the simplified transfer model is able to explain this curve. Take the saturated sample, for which relative humidity is 100%. As the vapour pressure at the drying face is the saturation pressure for that face's temperature ($p_{sat,s}(\theta_s)$), the drying flux to the chamber becomes:

$$g_{vd} = \beta \left(p_{sat,s} - \phi p_{sat} \right) \tag{2.163}$$

with β the surface film coefficient for diffusion, ϕ the relative humidity in the chamber and p_{sat} the saturation pressure at chamber temperature.

Since drying means evaporation, the drying flux carries the heat of evaporation, which reduces the temperature and saturation pressure at the drying face below the conditions in the chamber. To simplify, the sample is attributed an infinite thermal conductivity, while at the drying face no heat transfer by radiation is assumed. The face's saturation pressure then equals that related to the temperature in the chamber. The drying flux can thus be rewritten as:

$$g_{vd} = \beta p_{sat}(1 - \phi) \tag{2.164}$$

Whatever the material may be, drying apparently only depends on the conditions in the chamber: air temperature, relative humidity and, via the surface film coefficient

2.4 Moisture

for diffusion, air velocity. The warmer and drier the air and the more intense its movement, the faster the sample dries. People apply this principle unknowingly. Laundry is hung outside when the weather is sunny (= high temperature) and dry (= low relative humidity). If windy, drying accelerates.

The result is a drop in moisture content at the drying face. As a consequence, a gradient initiating moisture flow develops in the sample, while at any moment in time the following applies:

$$x=0, \quad -D_w(dw/dx)_{x=0} = \beta p_{sat}(1-\phi)$$

A capillary water flux to the drying face, evaporation there and the moisture dropping in the sample defines the first drying stage (see Figure 2.62), which continues until the moisture content at the drying face turns critical. Then, moisture diffusivity there becomes zero, the moisture front shifts into the sample and moisture transfer past the drying front becomes vapour only (Figure 2.62), changing the drying flux into:

$$g_{vd} = \frac{p_{sat}(1-\phi)}{1/\beta + \mu N x} \tag{2.165}$$

with μ the vapour resistance factor of the 'dry' material and x the distance between the moisture front and the drying face. The second drying stage has begun. As the distance to the drying face increases, drying slows down. That second stage lasts until the sample regains hygroscopic equilibrium with the ambient. The sample anyhow quickly appears superficially dry, which is too often taken as proof that a material is dry, while in fact it still contains quite a lot of moisture.

The first drying stage can be described using mass equilibrium with, as starting and boundary conditions: a surface at $x=0$ that allows drying, a vapour-tight surface at $x=d$, with, until time zero, a constant moisture content over the whole thickness ($t \le 0$, $0 \le x \le d$: $w = w_o$); and beyond time zero ($t \ge 0$) a gradient in moisture content equal to $dw/dx = -g_{vd}/D_w$ at surface $x=0$, and a gradient in moisture content equal to $dw/dx = 0$ at surface $x=d$. The solution is:

$$w = w_o - \frac{g_{vd}t}{d} - \frac{g_{vd}d}{D_w}\left\{\frac{3x^2-6dx+2d^2}{6d^2} + \underbrace{\frac{2}{\pi^2}\sum\left[\frac{(-1)^n}{n^2}\exp\left(-D_w\frac{n^2\pi^2 t}{d^2}\right)\cos\left(\frac{n\pi(d-x)}{d}\right)\right]}\right\}$$

The term above the brace describes how the moisture profile change moves from the drying face to the rear. Once there the change halts and the sample is left with a parabolic moisture profile:

$$w = w_o - \frac{g_{vd}t}{d} - \frac{g_{vd}d}{D_w}\left(\frac{3x^2-6dx+2d^2}{6d^2}\right) \tag{2.166}$$

with lowest moisture content at the drying face ($x=0$):

$$w = w_o - \frac{g_{vd}t}{d} - \frac{g_{vd}d}{3D_w} \tag{2.167}$$

The average value at any moment touches:

$$w = w_o \left(1 - \frac{g_{vd}t}{w_o d}\right) \qquad (2.168)$$

The second drying stage starts once the drying face turns critically moist, which requires a time span ($t_{second} - t_{start}$) given by:

$$t_{second} - t_{start} = d\left[\frac{w_o - w_{cr}}{g_{vd}} - \frac{d}{3D_w}\right] \qquad (2.169)$$

Moisture at that moment left in the sample fixes the transition moisture content (w_{tr}):

$$w_{tr} = w_{cr} + g_{vd}d/(3D_w) \qquad (2.170)$$

The difference between critical and transition moisture content clearly increases at higher drying fluxes. Hence, faster drying during the first stage does not guarantee sooner dryness. The second stage simply starts at a higher transition moisture content. At equal drying flux the difference between transition and critical is smaller for highly capillary materials, which have a high moisture diffusivity compared with less capillary ones. More generally, a low diffusion resistance factor, a high moisture diffusivity and a low critical moisture content will favour drying – a good example is brick. Less capillary materials with critical moisture content near capillary, such as concrete, show a very short first and long second drying stage. They dry very slowly.

Modelling the second drying stage proceeds as follows. Assume the transition and critical moisture contents are close, a hypothesis that presumes a low drying flux during the first stage. Then, with the withdrawing, critically wet moisture front at x and the drying face at $x = 0$, mass equilibrium gives:

$$\frac{p_{sat}(1-\phi)}{1/\beta + \mu N x} dt = (w_{cr} - w_H) dx \qquad (2.171)$$

where w_H stands for hygroscopic equilibrium with the ambient conditions. In steady state and with the vapour resistance factor constant, solving that mass balance gives as the position of the moisture front:

$$x = \frac{1}{\mu N \beta}\left(\sqrt{1 + \frac{2p_{sat}(1-\phi)\mu N \beta^2}{w_{cr} - w_H}\sqrt{t}} - 1\right) \qquad (2.172)$$

As soon as the vapour resistance $\mu N x$ outranges the surface resistance $1/\beta$, the equation simplifies to:

$$x = \sqrt{\frac{2p_{sat}(1-\phi)}{\mu N(w_{cr} - w_H)}}\sqrt{t} \qquad (2.173)$$

Or, drying results in a square root relationship between the position of the moisture front and time. Multiplying left and right with ($w_{cr} - w_H$) gives as quantity of

2.4 Moisture

moisture that dried:

$$m_{vd} = \sqrt{\frac{2p_{sat}(1-\phi)(w_{cr}-w_H)}{\mu N}}\sqrt{t} = A'\sqrt{t} \quad (kg/m^2) \quad (2.174)$$

When both the critical and hygroscopic moisture contents are known, this equation allows the calculation of the vapour diffusion factor from a drying test on a sample with a known drying surface. The higher the value found, the slower drying proceeds.

Despite all the simplifications, the model is quite usable at low drying fluxes and ready heat supply. At high fluxes, the isothermal character gets lost and the curve will first bend towards the $\sqrt{t}$-axis and then slowly become a straight line. The reason is the heat of evaporation, which lowers the temperature at the drying front, thereby reducing the flux below the straight line. As drying proceeds and the front shifts deeper in the material, the lowering drying flux requires less heat of evaporation, which is why the curve slowly becomes linear with the square root of time.

The drying model is easily extended to composite assemblies with a wet interface (j) or a more than critical moist layer between non-capillary layers. If <equivalent> diffusion is the only driver, then the steady-state vapour fluxes between interface j, or the interfaces k and l bounding the wet layer, and the ambient at both sides equals:

$$\text{Wet interface} \quad g_{md} = -\left(\frac{p_{sat,j}-p_1}{Z_1} + \frac{p_{sat,j}-p_2}{Z_2}\right) = \frac{p_2-p_{sat,j}}{Z_2} - \frac{p_{sat,j}-p_1}{Z_1}$$

(2.175)

$$\text{Wet zone} \quad g_{md} = -\left(\frac{p_{sat,k}-p_1}{Z_1} + \frac{p_{sat,l}-p_2}{Z_2}\right) = \frac{p_2-p_{sat,l}}{Z_2} - \frac{p_{sat,k}-p_1}{Z_1} \quad (2.176)$$

with $p_{sat,j}$, $p_{sat,k}$ and $p_{sat,l}$ the vapour saturation pressures in the interfaces j, k and l respectively. In the wet zone, relative humidity is 100% and the vapour pressure is saturated. Z_1 and Z_2 are the diffusion resistances between j and the ambient at both sides, or, between k and the cold ambient 1, vapour pressure p_1, and l and warm ambient 2, vapour pressure p_2.

In the [Z, p] plane, the flux terms equal the slopes of the lines connecting the saturation pressure in the wet interface j with the points [0, p_1] and [Z_T, p_2] or the saturation pressure in wet interface k with the point [0, p_1] and the saturation pressure in the wet interface l with the point [Z_T, p_2]. The line to the cold side 1 can intersect the saturation curve. Part of the vapour diffusing from j or k then condenses somewhere there. Finding where and how to correct the vapour pressure requires tracing the tangents from the saturation pressures $p_{sat,j}$ or $p_{sat,l}$ and the vapour pressure p_1 to the saturation curve, while the amounts condensing follow from the difference in slope between both tangents (Figure 2.63).

What is termed drying condensation could be a problem in any assembly that contains construction moisture, where one or more layers absorbed rain, or wherein vapour produced inside condenses in two or more interfaces or zones. When

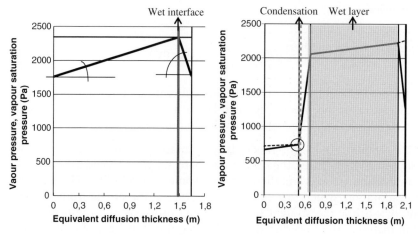

Fig. 2.63 Left: drying of a moist interface; right: drying of a moist layer in an assembly with condensation in an interface elsewhere as a result

convection supplements diffusion, drying proceeds more rapidly, but the condensation deposit in an interface elsewhere, if it takes place, will accelerate.

Interstitial condensation, limit state
The question when interstitial condensation happens is: does it end? Consider a flat assembly made of an insulating, capillary, vapour-permeable material with diffusion resistance Z_2, finished at the cold side by a vapour-retarding, air-tight layer with diffusion resistance Z_1. The yearly mean ambient temperature and vapour pressure at the warm side $(\theta_2, p_2, Z = Z_1 + Z_2)$ gives condensation behind the finish $(Z = Z_1)$ with as flux (g_{wc}):

$$g_{wc} = g_{v2} - g_{v1} = \frac{p_2 - p_{sat,Z_1}}{Z_2} - \frac{p_{sat,Z_1} - p_1}{Z_1}$$

where p_1 is the vapour pressure at the cold side and p_{sat,Z_1} is the vapour saturation pressure in the condensation interface. Being capillary, the insulating material sucks the condensate back, giving as water flux:

$$g_w = D_w \frac{w - w_{cr}}{x} = \frac{p_2 - p_{sat,Z_1}}{Z_2} - \frac{p_{sat,Z_1} - p_1}{Z_1}$$

A wet zone thus develops in the capillary layer, extending slowly to the warm side. In it, vapour pressure equals saturation, while moisture content stays slightly above critical except at the moving moisture front, where it is critical. Simultaneously, the incoming vapour pressure line rotates from being the tangent to the saturation pressure in the interface Z_1 with the finish to contacting the saturation pressure at the moisture front ($p_{sat,front}$) (Figure 2.64). As a result, the incoming vapour flux g_{v2} drops. Once the moisture front arrives where incoming equals outgoing (g_{v1}), condensation stops and the limit state is reached. In the wet zone, a capillary water flux from cold to warm persists in equilibrium with an identical vapour flux from warm to cold.

2.4 Moisture

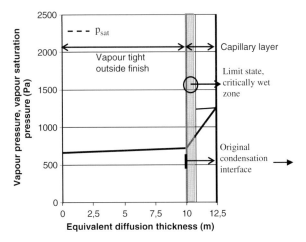

Fig. 2.64 Annually accumulating condensate: limit state in the case a capillary, vapour-permeable material has a non-capillary, vapour-retarding finish at the cold side

Ingoing is:

$$g_{v2\infty} = g_{v1} = \frac{p_2 - p_{\text{sat},f}}{Z_2 - \mu N d_w} = \frac{p_{\text{sat},f} - p_{\text{sat},Z_1}}{\mu_w N d_w} - D_w \frac{w_{Z1} - w_{cr}}{d_w} \qquad (2.177)$$

with w_{Z1} the moisture content in the interface Z_1, $p_{\text{sat},f}$ the vapour saturation pressure at the moisture front, μ and μ_w the vapour resistance factors of the dry and critically wet insulating material, and d_w the thickness of the wet zone. Solving the equation for w_{Z1} gives:

$$w_{Z1} = w_{cr} + \underbrace{\frac{1}{D_w}\left(\frac{p_{\text{sat},f} - p_{\text{sat},Z_1}}{\mu_w N} - g_{v1} d_w\right)} \qquad (2.178)$$

The term above the brace is so small that it can be set to zero ($w_{Z1} \approx w_{cr}$). Or, in the limit state, a zone, wherein the average moisture content hardly differs from critical, extends over a thickness d_w in the capillary material between the finish and somewhere closer to the warm side. That thickness increases with a greater difference in vapour pressure between cold and warm. An approximation of the front's final position gives the intersection with the original saturation curve of a parallel to the outgoing tangent through the warm side vapour pressure. The time taken to reach the limit state follows from:

$$w_{cr} dx_w = \left(\frac{p_2 - p_{\text{sat},f}}{Z_2 - \mu N x} - \frac{p_{\text{sat},Z_1} - p_1}{Z_1}\right) dt \qquad (2.179)$$

with x the thickness of the wet zone. To solve this equation, the relationship between saturation pressure ($p_{\text{sat},f}$) and temperature at the moisture front, or between $p_{\text{sat},f}$ and x, is needed. An approximate solution is found by assuming:

$$p_{\text{sat},f} = aZ + p_{\text{sat},Z_1}, \quad p_{\text{sat},Z_1} = C^{te}, \quad Z = \mu N x$$

with as a result:

$$(\mu N/w_{cr})t = a_1 Z - a_2 \ln(1 - a_3 Z) \tag{2.180}$$

where:

$$a_1 = \frac{1}{a - g_{v1}}, \quad a_2 = \frac{a - g_{v1}}{Z_2(g_{v2} - g_{v1})}, \quad a_3 = \frac{Z_2}{(a - g_{v1})},$$

$$g_{v1} = \frac{p_{sat,Z_1} - p_1}{Z_1}, \quad g_{v2} = \frac{p_2 - p_{sat,Z_1}}{Z_2}$$

The diffusion resistance (Z_w) and thus thickness of the wet zone (x_w) in the capillary layer are thereby known as implicit functions of time. The condensation deposit in kg/m² after t days is:

$$m_c = 86,400 \, w_{cr} Z(t)/(\mu N)$$

Table 2.6 lists the limit states for other layer combinations.

Whether interstitial condensation will be acceptable depends not only on the nature of the material in or against which condensation occurs and the quantity deposited, but also on the possible limit state. Case (4) is most negative. It occurs each time a capillary layer sits between a non-capillary thermal insulation at the warm side and a vapour-retarding finish at the cold side. An example in cold climates is the deck of a ventilated low-slope roof. An example for warm, humid climates would be a mineral-wool insulated timber-frame wall with vinyl paper finish on the inside gypsum board and a vapour-permeable finish outside.

2.5 Problems and solutions

Problem 2.1

The monthly average temperature outdoors is 0 °C, the dew point outdoors −2 °C and the temperature indoors 22 °C. The lowest temperature factor on the envelope is 0.4, while the daily mean vapour release indoors reaches 20 kg. What outside air ventilation in m³/h is needed to avoid mould (which means a surface relative humidity below 80%)?

Solution

The vapour pressure outdoors equals:

$$p_e = 611 \exp\left(\frac{22.44 \, \theta_{d,e}}{272.44 + \theta_{d,e}}\right) = 518 \, \text{Pa}$$

$\theta_{d,e}$ being the dew point. The lowest inside surface temperature thus is: $\theta_s = 0 + 22 \times 0.4 = 8.8$ °C. Allowable vapour pressure inside becomes:

$$p_i = 0.8 \times 611 \exp\left(\frac{17.08 \, \theta_s}{234.18 + \theta_s}\right) = 907 \, \text{Pa}$$

The ventilation flow needed is:

$$\dot{V}_a = \frac{R_v T_i G_{v,P}}{p_i - p_e} = \frac{462 \times 295.15 \times 20/24}{907 - 518} = 292 \, \text{m}^3/\text{h}$$

2.5 Problems and solutions

Table 2.6 Limit state in the case of annual accumulation of condensate

Cold side	Warm side	Limit state
Case (2) Vapour-retarding, non-capillary layer	Insulating, vapour-permeable, non-capillary layer	The layer at the warm side becomes saturated from the interface in between, towards the inside face, over a thickness (d_w) approximately equal to the distance between this interface and the intersection with the original saturation curve of a line parallel to the outgoing tangent across the vapour pressure at the warm side.
Case (3) Vapour-permeable capillary layer	Insulating, vapour, non-capillary layer	The layer at the cold side turns critically wet from the interface between the two outward over a thickness (d_w) approximately equal to the distance between this interface and the intersection with the original saturation curve of a line parallel to the ingoing tangent across the vapour pressure at the cold side
Case (4) Vapour-permeable, capillary layer, finished at its cold side with a vapour-retarding lining	Insulating, vapour-permeable, non-capillary layer	Interstitial condensation starts in the capillary layer itself and at its interfaces, one with the vapour-retarding lining, the other with the non-capillary layer at its warm side. At the limit state, the capillary layer ends above capillary wet, while water runs off in the interface with the non-capillary layer at its warm side.
Case (5) Vapour-permeable capillary layer	Insulating, vapour-permeable capillary layer	Both layers turn critically wet by back-sucking condensate from the interface in between. At the limit state, thickness of the critically moist part in both is such that the incoming and outgoing vapour pressure lines in the $[Z, p]$ plane have the same slope. The limit state cannot be determined graphically, but assume that: $$d_1 = \frac{1}{w_{kr,1}}\left(\frac{A_1^2}{A_1^2+A_2^2}\right), d_2 = \frac{1}{w_{kr,2}}\left(\frac{A_2^2}{A_1^2+A_2^2}\right)$$

If surface condensation was the risk to avoid, then a flow of 184 m³/h should suffice, (i.e. only 63% of what is needed to exclude mould).

Problem 2.2

Assume the ventilation flow calculated in Problem 2.1 is delivered by an extractor fan. The designer forgot to provide the necessary air inlets. As a result, air leakage through the envelope has to do the job. The envelope air permeance equals $0.083\,\Delta P_a^{-0.37}$ kg/(s.Pa). What air pressure excess indoors will generate the flow required?

Solution

The relation between airflow in kg/s and air pressure difference is given by:

$$G_a = K_a \Delta P_a = 0.083\,\Delta P_a^{0.63}$$

A volumetric flow of 292 m³/h means a weight-related flow of 345.3 kg/h or 0.095922 kg/s. The air pressure excess required indoors equals:

$$\Delta P_a = (0.095922/0.083)^{1/0.67} = 1.24\,\text{Pa}$$

a result that shows that the envelope must be highly air-permeable. In fact, at a pressure difference of 50 Pa, one gets a volumetric air flow of 3469 m³/h, which is more than a lot.

Problem 2.3

Repeat Problem 2.2 where the air-tightness of the envelope is 10 times higher, albeit with the same air pressure difference exponent. What is your conclusion?

Solution

Pressure excess needed is 38.6 Pa. Therefore, quite an air-tight envelope demands unrealistically high pressure differences before infiltration could deliver the ventilation needed.

Problem 2.4

Outdoors, the temperature reaches 34 °C for 80% relative humidity. Indoors, air conditioning provides 24 °C at 60% relative humidity. The building has a volume of 600 m³. The ventilation rate is $0.6\,\text{h}^{-1}$ and the daily mean vapour release indoors 15 kg. How much vapour must be removed by the system (in g/h) to maintain this 60%? Could that quantity be influenced by selecting appropriate glazing?

Solution

The formula describing the indoor/outdoor vapour pressure difference is:

$$\Delta p_{i,e} = \frac{RT_i X}{nV}$$

2.5 Problems and solutions

with X the vapour flow to be removed from the outdoor air. In this case, that difference equals:

$$611\left[0.6\times\exp\left(\frac{17.08\times 24}{234.18+24}\right)-0.8\times\exp\left(\frac{17.08\times 34}{234.18+34}\right)\right]=-2468\,\text{Pa}$$

which gives 6471 g/h vapour to be removed from the outside air, added to the 625 g/h of vapour released indoors.

The glazing system chosen has no impact. In fact, as the outdoor temperature surpasses the indoor one, the inside surface temperature will be higher than the indoor air temperature. Condensation on the inside surface of the glass is prohibited this way.

Problem 2.5

A dwelling has a bathroom which is directly accessible from the parents' bedroom. The bathroom has a volume of 15 m³, while the daily vapour release in it reaches 2 kg. In the bedroom, with a volume of 50 m³, 1 kg a day is produced. What average vapour pressure will be measured in the bedroom when the ventilation air entering the bathroom at a rate of 1.7 ach linked to its air volume moves to the bedroom to mix there with an outside air flow of 30 m³/h? Outdoors, it is $-10\,°C$ with 90% relative humidity. The temperature in the bathroom is 24 °C, in the bedroom 18 °C.

Solution

First calculate the vapour flow that moves from the bathroom to the bedroom. Vapour pressure in the bathroom ($p_{i,bath}$) is given by:

$$p_{i,bath}=p_e+RT_i\left(\frac{G_{v,P}}{(nV)}\right)_{bathroom}\quad\text{with}\quad p_e=0.9\times 611\exp\left(\frac{22.44\times-10\theta_{d,e}}{272.44-10}\right)=234\,\text{Pa}$$

With a vapour flow from the bathroom to the bedroom equal to $2/24 = 0.127$ kg/h, the vapour pressure value in the bathroom equals:

$$p_{i,bath}=234+462\times 297.15\left(\frac{2/24}{1.7\times 15}\right)=683\,\text{Pa}$$

The vapour balance in the bedroom is then:

$$0.127+\frac{30\times 234}{462\times 291.15}+1/24-\frac{p_{i,bed}(30+1.7\times 15)}{462\times 291.15}=0$$

giving as vapour pressure:

$$p_{i,bed}=234\left(\frac{30}{30+1.7\times 15}\right)+462\times 291.15\left(\frac{0.127+1/24}{30+1.7\times 15}\right)=535\,\text{Pa}$$

Or, between bathroom and bedroom, the vapour pressure drops. The reason is the additional ventilation in the bedroom.

Problem 2.6

Repeat Problem 2.5 for the case where the only ventilation in the bedroom comes from the airflow leaving the bathroom. Conclusions?

Solution

The vapour pressure in the bedroom is 902 Pa, which is now higher than the vapour pressure in the bathroom.

Problem 2.7

A cavity wall has as section:

Layer	Thickness, m	λ value, W/(m.K)	Air permeance, $m^3/(m^2.s.Pa)$
Inside leaf (no fines blocks)	0.14	1.00	$33 \times 10^{-4} \Delta P_a^{-0.42}$
Cavity fill (mineral fibre)	0.10	0.04	0.00 081
Air layer	0.01	0.067	∞
Brick veneer	0.09	1.00	$0.32 \times 10^{-4} \Delta P_a^{-0.19}$

Calculate the air flux across the wall, knowing that wind induces a 6 Pa higher average air pressure outdoors than indoors and assuming the flux develops normal to the wall.

Solution

The air flux equals:

$$g_a = \frac{6}{g_a^{\frac{1}{0.58}-1}/(33 \times 10^{-4})^{\frac{1}{0.58}} + 1/0.00081 + g_a^{\frac{1}{0.81}-1}/(0.32 \times 10^{-4})^{\frac{1}{0.81}}}$$

This formula requires iteration, starting from an assumed air flux at the right-hand side. As a starting value, the flux in the case that the wall consisted of the fill only is used:

$$g_a = 6 \times 0.00081 = 0.00486 \ m^3/s$$

Successive iterations give as the final result:

$$g_a = 0.000133 \ m^3/(m^2.s)$$

As the top graph in the figure shows, a good approximation only demands a few iterations. Thus, 0.48 m^3 of air per hour passes through each square metre of cavity wall.

Air pressure differences across the successive layers are found by introducing this air flux in the three flux equations ($g_a = K_a \Delta P_a$). The air pressures in the interfaces are then:

$$P_a = P_{a,o} - \sum_{i=1}^{5} \Delta P_a$$

2.5 Problems and solutions

See the left-hand graph in the figure below. Apparently the veneer bears the greatest pressure difference, which means that excessive wind load may threaten its stability. Rain will also seep easily across the head joints and run off at the backside of the veneer. In other words, try to avoid!

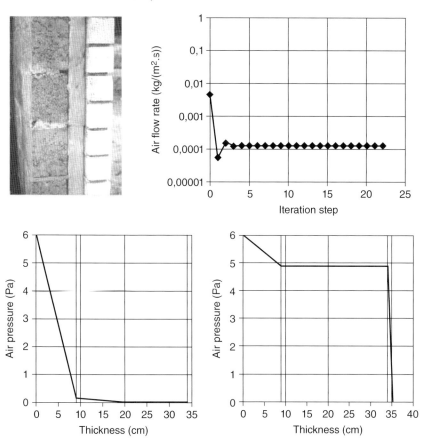

A better choice consists of having the greatest pressure difference over the inside leaf. This is realized by applying an inside plaster, which gives as the adapted section:

Layer	Thickness, m	λ value, W/(m.K)	Air permeance, m³/(m².s.Pa)
Plaster	0.01	0.3	$0.1 \times 10^{-4} \Delta P_a^{-0.23}$
Inside leaf (no fines blocks)	0.14	1.00	
Cavity fill (mineral fibre)	0.10	0.04	0.00081
Air layer	0.01	0.067	∞
Brick veneer	0.09	1.00	$0.32 \times 10^{-4} \Delta P_a^{-0.19}$

The result is a 75% decrease in infiltration rate, down to $0.12 \, m^3/(m^2.h)$. Also the air pressures in the interfaces change substantially as the graph at the right in the figure above shows. In fact, the plaster faces the greatest difference now, while the veneer is left with some 20% of the total. This creates a much better situation, with less wind load and less rain seeping across the veneer's head joints.

Problem 2.8

Return to the cavity wall in Problem 2.7. Outdoors it is $-10\,°C$, and indoors $21\,°C$. Calculate the temperatures in the wall and evaluate the heat loss by conduction at the inside surface when the wall faces a $4\,m/s$ wind, giving $6\,Pa$ overpressure on its outside face. The surface film coefficient inside is $8\,W/(m^2.K)$, and outside $22.2\,W/(m^2.K)$. To simplify things, stack is not considered.

Solution

The temperature difference over the wall in this windy weather leads to a combined heat and enthalpy flow that, without stack, remains homogeneously distributed over the wall's surface, with as temperature course across the assembly:

$$\theta = \theta_e + (\theta_i - \theta_e) F_1(R) \quad \text{with} \quad F_1(R) = \frac{1 - \exp(c_a g_a R)}{1 - \exp(c_a g_a R_T)}$$

The original cavity wall gave an infiltration rate of $0.000\,133\,m^3/(m^2.s)$ or $0.00\,016\,kg/(m^2.s)$, causing a unit enthalpy flow of $0.1602\,W/(m^2.K)$ ($c_a = 1008\,J/(kg.K)$). Temperatures in the interfaces thus become:

Layer	Thickness m	λ value W/(m.K)	$R\,m^2$. K/W	$\Sigma R\,m^2$. K/W	$F_1(R)$	Temp. °C
						21.0
Surface film resistance (R_i)			0.125	0.125	0.0514	19.4
Inside leaf (no fines blocks)	0.14	1.00	0.14	0.265	0.1077	17.7
Cavity fill (mineral fibre)	0.10	0.04	2.5	2.765	0.9260	−7.7
Air layer	0.01	0.067	0.15	2.915	0.9654	−8.9
Brick veneer	0.09	1.00	0.09	3.005	0.9885	−9.6
Surface film resistance (R_e)			0.045	3.05	1	−10

2.5 Problems and solutions

As a figure:

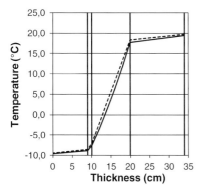

The solid line gives the results with infiltration, and the dashed line for conduction only. Clearly, infiltration cools down the wall somewhat. The heat flux by conduction at the inside face is:

$$q = F_2(\theta_i - \theta_e) \quad \text{with} \quad F_2 = \frac{c_a g_a \exp(c_a g_a R_T)}{1 - \exp(c_a g_a R_T)},$$

$$R_T = 3.05 \text{ m}^2.\text{K/W}, \quad c_a g_a = 0.1602 \text{ W/(m}^2.\text{K)}$$

giving 12.9 W/m². Without infiltration, the result would have been 10.2 W/m². Or, infiltration lifts the conductive losses at the inside face a little. Simultaneously, an enthalpy flow of 3.36 W/m² passes across the wall.

Problem 2.9

Return to the cavity wall of Problem 2.7, but now plastered inside. Calculate the temperatures across and heat flux by conduction at the inside surface for the same indoor and outdoor conditions as in Problem 2.8. What are the conclusions?

Solution

Layer	Thickness m	λ value W/(m.K)	R m². K/W	ΣR m². K/W	$F_1(R)$	Temp °C
					0.000	21.0
Surface film resistance (R_i)			0.125	0.125	0.043	19.7
Plaster	0.01	0.3	0.035	0.16	0.055	19.3
Inside leaf (no fines blocks)	0.14	1.00	0.14	0.30	0.103	17.8
Cavity fill (mineral fibre)	0.10	0.04	2.50	2.80	0.913	−7.3
Air layer	0.01	0.067	0.15	2.95	0.959	−8.7
Brick veneer	0.09	1.00	0.09	3.04	0.986	−9.6
Surface film resistance (R_e)			0.045	3.09	1.000	−10.0

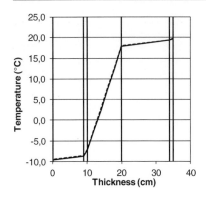

The conductive heat flux at the inside surface of the plastered cavity wall equals 10.7 W/m², whereas pure transmission should have given 10.0 W/m². So the plaster is quite effective in ensuring acceptable air-tightness.

Problem 2.10

The cavity wall of Problem 2.7 is 2.7 m high. Outdoors it is −10 °C, and indoors 21 °C. Given that the wall is air-permeable, thermal stack flow develops. What do stack and the related heat fluxes look like for a windowless one-storey high wall?

Solution

In the case considered, the neutral plane for stack sits at mid-height, resulting in a triangular pressure profile. Just above the floor an underpressure of 1.8 Pa forces outside air to infiltrate. Above mid-height, inside air exfiltrates, with a maximum just below the ceiling where the overpressure touches 1.8 Pa.

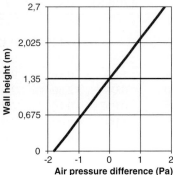

2.5 Problems and solutions

The air flux at different heights is calculated in the same way as for Problem 2.7. For the results, see the table and figure below.

Height, m	Airflow rate, $m^3/(m^2.h)$
2.7	0.180
2.4	0.147
2.1	0.112
1.8	0.074
1.5	0.030
1.2	−0.030
0.9	−0.074
0.6	−0.112
0.3	−0.147
0	−0.180

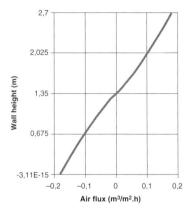

The air flux profile generated by stack is close to linear. The total infiltrating and exfiltrating air equals a moderate $0.13\,m^3/h$ per metre run. As a consequence, the heat flux by conduction at the inside face will vary somewhat: highest just above the floor, pure conduction at mid-height, and lowest under the ceiling (see the figure below). Calculating those fluxes does not differ from the method used in Problem 2.8.

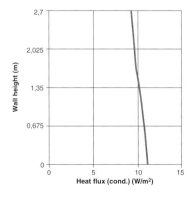

Problem 2.11

Given a timber-frame wall with section:

Layer	Thickness m	Density kg/m^3	λ value W/(m.K)	μ value	Air permeability m^3/(m^2.s.Pa)
Gypsum board lining	0.012		0.1	12	$3.1 \times 10^{-5} \Delta P_a^{-0.19}$
Insulation (mineral fibre)	0.15	20	0.04	1.2	0.00 081
OSB sheathing	0.01	400	0.13	15	$3.5 \times 10^{-4} \Delta P_a^{-0.31}$
Building paper	0.0005		0.2	200	$4.2 \times 10^{-4} \Delta P_a^{-0.4}$
Cavity	0.03		0.18	0	∞
Brick veneer	0.09	1600	0.9	5	$0.32 \times 10^{-4} \Delta P_a^{-0.19}$

Will this design suffer from unacceptable interstitial condensation? If so, what vapour retarder should be added where? Looking north-east, the boundary conditions to be considered are:

	J	F	M	A	M	J	J	A	S	O	N	D
Outdoors												
Air temp. (°C)	2.3	3.3	5.9	8.8	13.5	16.1	16.8	15.7	13.5	8.8	5.5	2.9
Eq. temp. (°C)	3.1	4.2	7.2	10.6	16.1	19.1	19.9	18.6	16.1	10.6	6.7	3.8
Vapour press. (Pa)	619	675	825	997	1277	1427	1470	1406	1277	997	804	653
Indoors												
Air temp. (°C)	20.0	20.3	21	21.8	23.1	23.8	24	23.7	23.1	21.8	20.9	20.2
Vapour press. (Pa)	1224	1253	1330	1418	1561	1638	1660	1627	1561	1418	1319	1242

The surface film resistance inside is 0.13 m^2.K/W, outside 0.04 m^2.K/W. Calculations use the 'equivalent temperature' (the Eq. temp. row in the boundary conditions table), a value accounting for solar gains, undercooling and the exponential relationship between temperature and vapour saturation pressure.

Solution

Assume first that <equivalent> diffusion is the only driving force. To evaluate whether interstitial condensation happens, a Glaser analysis is applied using the climate data for the coldest month, in this case January.

Step 1: Temperatures in all interfaces.

Thickness m	R m^2.K/W	ΣR m^2.K/W	Temperature °C
0	0	0	20.0
0	0.13	0.13	19.5
0.012	0.12	0.25	19.0
0.162	3.75	4.00	3.9
0.172	0.08	4.08	3.6
0.172	0	4.08	3.6
0.2025	0.17	4.25	2.9
0.2925	0.1	4.35	2.5
0.2925	0.04	4.39	2.3

2.5 Problems and solutions

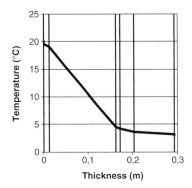

Step 2: Vapour saturation pressure in all interfaces.

Temperature °C	P_{sat} Pa
20.0	2348
19.5	2273
19.0	2206
3.9	808
3.6	791
3.6	791
2.9	754
2.5	732
2.3	724

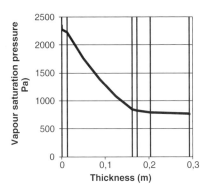

Step 3: Vapour pressure in all interfaces.

Thickness m	μd m	$\Sigma \mu d$ m	Vapour pressure Pa
0	0	0	1224
0	0.007	0.007	1220
0.012	0.144	0.151	1136

(*continued*)

(Continued)

Thickness m	µd m	Σµd m	Vapour pressure Pa
0.162	0.180	0.331	1030
0.172	0.150	0.481	942
0.172	0.100	0.581	884
0.2025	0.000	0.581	884
0.2925	0.450	1.031	620
0.2925	0.001	1.032	619

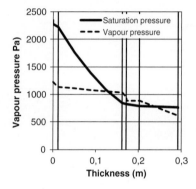

Step 4: The vapour saturation and vapour pressure curves intersect, so condensate will deposit. To get the condensation interface, the correct vapour pressure curve and the deposit, the wall is redrawn in the [vapour diffusion resistance versus vapour pressure] plane and the tangents to the saturation line are traced. For the result, see the figure below. Condensation occurs at the back of the OSB sheathing and is wetting it stepwise.

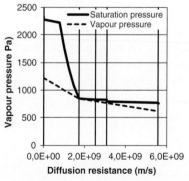

The deposit in January totals:

$$g_c = \left(\frac{1224 - 842}{1.72 \times 10^9} - \frac{842 - 619}{5.62 \times 10^9 - 1.72 \times 10^9} \right) \times 3600 \times 24 \times 31 = 0.41 \text{ kg/m}^2$$

2.5 Problems and solutions

As the 1 cm thick OSB sheathing has a density of 400 kg/m³, the result is a 10% increase in moisture ratio by weight. The condensation/drying cycle over a whole year becomes (in kg/m²):

	J	F	M	A	M	J	J	A	S	O	N	D
Deposit	0.41	0.33	0.16	−0.16	−0.95	−1.48	−1.72	−1.43	−0.92	−0.17	0.18	0.38
Cumulated	0.98	1.31	1.46	1.30	0.35	0	0	0	0	0	0.18	0.56

As a figure:

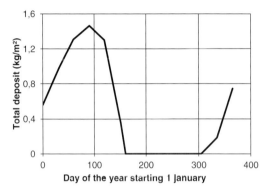

The deposit accumulates a moisture ratio by weight in the OSB of up to 36.5%, which is unacceptable for a timber-based material. Drying in the spring is rapid. In summer, little happens, at least as long as hygroscopicity and wind-driven rain is not accounted for.

As the deposit is far too high, which vapour retarder is needed to achieve acceptability? A rule of the thumb says that in timber-based materials the deposit should not give a moisture ratio increase beyond 3% kg/kg, a limit allowing an end of winter maximum of 0.12 kg/m². An iterative calculation shows that a retarder with vapour permeance below 3.3×10^{-10} s/m on the inside of the thermal insulation is sufficient: this is not too demanding. Expressed in terms of equivalent diffusion thickness, 0.56 m is needed, a value that a vapour-retarding paint already offers.

Does the result reflect reality? Three elements are overlooked: advection, hygric inertia, and wind-driven rain sucked by the capillary veneer.

Correction 1: Assume that the difference in monthly mean indoor and outdoor air temperature is the main cause of air egress across the upper part of the 2.7 m high wall. For reasons of simplicity, the neutral plane is fixed at mid-height. The air pressure difference under the ceiling is:

	J	F	M	A	M	J	J	A	S	O	N	D
ΔP_a (Pa)	0.41	0.33	0.16	−0.16	−0.95	−1.48	−1.72	−1.43	−0.92	−0.17	0.18	0.38

Steady-state advection across the flat composite assembly gives as the cumulated condensation deposit per month (in kg/m²):

	J	F	M	A	M	J	J	A	S	O	N	D
Deposit	0.54	0.43	0.25	−0.10	−0.92	−1.47	−1.71	−1.42	−0.89	−0.10	0.28	0.50
Cumulated	1.32	1.75	2.00	1.90	0.98	0	0	0	0	0	0.28	0.78

As a figure:

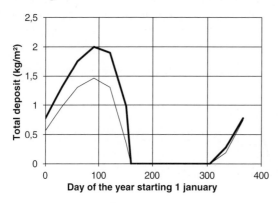

Although the gypsum board takes care of air-tightness, exfiltration remains strong enough to increase accumulation in the OSB sheathing on top of the wall by 21%, resulting in an increase in moisture ratio by weight of 44%, thus far too high. Improving air-tightness is consequently the first thing to do, in combination with a better vapour resistance at the inside of the thermal insulation as calculated above. At mid-height, diffusion remains the only driving force, while just above floor level, infiltration will reduce the amounts deposited.

Correction 2: Calculating the impact of sorption/desorption as correctly as possible demands an appropriate software tool, such as Wufi, Match or others. A first approximation anyhow is possible, assuming gypsum board and OSB have constant material properties, included a constant specific moisture content, and by modelling the timber frame wall as a serial network of two capacitances, coupled by resistances, whereby the cavity between the building paper and the brick veneer is supposed to be so well vented that the air passing through has the same vapour pressure as outdoors. The equations are (forward differences):

Capacitance 1: gypsum board

$$\frac{p_i^t - p_{sat,1}^t \phi_1^t}{\mu_{gyps} N d_{gyps}/2 + 1/\beta_i} + \frac{p_{sat,2}^t \phi_2^t - p_{sat,1}^t \phi_1^t}{\mu_{gyps} N d_{gyps}/2 + \mu_{insul} N d_{insul} + \mu_{OSB} N d_{OSB}/2}$$
$$= \rho_{gyps} \xi_{gyps} d_{gyps} \frac{\phi_1^{t+\Delta t} - \phi_1^t}{\Delta t}$$

2.5 Problems and solutions

Capacitance 2: OSB

$$\frac{p^t_{sat,1}\phi^t_1 - p^t_{sat,2}\phi^t_2}{\mu_{gyps}Nd_{gyps}/2 + \mu_{insul}Nd_{insul} + \mu_{OSB}Nd_{OSB}/2} + \frac{p^t_{sat,e}\phi^t_e - p^t_{sat,2}\phi^t_2}{\mu_{OSB}Nd_{OSB}/2 + \mu_{bp}Nd_{bp} + 1/\beta_{cavity}}$$
$$= \rho_{OSB}\xi_{OSB}d_{OSB}\frac{\phi^{t+\Delta t}_2 - \phi^t_2}{\Delta t}$$

Solving this system of two equations demands knowledge of the boundary conditions. See the climate data above. Each month is modelled as a step function, with the temperatures and saturation pressures invariable. If, in such a case, the steady-state vapour pressure is reached before the month ends, Glaser reflects vapour addition in winter quite well.

The time interval used is one day. The additional material properties needed are:

Layer	$\rho\xi d$, kg/m²
Gypsum board lining	0.46
Insulation (mineral fibre)	0
OSB sheathing	0.93, 2.83 beyond 90% RH
Building paper	0

The starting conditions are: 80% relative humidity in the OSB and 50% relative humidity in the gypsum board. Calculation starts on 1 January. The result shows that hygroscopic uptake, adsorbing in total 0.43 kg/m², progresses until day 29 – see the figure below.

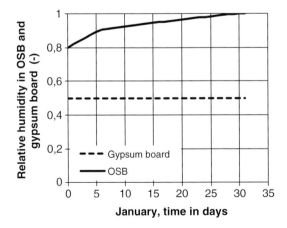

Then condensation at the backside of the OSB begins, depositing 0.048 kg/m² during the two days left. This deposit forms droplets that end in runoff towards the sill plate.

In February and March the whole deposit runs off, in total 0.485 kg/m². In April, drying begins.

Correction 3: Inserting an air and vapour retarder at the inside of the thermal insulation is a valid solution, as long as wind-driven rain does not wet the brick veneer. Take a south-west orientated wall – the dominant rain direction in western Europe. In a cool, humid climate, drying of a brick veneer, wetted by successive rain events, proceeds so slowly that it stays at 100% relative humidity from November till March, but many other times all year round. Assume that the building paper, equivalent diffusion thickness 0.1 m, is replaced by a high-permeance wrap with an equivalent diffusion thickness of 0.01 m. The boundary conditions are:

	J	F	M	A	M	J	J	A	S	O	N	D
Outdoors												
Air temp. (°C)	2.3	3.3	5.9	8.8	13.5	16.1	16.8	15.7	13.5	8.8	5.5	2.9
Eq. temp. (°C)	5.5	7.0	9.2	12.3	17.3	20.0	20.8	19.7	17.3	12.3	8.8	6.1
Vap. pres. (Pa)	619	675	825	997	1277	1427	1470	1406	1277	997	804	653
Indoors												
Air temp. (°C)	20.0	20.3	21	21.8	23.1	23.8	24	23.7	23.1	21.8	20.9	20.2
Vap. pres. (Pa)	1224	1253	1330	1418	1561	1638	1660	1627	1561	1418	1319	1242

In winter, things hardly change. A Glaser calculation gives some condensation at the backside of the OSB with an accumulated maximum of 0.47 kg/m². In reality, this amount mainly consists of hygroscopic wetting, followed by a little deposit. In applying the method, the brick veneer is kept at vapour saturation pressure from November till March.

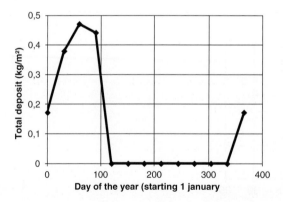

In summer, when the veneer turns first wet by rain and is heated by the sun after, solar-driven inward vapour flow will develop. As a result, condensate deposits on the wrap. On a sunny day, the amount may reach 0.16 kg/m², which is close to the vapour adsorbed by the OSB during the whole month of December! Without a wrap, the condensate wets the OSB. That the wet veneer induces very high vapour saturation pressures is underlined by the figure below.

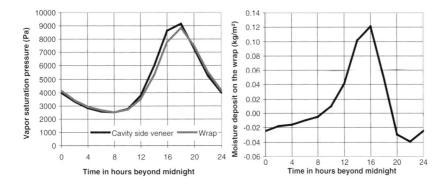

Problem 2.12

Given a pitched roof assembly with as section:

Layer	Thickness m	Density kg/m^3	λ value W/(m.K)	μ value	Air permeability m^3/(m^2.s.Pa)
Lathed ceiling	0.01	450	0.14	85	$2 \times 10^{-4} \Delta P_a^{-0.32}$
Insulation (mineral fibre)	0.15	20	0.04	1.2	0.00 081
Capillary underlay (fibre cement)	0.0032	–	0.2	45	$4.2 \times 10^{-4} \Delta P_a^{-0.4}$
Cavity	0.06	–	$\lambda_{eq} = 0.33$	0	∞
Tiles	0.012	1800	1	20	$1.6 \times 10^{-2} \Delta P_a^{-0.5}$

Will the design suffer from unacceptable interstitial condensation? If so, what vapour retarder should be added where? Boundary conditions to be considered are (slope: 30°, SW):

	J	F	M	A	M	J	J	A	S	O	N	D
Outdoors												
Air temp. (°C)	2.3	3.3	5.9	8.8	13.5	16.1	16.8	15.7	13.5	8.8	5.5	2.9
Eq. temp. (°C)	3.9	5.2	8.7	12.6	19.1	22.5	23.5	22.1	19.1	12.6	8.2	4.7
Vap. pres. (Pa)	619	675	825	997	1277	1427	1470	1406	1277	997	804	653
Indoors												
Air temp. (°C)	20.0	20.3	21	21.8	23.1	23.8	24	23.7	23.1	21.8	20.9	20.2
Vap. pres. (Pa)	1224	1253	1330	1418	1561	1638	1660	1627	1561	1418	1319	1242

The surface resistances for heat are 0.13 m².K/W indoors and 0.04 m².K/W outdoors. Stack height above the neutral plane is 3 m. The underlay is capillary. The tiled roof cover is normally wind washed permanently, but neglect this in the <equivalent> diffusion mode, that is consider the tiles and cavity as two separate layers with properties as listed in the table. Account for it when looking to the impact of stack by replacing the cavity and tiled cover by a surface resistance 0.1 m².K/W and zero vapour resistance. Calculations are done using the equivalent temperature (see Problem 2.11).

Solution

If <equivalent> diffusion was the only driving force, no interstitial condensation should be noted against the backside of the underlay at the end of January (see the first figure below). Anyhow, with the air outflow equal to $1.49 \times 10^{-4} \Delta P_a^{0.68}$ m³/(m². s), convection is added, causing condensation at the backside of the underlay. The amounts reach:

End of	J	F	M	A	M	J	J	A	S	O	N	D
Stack (Pa)	2.28	2.18	1.91	1.62	1.17	0.93	0.87	0.97	1.17	1.62	1.95	2.22
Airflow (kg/(m².h))	1.13	1.10	1.00	0.90	0.72	0.61	0.58	0.63	0.72	0.90	1.02	1.11
Cond. (kg/m²)	3.26	4.39	4.55	3.35	0.00	0.00	0.00	0.00	0.00	0.00	0.32	1.69

Interface	Temperature °C	p_{sat} Pa	p Pa
	20	2343	1224
$1/h_i$	19.5	2271	1221
Lathed ceiling	19.2	2233	859
Insulation (mineral fibre)	4.9	865	783
Underlay	4.8	861	722
Cavity	4.1	820	722
Tiles	4.1	817	620
$1/h_e$	3.9	808	619

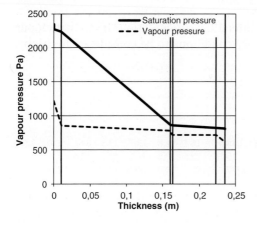

2.5 Problems and solutions

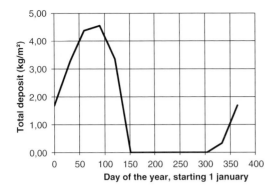

A maximum of 4.55 kg/m² at the end of March looks really problematic. So, not a vapour but an air retarder at the warm side of the insulation is needed. In reality, however, these amounts will not cause problems. In fact, the capillary underlay will become wet over a thickness such that the condensate sucked will evaporate and diffuse to the wind-washed air layer under the tiled deck at a rate equal to the vapour flux coming from indoors. Anyhow, an air retarder still remains a positive addition as it will add air-tightness to the building enclosure.

Problem 2.13

Back to the timber-frame wall of Problem 2.11. Assume the design was focusing on a cold climate with, as air and vapour retarder, a gypsum board lining, finished with vinyl wallpaper. The assembly is applied now in a hot and humid climate, where cooling and dehumidification is needed to keep the building comfortable. The indoors is cooled to an air temperature not exceeding 20 °C. What problems could be expected in case the veneer stays dry or is wetted by wind-driven rain each day? The layer properties are:

Layer	Thickness m	Density kg/m³	λ value W/(m.K)	μ value	μd value m
Vinyl paper	–	–	–		1
Gypsum board lining	0.012	–	0.1	12	
Insulation (mineral fibre)	0.15	20	0.04	1.2	
OSB sheathing	0.01	400	0.13	15	
Building paper	0.0005	–	0.2	200	
Cavity	0.03	–	0.18		0
Brick veneer	0.09	1600	0.9	5	

The maximum amount of moisture that can stick on the building paper without running off is 100 g/m², while the capillary moisture content of gypsum board equals

150 kg/m³. The surface film resistances are 0.13 m².K/W indoors and 0.04 m².K/W outdoors. The boundary conditions to be considered are (orientated north)

	J	F	M	A	M	J	J	A	S	O	N	D
Outdoors												
Air temp. (°C)	22.5	23.2	25.3	28.1	30.9	32.9	33.6	32.9	30.9	28.1	25.3	23.2
Vap. pres. (Pa)	2185	2279	2585	3047	3579	4008	4168	4008	3579	3047	2585	2279
Indoors												
Air temp. (°C)	20	20	20	20	20	20	20	20	20	20	20	20
Vap. pres. (Pa)	1910	1913	1895	1808	1656	1512	1455	1512	1656	1808	1895	1913

Solution

In the case of <equivalent> diffusion only without rain, interstitial condensate will accumulate year round at the backside of the gypsum board with the following quantities at the end of each month during the first year:

End of:	J	F	M	A	M	J	J	A	S	O	N	D
Cond. (kg/m²)	0.00	0.00	0.00	0.071	0.217	0.460	0.749	1.00	1.141	1.146	1.016	0.773

As a graph, covering 3 years:

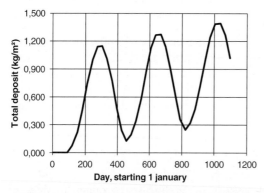

When additionally abundant wind-driven rain keeps the veneer wet, 100% relative humidity will be maintained at its backside with quite devastating consequences for the wall as a whole (see the Glaser analysis in the figure below).

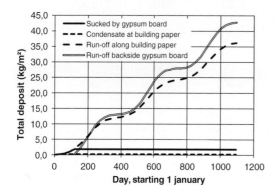

Problem 2.14

Given are 3 m high masonry cavity walls with, as layer properties:

Layer	Thickness m	Density kg/m³	λ value W/(m.K)	μ value	μd value m
Gypsum plaster	0.01	980	0.3	7	
Inside leaf (fast bricks)	0.14	1400	0.5	5	
Insulation (mineral fibre)	0.12	30	0.04	1.2	
Cavity	0.03	400	0.18		0
Brick veneer	0.09	1600	0.9	5	

The cavities are vented by two open head joints per metre run at the bottom and top, all with section $1 \times 6.5 \text{ cm}^2$ and a depth of 9 cm. In some, the insulation is correctly mounted. In others, it sits centrally in the cavity with a 1.5 cm wide air layer left at both sides. In a few, the insulation touches the brick veneer. The tray at the bottom guarantees an open contact between the cavity or the remaining air layers and the open head joints, while at the top, the insulation stops just below these head joints. That results in some thermal bridging, which is not considered. The question is, to what extent is the thermal transmittance affected by the venting flow when, on average, a 4 Pa air pressure difference exists between the open head joints at the top (highest air pressure) and the bottom (lowest air pressure), in the three cases? The surface film coefficients are $25 \text{ W}/(\text{m}^2.\text{K})$ outdoors and $8 \text{ W}/(\text{m}^2.\text{K})$ indoors. Indoors, it is $20\,°\text{C}$, and outdoors $-10\,°\text{C}$.

Solution

Only the case with the insulation centrally in the cavity is considered. For the other two cases, it is up to the reader to find the solution.

The two open head joints per metre run at the top and bottom fix the airflow. Their hydraulic resistance in fact outweighs the one of the two 1.5 cm thick air layers at both sides of the insulation. The equation thus is:

$$\Delta P_a = 2\left[1.5\frac{\rho_a}{2}\left(\frac{G_a}{\rho_a 2A_{\text{head joint}}}\right)^2 + 0.42\frac{96\,\nu\rho_a 2A_{\text{head joint}}}{G_a d_H}\frac{L}{d_H}\rho_a\left(\frac{G_a}{\rho_a 2A_{\text{head joint}}}\right)^2\right]$$

$$\approx 0.32\left(\frac{G_a}{0.00065}\right)^2 + 0.08\left(\frac{G_a}{0.00065}\right) = 757,400\,G_a^2 + 12.3\,G_a$$

For $\Delta P_a = 4$ Pa, the result is an airflow from top to bottom equal to 0.0023 kg/s. As the air layers at both sides are of equal thickness, this flow splits into two, one washing the air layer in front, the other the air layer behind the insulation. As the insulation has a thermal resistance of $3 \text{ m}^2.\text{K/W}$, for reasons of simplicity the temperature behind the brick veneer is set equal to the temperature outdoors. This way the problem does not differ from calculating the equivalent thermal transmittance of a wall with a vented cavity between two leafs, the one consisting of the insulation with thermal resistance $R_1 = 3.07$ m².K/W, the other consisting of the plastered inside leaf and the inside surface resistance with thermal resistance $R_2 = 0.44$ m².K/W.

For all the constants advanced below, see Section 2.2.7.7 on 'Vented cavity'. The constant D equals:

$$D = (3.33 + 4.45 + 1/3.07) \times (3.33 + 4.45 + 1/0.438) - 4.45^2 = 61.9$$

In this sum, 3.33 W/(m².K) is the convective surface film coefficient at each cavity face and 4.45 W/(m².K) is the radiant surface film coefficient across the cavity. In fact, as for a 1.5 cm wide air layer, the Nusselt number in the cavity is 1 and the convective surface film coefficient becomes: $0.025/0.015 \times 2 = 3.33$ W/(m².K). The other constants are: $A_1 = 0.0531$, $A_2 = 0.0235$, $B_1 = 0.1643$, $B_2 = 0.2993$, $C_1 = 0.7826$, $C_2 = 0.6773$. The bounding face temperatures of the cavity thus are:

$$\theta_{s1} = 2.8 + 0.783\,\theta_{cav} \qquad \theta_{s1} = 5.7 + 0.677\theta_{cav}$$

For the temperatures along the cavity, see the figure below.

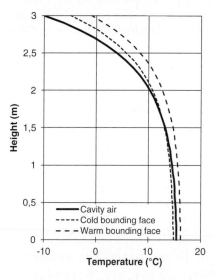

It is interesting to see how radiation impacts the bounding face temperatures. In fact, just below the air inlet, the warm and cold faces are warmer than the air in the cavity.

The equivalent thermal transmittance (U_{eq}) becomes:

$$U = U_o \left\{ 1 + \frac{C_2 b_1 R_1}{L R_2} \left[1 - \exp\left(-\frac{L}{b_1}\right) \right] \right\} \quad \text{with} \quad b_1 = \frac{1008 G_a}{h_c(2 - C_1 - C_2)}$$

Thus calculated it is 0.55 W/(m².K). Without the wind-washed 1.5 cm thick air layer behind the thermal insulation, the clear wall value should be 0.27 W/(m².K), or the increase by wind-washing exceeds 100%, which is unacceptable.

Further reading

Abuku, M. (2009) Moisture stress of wind-driven rain on building enclosures. PhD thesis, KU-Leuven.

Arfvidsson, J. (1998) Moisture transport in porous media, modelling based on Kirchhoffs potentials. Doctoral thesis, Lund University.

ASHRAE (2017) *ASHRAE Handbook of Fundamentals*, SI ed., Published by ASHRAE, 1791, Tullie Circle, Atlanta, GA.

Blocken, B. (2004) Wind-driven rain on buildings, PhD thesis, KU-Leuven.

Blocken, B., Hens, H. and Carmeliet, J. (2002) Methods for the quantification of driving rain on buildings. *ASHRAE Transactions*, **108**, 338–350.

Bomberg, H. (1971) Waterflow through porous materials, Part 1, Methods of water transport measurements. Report 19, Lund Institute of Technology, Section of Building Technology.

Bomberg, H. (1971) Waterflow through porous materials, Part 2, Relative suction model. Report 20, Lund Institute of Technology, Section of Building Technology.

Brocken, H. (1998) Moisture transport in brick masonry: the grey area between bricks. Doctoral thesis, TU/e, Eindhoven.

Brunauer, S., Emmett, P.H. and Teller, E. (1938) Adsorption of gases in multimolecular layers. *Journal of the American Chemical Society*, **60**, 309–319.

Cammerer, J.S. (1962) *Wärme- und Kälteschutz in der Industrie*, Springer Verlag, Berlin, [in German].

Carmeliet, J., Hens, H. and Vermeir, G. (eds) (2003) *Research in Building Physics*, Balkema, Lisse.

Chaddock, J.B. and Todorovic, B. (1991) *Heat and Mass Transfer in Building Materials and Structures*, Hemisphere Publishing Corporation, New York.

Cranck, J. (1956) *The Mathematics of Diffusion*, Clarendon Press, Oxford.

De Grave, A. (1957) *Bouwfysica 1*, Uitgeverij SIC, Brussels, [in Dutch].

Devries, D.A. (1958) Simultaneous transfer of heat and moisture in porous media. *Transactions of the American Geophysical Union*, **39**, 909–916.

de Wit, M.H. (1995) *Warmte en Vocht in Constructies (Heat and Moisture in Building Constructions)*, TU/e, Eindhoven, [in Dutch].

Feynman, R., Leighton, R. and Sands, M. (1977) *Lectures on Physics*, vol. **1**, Addison-Wesley Publishing Company, Reading, MA.

Garrecht, H. (1992) Porenstrukturmodelle für den Feuchtehaushalt von Baustoffen mit und ohne Salzbefrachtung und rechnerische Anwendung auf Mauerwerk. Doctoral thesis, Universität Karlsruhe [in German].

Glaser, H. (1958) Wärmeleitung und Feuchtigkeitsdurchgang durch Kühlraumisolierungen. Kältetechnik, no. 3 [in German].

Glaser, H. (1958) Temperatur- und Dampfdruckverlauf in einer homogenen Wand bei Feuchtigkeitsausscheidung. Kältetechnik, no. 6 [in German].

Glaser, H. (1958) Vereinfachte Berechnung der Dampfdiffusion durch geschichtete Wände bei Ausscheiden von Wasser und Eis. Kältetechnik, no. 11 & no. 12 [in German].

Glaser, H. (1959) Grafisches Verfahren zur Untersuchung von Diffusionsvorgängen. Kältetechnik, no. 10 [in German].

Gösele, K. and Schüle, W. (1973) *Schall, Wärme, Feuchtigkeit*, 3rd edn, Bauverlag GMBH, Wiesbaden-Berlin, [in German].

Hagentoft, C.-E. (2001) *Introduction in Building Physics*, Studentlitteratur, Lund.

Hall, C. and D'Hoff, W. (2002) *Water Transport in Brick, Stone and Concrete*, Spon Press, London.

Harmathy, T.Z. (1967) *Moisture Sorption of Building Materials.* Technical Paper no. 242, NRC, Division of Building Research, Ottawa.

Häupl, P. and Roloff, J. (eds) (1999) Proceedings of the 10th Bauklimatisches Symposium, Band 1 & 2. T.U. Dresden, Institut für Bauklimatik.

Häupl, P. and Roloff, J. (eds) (2002) Proceedings of the 11th Bauklimatisches Symposium, Band 1 & 2. T.U. Dresden, Institut für Bauklimatik.

Hens, H. (1975) Theoretische en experimentele studie van het hygrothermisch gedrag van bouw- en isolatiematerialen bij inwendige condensation en droging, met toepassing op de platte daken. Doctoral thesis, KU-Leuven [in Dutch].

Hens, H. (1978, 1981) Bouwfysica, Warmte en Vocht, Theoretische grondslagen, 1^e en 2^e uitgave. ACCO, Leuven [in Dutch].

Hens, H. (1992, 1997, 2000) Bouwfysica 1, Warmte en Massatransport, 3^e, 4^e en 5^e uitgave. ACCO, Leuven [in Dutch].

Hens, H. (1996) *Modelling*. Vol. 1 of the Final Report Task 1, IEA-Annex 24. ACCO, Leuven.

Holm, A. (2001) Ermittlung der Genauigkeit von instationären hygrothermischen Bauteilberechnungen mittels eines stochastischen Konzeptes. Doctoral thesis, Universität Stuttgart [in German].

IEA-Annex 14 (1990) *Condensation and Energy: Guidelines and Practice*, ACCO, Leuven.

Janssen, H. (2002) The influence of soil moisture transfer on building heat loss via the ground. Doctoral thesis, KU-Leuven.

Janssens, A. (1998) Reliable control of interstitial condensation in lightweight roof systems. Doctoral thesis, KU-Leuven.

Kalagasidis, A. (2004) HAM-tools, an integrated simulation tool for heat, air and moisture transfer analysis in building physics. PhD thesis, Chalmers University of Technology.

Kiessl, K. (1983) Kapillarer und dampfförmiger Feuchtetransport in mehrschichtigen Bauteilen. Dissertation, Essen [in German].

Klopfer, H. (1974) *Wassertransport durch Diffusion in Feststoffen*, 1st edn, Bauverlag GMBH, Wiesbaden-Berlin [in German].

Kohonen, R. and Ojanen, S. (1985) Coupled convection and conduction in two dimensional building structures. Proceedings of the 4th Conference on Numerical Methods in Thermal Problems, Swansea.

Kohonen, R. and Ojanen, T. (1987) Coupled diffusion and convection heat and mass transfer in building structures. Building Physics Symposium, Lund.

Kreith, F. (1976) *Principles of Heat Transfer*, Harper & Row Publishers, New York.

Krischer, O. (1963) *Die wissenschaftlichen Grundlagen der Trocknungstechnik*, 2nd edn, Springer Verlag, Berlin [in German].

Krus, M. (1995) Feuchtetransport- und Speicherkoeffizienten poröser mineralischer Baustoffe. Theoretische Grundlagen und Neue Messtechniken. Doctoral thesis, Universität Stuttgart [in German].

Künzel, H.M. (1994) Verfahren zur ein- und zweidimensionalen Berechnung des gekoppelten Wärme- und Feuchtetransports in Bauteilen mit einfachen Kennwerten. Doctoral thesis, Universität Stuttgart [in German].

Langmans, J. (2013) Feasibility of exterior air barriers in timber frame construction. PhD thesis, KU-Leuven.

Langmans, J., Klein, R. and Roels, S. (2012) Hygrothermal risks of using exterior air barrier systems for highly insulated light weight walls: a laboratory investigation. *Buildings and Environment*, **58**, 209–218.

Langmuir, I. (1918) The adsorption of gases on plane surfaces of glass, mica and platinum. *Journal of the American Chemical Society*, **40**, 1361–1403.

Luikov, A. (1966) *Heat and Mass Transfer in Capillar Porous Bodies*, Pergamon Press, Oxford.

Lutz, P., Jenisch, R., Klopfer, H., Freymuth, H. and Krampf, L. (1989) *Lehrbuch der Bauphysik*, B.G. Teubner Verlag, Stuttgart [in German].

Nevander, E.L. and Elmarsson, B. (1981) *Fukt-Handbok*, Svensk Byggtjänst, Stockholm, [in Swedish].

Pedersen, C.R. (1990) Combined heat and moisture transfer in building constructions. PhD thesis, Technical University of Denmark.

Roels, S. (2000) Modelling unsaturated moisture transport in heterogeneous limestone. Doctoral thesis, KU-Leuven.

Rose, W.B. (2005) *Water in Buildings*, John Wiley and Sons, Inc., Chichester.

Sedlbauer, K. (2001) Vorhersage von Schimmelpilzbildung auf und in Bauteilen. Doctoral thesis, Universität Stuttgart [in German].

Taveirne, W. (1990) *Eenhedenstelsels en groothedenvergelijkingen: overgang naar het SI*, Pudoc, Wageningen [in Dutch].

Time, B. (1998) Hygroscopic moisture transport in wood. Doctoral thesis, NUST Trondheim.

Trechsel, H.R. (ed.) (1994) *Moisture Control in Buildings*, ASTM Manual Series, MNL 18, Published by ASTM.

TU-Delft (1975 1985) Faculteit Civiele Techniek, Vakgroep Utiliteitsbouw-Bouwfysica. *Bouwfysica, naar de colleges van Prof A.C.Verhoeven* [in Dutch].

Van der Kooy, J. (1971) *Moisture Transport in Cellular Concrete Roofs*, Uitgeverij Waltman, Delft.

Van Mook, J.R. (2003) Driving rain on building envelopes. *TU/e Bouwstenen*, **69**, 198.

Vos, B.H. and Coelman, E.J.W. (1967) *Condensation in structures*. Report No. BI-67-33/23 TNO-IBBC, Rijswijk.

Vos, B.H. and Tammes, E. (1969) *Moisture and Moisture Transfer in Porous Materials*. Report No. B1-69-96/03.1.001. TNO-IBBC, Rijswijk.

Welty, J., Wicks, C. and Wilson, R. (1969) *Fundamentals of Momentum, Heat and Mass Transfer*, John Wiley & Sons, New York.

Woloszyn, M. and Rode, C. (2008) *Modelling Principles and Common Exercises*. Final Report IEA ECBCS, Annex 41. ACCO, Leuven.

Zillig, W. (2009) Moisture transport in wood using a multiscale approach. PhD thesis, KU-Leuven.

3 Combined heat, air and moisture flow

3.1 Introduction

With the exception of the paragraphs on combined air and heat transfer, combined vapour diffusion and convection, interstitial condensation and drying, heat, air and moisture were mostly treated separately. In reality, they intertwine. In fact, mass transfer means enthalpy flow, and thus heat transfer, as explained in Chapter 2. Temperature differences induce mass displacement, such as natural convection, vapour diffusion and water movement in the pores due to changes in surface tension. The thermal characteristics termed volumetric specific heat capacity and thermal conductivity are a function of moisture content, while the moisture transfer characteristics are a function of temperature.

These interrelationships are the reason why a separate analysis of the three factors oversimplifies reality, resulting in a loss of information and understanding. Therefore, in this last chapter, combined heat, air and moisture transfer is discussed.

3.2 Material and assembly level

3.2.1 Assumptions

An exact mathematical treatment of combined heat, air and moisture transfer is not possible. Precipitation of soluble substances, crystallization and hydration of salts, diffusion of solid substances, will all alter the material matrix, modify pores in form and equivalent diameter, and change the specific pore surface. These effects are overlooked. A first assumption, therefore, is that the material matrix is invariable. Consequently, material characteristics such as the dry density, the pore distribution and the specific pore surface do not change. Furthermore, each porous material layer is assumed to be composed of a large number of infinitely small representative volumes (V_{rep}), each possessing all characteristics of the layer as a whole. In reality, this is untrue. Although representative volumes are finite in nature, a continuum approach is adopted, assuming this mirrors a calculation of average values per representative volume of all quantities involved.

3.2.2 Solution

Any combined heat, air and moisture problem is considered solved when the temperatures, air pressures, moisture potential values, heat fluxes, air fluxes and moisture fluxes in each point in the solution space are known, here a material layer or assembly. While potentials are scalars, fluxes are vectors. Consequently, a system of three scalar and three vector equations is needed, completed with the necessary equations of state, which describe the thermodynamic equilibrium among the separate potentials and the relations with certain material characteristics. In addition, solving that system demands a perfect knowledge of the geometry and the initial, boundary and contact conditions.

Fundamental to any solution is the conservation of mass, energy and momentum equations that govern combined heat, air and moisture transfer. For the mass flows, empirical equations substitute the momentum balances, although convection in air and mass movement in highly permeable materials may necessitate solving conservation of momentum, reflected by the Navier-Stokes equations in combination with a turbulence model.

3.2.3 Conservation of mass

Below $0\,°C$ the pores in a moist material contain air, liquid, vapour, ice, often dissolved salts and other substances. Above $0\,°C$, ice is gone but the others remain. To simplify things, salts and other substances are assumed absent. For liquid water, vapour, ice and air the quantities per unit volume of material are described by the water, vapour, ice and air content:

	$\theta < 0$	$\theta \geq 0$
Water	$w_l = \frac{m_l}{V}$	$w_l = \frac{m_l}{V}$
Vapour	$w_v = \frac{m_v}{V}$	$w_v = \frac{m_v}{V}$
Ice	$w_i = \frac{m_i}{V}$	
Air	$w_a = \frac{m_a}{V}$	$w_a = \frac{m_a}{V}$

The moisture content bundles together liquid water, vapour and ice. Most of the time, vapour can be neglected and the following applies:

$$w = w_l + w_i$$

The air content can be written as:

$$w_a = \Psi_{of} \rho_a$$

where Ψ_{of} is the open pore volume not containing liquid or ice, and ρ_a the air density. For vapour:

$$w_v = x_v w_a$$

with x_v the vapour ratio in the pore air in kg/kg. Applying mass conservation to each of these components gives for the liquid:

$$\text{div } \mathbf{g}_w \pm G'_1 = -\frac{\partial w_l}{\partial t} \qquad (3.1)$$

with $\mathbf{g}_w$ the water flux and G'_1 a water source due to condensation or a local water supply, or a water sink due to evaporation. For vapour, the balance is:

$$\text{div } \mathbf{g}_v \pm G'_2 = -\frac{\partial w_v}{\partial t} \qquad (3.2)$$

where $\mathbf{g}_v$ is vapour flux and G'_2 a vapour source linked to evaporation or a vapour sink caused by condensation. For hygroscopic materials, (de)sorption moisture content w_H replaces the vapour content. Buffering is then given by the related derivative versus time. Above freezing, summing up the water and vapour sources and sinks gives zero. Condensation in fact is a water source but a vapour sink, while evaporation is the reverse. For ice, the balance becomes:

$$\text{div } \mathbf{g_i} \pm G'_3 = -\frac{\partial w_i}{\partial t}$$

with $\mathbf{g_i}$ the ice flux and G'_3 an ice source due to freezing of water and vapour or an ice sink because of ice melting and sublimating. In porous materials, ice formation in the largest pores starts at $0\,°C$. As temperature drops, water in smaller pores freezes. Because ice is a solid, the related flux is quasi zero, changing the equation into:

$$\pm G'_3 = -\frac{\partial w_i}{\partial t} \tag{3.3}$$

Summing up the liquid, vapour and ice sources and sinks gives:

$$G'_1 + G'_2 + G'_3 = 0$$

Combining the three conservation equations with w, moisture content in the material, as storage variable, thus gives:

$$\text{div}(g_w + g_v) = -\frac{\partial w}{\partial t} \tag{3.4}$$

Air obeys:

$$\text{div } \mathbf{g_a} = -\frac{\partial w_a}{\partial t} = -\Psi_r \frac{\partial \rho_a}{\partial t}$$

with $\mathbf{g_a}$ the air flux. Air faces neither sources nor sinks, while, compared with moisture and heat, the air content changes so fast that related inertia can be set to zero, or:

$$\text{div } \mathbf{g_a} = 0 \tag{3.5}$$

3.2.4 Conservation of energy

In the discussion about heat conduction, the conservation law was applied to media free of mass flow. Air ingress and egress introduced air transfer carrying enthalpy. Here, the law is extended to porous media with air, liquid and vapour migrating. All induce enthalpy flows:

$$q_j = g_j(h_j - h_{0,j}) \tag{3.6}$$

with h_j the specific enthalpy for mass component j, units J/kg, and $h_{0,j}$ the reference value at $0\,°C$, mostly set at zero. As the kinetic and frictional energy is too small to play a role, energy conservation becomes:

$$\text{div}\left[\mathbf{q} + \sum_{j=1}^{3}(\mathbf{g}_j h_j)\right] \pm \Phi' = -\frac{\partial}{\partial t}\left[\rho_0 c_0 \theta + \sum_{j=1}^{4}(w_j h_j)\right] \tag{3.7}$$

with $\rho_0 c_0$ the volumetric specific heat capacity of the material matrix. The term Φ' represents heat dissipation by other processes than the changes of state included in $\sum(w_j h_j)$, with $T_0 = 273.15\,K = 0\,°C$:

Water (1)	$h_1 = c_l(T - T_0) = c_l\theta$
Vapour (2)	$h_v = c_v(T - T_0) + h_{b0} = c_v\theta + h_{b0}$
Ice (3)	$h_i = c_i(T - T_0) + h_{m0} = c_i\theta + h_{m0}$
Dry air (4)	$h_a = c_a(T - T_0) = c_a\theta$

In these formulae, l_{b0} is the heat of evaporation and l_{m0} the heat of melting, both at 0 °C (2,500,000 and 330,000 J/kg respectively). In case the material matrix and all mass components have the same temperature per representative volume, an assumption that fits for fine pores and low flow velocities, rather than for air passing across apertures, cavities, cracks and fissures, the changes in enthalpy can be written as:

$$dh_j = c_j d\theta \quad \text{and} \quad \mathbf{grad}\, h_j = c_j \mathbf{grad}\, \theta$$

Energy conservation thus converts into:

$$-\text{div}\,\mathbf{q} = \sum_{j=1}^{3}\left(h_j \, \text{div}\, \mathbf{g_j} + \mathbf{g_j}\,\mathbf{grad}\,h_j\right) + \frac{\partial}{\partial t}\left\{\left[\rho_0 c_0 + \sum_{j=1}^{4}(c_j w_j)\right]\theta\right\}$$

$$+ \sum_{j=1}^{4}\left(h_j \frac{\partial w_j}{\partial t}\right) \pm \Phi'$$

a vector equation which can be reshuffled to:

$$-\text{div}\,\mathbf{q} - \underbrace{\sum_{j=1}^{3}\left(\mathbf{g_j} c_j \,\mathbf{grad}\,\theta\right) = \sum_{j=1}^{3}\left(h_j\,\text{div}\,\mathbf{g_j}\right) + \sum_{j=1}^{4}\left(h_j \frac{\partial w_j}{\partial t}\right)}_{(B)}$$

$$+ \underbrace{\frac{\partial}{\partial t}\left\{\left[\rho_0 c_0 + \sum_{j=1}^{4}(c_j w_j)\right]\theta\right\}}_{(A)} \pm \Phi' \qquad (3.8)$$

Term (A) in the equation includes the volumetric heat capacity of the matrix and the three or four mass components. As the values for vapour and air are negligible, the term simplifies to:

$$\left[\rho_0 c_0 + \sum_{j=1}^{4}(c_j w_j)\right] = (\rho_0 c_0 + c_1 w_1 + c_i w_i) = \rho_0 c'$$

with:

$$c' = c_0 + \frac{c_1 w_1 + c_i w_i}{\rho_0} \qquad (3.9)$$

3.2 Material and assembly level

As for ice, the flux (g_i) is zero, so term (B) resets as:

$$\sum \left[h_j \underbrace{\left(\text{div } \mathbf{g_j} + \frac{\partial w_j}{\partial t} \right)} \right]$$

The expression above the brace stands for mass conservation per component without source or sink, so equals:

$$\sum_{j=1}^{3} h_j G'_j \tag{3.10}$$

Sum 1 to 3, not 1 to 4, indicates that changes of state are excluded for dry air (4). For ice, which is 3, the source term (G'_3) is zero beyond $0\,°C$, or:

$$G'_1 = -G'_2$$

turning the sum into:

$$\sum_{j=1}^{3} h_j G'_j = G'_1(c_1\theta - l_{b0} - c_v\theta) = -G' l_b(\theta)$$

with

$$l_b(\theta) = l_b(0) + (c_v - c_1)\theta$$

In freezing conditions, $G'_1 + G'_2 + G'_3$ becomes zero, changing the sum equation into:

$$\sum_{j=1}^{3} h_j G'_j = -G'_1 l_b(\theta) - G'_3 l_m(\theta).$$

Inserting both relations in the reshuffled conservation equation gives:

$$\text{Above } 0\,°C: -\underbrace{\text{div }\mathbf{q}}_{(1)} - \underbrace{\sum_{j=1}^{3} \left(\mathbf{g_j} c_j \text{ grad } \theta \right)}_{(2)} = \underbrace{-G'_1 l_b}_{(3)} + \underbrace{\frac{\partial}{\partial t}(\rho_0 c'\theta)}_{(4)} \pm \underbrace{\Phi'}_{(5)} \tag{3.11}$$

$$\text{Below and at } 0\,°C: -\underbrace{\text{div }\mathbf{q}}_{(1)} - \underbrace{\sum_{j=1}^{3} \left(\mathbf{g_j} c_j \text{ grad } \theta \right)}_{(2)} = \underbrace{-G'_1 l_b - G'_3 l_m}_{(3)} + \underbrace{\frac{\partial}{\partial t}(\rho_0 c'\theta)}_{(4)} \pm \underbrace{\Phi'}_{(5)}$$

$$\tag{3.12}$$

Term (1) in these equations represents the heat transfer by 'equivalent' conduction at the micro-scale – not only pure conduction along the matrix but also conduction and convection in the pore gas, radiation between the pore walls, conduction in all absorbed water layers and latent heat transfer by local evaporation and condensation in wetted pores. All this means that the material property 'thermal conductivity' is a function of temperature, moisture content, thickness, and so on. Term (2) quantifies the sensible enthalpy transfer by mass fluxes, especially air movement, whereas the much smaller liquid and vapour fluxes have much less impact. Term (3) stands for the latent enthalpy transfer with the inter-porous changes of state between liquid, vapour and ice as drivers, where evaporation and condensation dominate. Term (4) concerns the heat stored, wherein the material matrix and the water content play the leading roles. Finally, term (5) considers all other heat sources and sinks, for example due to chemical processes.

3.2.5 Flux equations

The flux equations, which replace conservation of momentum, are empirical in nature.

3.2.5.1 Heat

Heat conduction is, as explained previously, described by Fourier's law:

$$\mathbf{q} = -\lambda \operatorname{grad} \theta$$

with λ being the (equivalent) thermal conductivity (W/(m.K)), which for anisotropic materials is a tensor. Air layers demand a specific approach (see Sections 1.4.4.2 and 1.6.2.4).

3.2.5.2 Mass, air

In open-porous materials, as discussed, Poseuille's law applied to air movement gives:

$$\mathbf{g}_a = -k_a \operatorname{grad} P_a$$

with P_a the air pressure (Pa) included stack $(-\rho_a \mathbf{g} \mathbf{z})$, and k_a the air permeability (s) equal to:

$$k_a = \frac{1}{W' \nu_a} = \frac{B}{\nu_a}$$

with ν_a the kinematic viscosity of air and B the penetration coefficient of the material (m²/s²), a characteristic of any porous system not impacted by the particular fluid or gas migrating. The air permeability again is a tensor for anisotropic materials, while its value nears zero for capillary wet materials. For air-permeable layers, apertures, joints, cracks, leaks and cavities, the flux or flow equations are:

Air permeable layers (per m²):	$g_a = -K_a \Delta P_a$	(G_a in kg/(m².s) and K_a in s/m)
Joints, cracks, cavities (per m):	$G_a = -K_a^\psi \Delta P_a$	(G_a in kg/(m.s) and K_a^ψ in s)
Leaks, voids, apertures (each):	$G_a = -K_a^\chi \Delta P_a$	(G_a in kg/s and K_a^χ in m.s)

3.2 Material and assembly level

with K_a^x the air permeance and ΔP_a the air pressure difference including stack along the flow direction. Most air permeances change with the pressure differential imposed:

$$K_a^x = a(\Delta P_a - \rho_a \mathbf{g} z)^{b-1}$$

Apertures and joints, but also accidental cracks, voids, leaks and unplanned air layers, whose locations are rarely known, assist in shaping the real geometry of an assembly, although the uncertainty about where they are and what they look like makes air and liquid flow troubling to simulate.

3.2.5.3 Mass, moisture

The driving forces behind vapour flow are equivalent diffusion and convection. Quantifying the first is done using Fick's law:

$$\mathbf{g}_a = -\delta \operatorname{grad} p$$

with δ the vapour permeability (units: s), a tensor for anisotropic materials. The property reflects the magnitude and deviousness of the pore system. Its value depends on moisture content and temperature. 'Equivalent' indicates that diffusive fluxes are, as explained, not purely Fickian. The airflow-driven vapour transfer is given by:

$$\mathbf{g}_v = \mathbf{g}_a x_v = \frac{0.622\, \mathbf{g}_a p}{P_a - p} \approx \frac{0.622\, \mathbf{g}_a p}{P_a} \approx 6.21 \times 10^{-6}\, \mathbf{g}_a p$$

with $\mathbf{g}_a$ the air flux (kg/(m².s)) and P_a the air pressure (Pa). The sign $\approx$ after the correct expression indicates that related simplifications only apply for temperatures below 50 °C and for very small air pressure differences.

For unsaturated water flow, an equation similar to Darcy's law for saturated flow applies:

$$\mathbf{g}_w = -k_w \operatorname{grad} s$$

with k_w the unsaturated water permeability (units: s), again a tensor for anisotropic materials, and s being suction. The unsaturated water permeability normally has very low values up to a given moisture content. Once beyond, it increases steeply.

3.2.6 Equations of state

3.2.6.1 Enthalpy and vapour saturation pressure versus temperature

See Section 2.1.6 and Section 2.3.1.3.

3.2.6.2 Relative humidity versus moisture content

As explained in Section 2.3.2, this relationship is known as the sorption/desorption curve with hysteresis in between. The derivative represents the relative humidity-related specific moisture content.

3.2.6.3 Suction versus moisture content

This is known as the moisture characteristic. In contact with water, the values that matter go from a moisture content of zero to capillary. The derivative, termed the suction-related specific moisture content, supersedes the relative humidity related moisture content beyond a relative humidity pivot ϕ_M, when capillary condensation starts to create continuous water paths in porous materials. Of course, linkage to relative humidity remains optional.

3.2.7 Starting, boundary and contact conditions

The starting conditions define the heat, air and moisture situation at time zero. They are assumed or follow from testing. Most often, construction moisture, whether measured or realistically estimated, figures as such. However, steady state and harmonic approaches do not require starting conditions.

The boundary conditions include the temperatures and heat fluxes, vapour pressures and vapour fluxes, suction, relative humidity, sometimes moisture content and liquid water fluxes, plus the air pressures and fluxes that act on the end faces of an assembly during the time span covered by the calculations. They can remain constant over time, a situation called steady state, or transient. Along surfaces, they are constant or variable. During the period considered, changes in the type of conditions are possible, like the case for wind-driven rain impinging on an outer wall at the time a water film forms. No heat, air and moisture problem can be solved without knowledge of the boundary conditions, but in most cases reality is poorly known, thus requiring us to schematize.

The contact conditions describe the situation in the interfaces between layers and material volumes. Most tools presume ideal contact with the driving forces, maintaining equality and continuity between fluxes, but this conflicts with what is observed. Thermally, for materials with high thermal conductivity such as metals, contact resistances intervene, and in most cases are unknown but have quite an impact. Across real free contacts, characterized by thin air gaps between successive layers, only vapour flow keeps equality. Liquid transfer, of course, can fill the air gap with water, which is either absorbed when the still-dry next layer is capillary active, or retained until gravity wins from friction and produced runoff. Airflows, in turn, could deviate and partly move parallel to the interface. Non-free contacts, such as glued or chemically active ones, add resistances. Often, real contacts mix free areas with locations where interaction creates extra resistance. As for the boundary conditions, the real contact conditions between material volumes and layers are mostly unknown.

3.2.8 Two examples of simplified models

3.2.8.1 Non-hygroscopic, non-capillary materials

Insertion of the air flux in the air balance gives:

$$\mathrm{div}(k_a \, \mathbf{grad} \, P_a) = 0$$

3.2 Material and assembly level

The overall air pressure P_a includes the actual air pressure $P_{a,o}$ and stack ($-\rho_a gz$). When air permeability is a constant, which it mostly is not, the balance simplifies to:

$$\nabla^2 (P_{a,o} - \rho_a gz) = 0 \tag{3.13}$$

In isothermal conditions, thermal stack ($-\rho_a gz$) is zero. In non-isothermal conditions, it differs from zero in all directions except the horizontal.

Inserting both the air-driven vapour movement and the flux by equivalent diffusion into the vapour balance yields:

$$\text{div}\left(\delta \, \text{grad} \, p - \frac{0.622 g_a}{P_a} p\right) \pm G'_2 = \frac{\partial w_v}{\partial t}$$

In non-hygroscopic materials, the storage term represents the water vapour concentration increase or decrease in the pore air:

$$\frac{\partial w_v}{\partial t} = \frac{\partial}{\partial t}\left(\frac{\Psi_0 p}{RT}\right)$$

As long as the relative humidity stays below 100%, which can be controlled thanks to the equation of state for the vapour saturation pressure ($p_{sat}(\theta)$), the source term G'_2 is zero, which simplifies the mass balance to:

$$\text{div}\left(\delta \, \text{grad} \, p - \frac{0.622 g_a}{P_a} p\right) = \frac{\Psi_0}{R} \frac{\partial}{\partial t}\left(\frac{p}{T}\right)$$

At 100% relative humidity, it changes to:

$$\text{div}\left(\delta \, \text{grad} \, p_{sat} - \frac{0.622 g_a}{P_a} p_{sat}\right) \pm G'_2 = \frac{1}{R}\frac{\partial}{\partial t}\left(\frac{\Psi_0 p_{sat}}{T}\right)$$

Compared with condensation and evaporation (G'_2), the right-hand storage term is negligible, or:

$$\text{div}(\delta \, \text{grad} \, p_{sat}) - \frac{0.622 g_a}{P_a} \text{grad} \, p_{sat} = \pm G'_2 \tag{3.14}$$

Energy conservation gives:

$$\text{div}(\lambda \, \text{grad} \, \theta) - \sum_{j=1}^{3}\left(g_j c_j \, \text{grad} \, \theta\right) = -G'_2 l_b + \frac{\partial}{\partial t}(\rho_0 c' \theta) \pm \Phi'$$

In the enthalpy transfer, only air plays a role of importance. Without condensation or drying, no heat of evaporation intervenes, while in the volumetric heat capacity, the material dominates. All this reduces the expression to:

$$\text{div}(\lambda \, \text{grad} \, \theta) - g_a c_a \, \text{grad} \, \theta = \rho_0 c_0 \frac{\partial \theta}{\partial t} \tag{3.15}$$

With condensation and drying, the equation changes into:

$$\text{div}(\lambda \, \text{grad} \, \theta) - g_a c_a \, \text{grad} \, \theta + l_b \, \text{div}(\delta \, \text{grad} \, p_{sat}) - l_b \frac{0.622 g_a}{P_a} \text{grad} \, p_{sat} = \rho_0 c' \frac{\partial \theta}{\partial t}$$

or:

$$\text{div}\left[\left(\lambda + l_b \delta \frac{dp_{\text{sat}}}{d\theta}\right) \mathbf{grad}\,\theta\right] - \mathbf{g_a}\left(c_a + l_b \frac{0.622}{P_a} \frac{dp_{\text{sat}}}{d\theta}\right) \mathbf{grad}\,\theta = \rho_0 c' \frac{\partial \theta}{\partial t} \qquad (3.16)$$

The heat, air and moisture transfer model for non-hygroscopic, non-capillary materials is ready for use now. Included the equation of state $p_{\text{sat}}(\theta)$, a system of four equations, the mass balance for air, the mass balance for vapour and the energy balance, with temperature θ, vapour pressure p, the condensation or evaporation rate G_2', and air pressure P_a' as unknowns has to be solved. The fluxes follow from the flow equations. The iteration needed starts from assuming a particular temperature distribution. Then the airflows are calculated, the temperatures recalculated, the airflows rechecked, and so on.

Without airflow, the model is much simpler. Has neither condensation nor drying an effect, the mass and heat balances are:

$$\text{div}(\delta\,\mathbf{grad}\,p) = \frac{\partial}{\partial t}\left(\frac{\Psi_0 p}{RT}\right) \qquad \text{div}(\lambda\,\mathbf{grad}\,\theta) = \rho_0 c_0 \frac{\partial \theta}{\partial t}$$

Only when the material properties depend on temperature are the two connected, although a solution assuming constant properties already gives valuable information. In the case condensation or drying takes place, the mass and heat balances change into:

$$\text{div}\left(\delta \frac{dp_{\text{sat}}}{d\theta}\mathbf{grad}\,\theta\right) = \pm G_c' \qquad \text{div}\left[\left(\lambda + l_b \delta \frac{dp_{\text{sat}}}{d\theta}\right)\mathbf{grad}\,\theta\right] = \rho_0 c_0 \frac{\partial \theta}{\partial t}$$

showing that temperature is the only driving potential to be calculated. The condensation or drying rates per unit volume (G_2') of course are also unknown.

3.2.8.2 Hygroscopic materials at low moisture content

Now, the material is only sorption-active, which means that vapour flow is the sole moisture transport mode. Mass conservation gives:

$$\text{div}\left(\delta\,\mathbf{grad}\,p - \frac{0.622\,\mathbf{g_a}}{P_a}p\right) = \frac{\partial w_H}{\partial t}$$

As air transfer is negligible in such fine-porous materials, the equation simplifies to:

$$\text{div}(\delta\,\mathbf{grad}\,p) = \partial w_H / \partial t$$

Introducing temperature and relative humidity as driving forces transforms this mass balance into:

$$\text{div}\left(\delta p_{\text{sat}}\,\mathbf{grad}\,\phi + \delta\phi \frac{dp_{\text{sat}}}{d\theta}\mathbf{grad}\,\theta\right) = \rho \xi_\phi \frac{\partial \phi}{\partial t} \qquad (3.17)$$

Conservation of energy yields:

$$\text{div}(\lambda\,\mathbf{grad}\,\theta) - \sum_{j=1}^{3}\left(\mathbf{g_j} c_j\,\mathbf{grad}\,\theta\right) = -G_1' l_b + \frac{\partial}{\partial t}(\rho_0 c'\theta) \pm \Phi'$$

The source or sink term Φ' remains zero, except when liquid water should be deposited by condensation somewhere in the assembly, or liquid deposit should evaporate. Diffusion only gives vapour fluxes too small to consider for related enthalpy displacement, while phase changes are restricted to moisture sorption and desorption. All this allows the expression of the sink or source term G'_1 as:

$$G'_1 = \rho \xi_\phi \frac{\partial \phi}{\partial t} = \mathrm{div}\left(\delta p_{\mathrm{sat}}\, \mathbf{grad}\, \phi + \delta \phi \frac{\mathrm{d}p_{\mathrm{sat}}}{\mathrm{d}\theta} \mathbf{grad}\, \theta\right)$$

Insertion into the energy balance gives:

$$\mathrm{div}\left[l_\mathrm{b} \delta p_{\mathrm{sat}}\, \mathbf{grad}\, \phi + \left(\lambda + l_\mathrm{b} \delta \phi \frac{\mathrm{d}p_{\mathrm{sat}}}{\mathrm{d}\theta}\right) \mathbf{grad}\, \theta\right] = \rho_0 c' \frac{\partial \theta}{\partial t} \tag{3.18}$$

Both balance equations, the one for energy and the other for vapour, with relative humidity (ϕ) and temperature (θ) as driving forces, allow the quantification of combined heat and moisture transfer in hygroscopic materials at low moisture content. A system with the material properties and coefficients variable is solved using CVM.

3.3 Whole building level

Combined heat, air and moisture transfer involves the building and building use as a whole. This is undoubtedly the case for thermal comfort, good indoor air quality (which demands a high enough fresh air inflow), and relative humidity. Too high a value of relative humidity at indoor surfaces favours mould, whose presence impacts the mental and sometimes physiological health of the inhabitants.

The thermal balance at the zone level was analyzed in Chapter 1, but only for harmonic boundary conditions. The section on mass transfer offered information about inter-zone airflows in buildings. Here, the vapour balance at zone level and whole building level is explored, with the inclusion of sorption.

3.3.1 Balance equations

3.3.1.1 Vapour

In Chapter 2, conservation of mass was applied to the vapour entering a zone with the supply and infiltrating air, the vapour released indoors by sources and surface drying or removed by sinks and surface condensation, the vapour leaving with the extract or exfiltrating air, and the vapour stored in the zone air. Here, we also include the vapour storage and release by sorption active surfaces, changing the vapour balance into (Figure 3.1):

$$\sum (x_j G_{\mathrm{a},ji}) + \sum (x_i G_{\mathrm{a},ij}) + G_{\mathrm{a,e}} x_\mathrm{e} + G_{\mathrm{vP},i} + \sum (G_{\mathrm{vc/d},ik}) + \sum (G_{\mathrm{vH},il}) = \rho_\mathrm{a} V_i \frac{\mathrm{d}x_i}{\mathrm{d}t}$$

$$\tag{3.19}$$

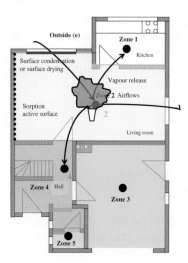

Fig. 3.1 Water vapour balance at the zone level

with x_j the vapour ratio in the airflows coming from the neighbouring zones and x_e the vapour ratio in the airflow from outdoors.

The term above the brace in the equation bundles all sorption-active surfaces, with the suffix $_{vH}$ denoting hygroscopic. As explained previously, surface condensation and drying obeys:

$$G_{vc/d,ik} = \beta_k A_k \left(p_{sat,s,k} - p_i \right) \qquad (3.20)$$

with p_i the vapour pressure in the zone and $p_{sat,s,k}$ the vapour saturation pressure on the surface with area A_k where vapour condenses or condensate evaporates. The equation governing the vapour exchange between a sorption active surface (A_l) and the zone air is:

$$G_{vH,il} = \beta_l A_l \left(p_{sat,s,l} \phi_{s,l} - p_i \right) \qquad (3.21)$$

with $\phi_{s,l}$ the relative humidity and $p_{sat,s,l}$ the vapour saturation pressure at that surface. Vapour inflow and outflow linked to the supply and extraction of air gives:

$$x G_a = \frac{R_l p}{R_v (P_a - p)} G_a \approx 6.21 \times 10^{-6} p G_a \qquad (3.22)$$

The simple expression on the right holds as long as the vapour pressure (p) is only a few per cent of the total air pressure (P_a).

Entering the three flow equations into the zone balance results in:

$$-p_i \left[6.21 \times 10^{-6} \sum G_{a,ij} + \sum_{k=1}^{m}(\beta_k A_k) + \sum_{l=1}^{n}(\beta_l A_l) \right] + 6.21 \times 10^{-6} \sum \left(p_j G_{a,ji} \right)$$
$$+ G_{vP,i} + \sum_{k=1}^{m}\left(\beta_k A_k p_{\text{sat},s,k} \right) + \sum_{l=1}^{n}\left(\beta_l A_l p_{\text{sat},s,l} \phi_{s,l} \right) = \frac{V_i}{RT} \frac{dp_i}{dt}$$

(3.23)

with R the gas constant of water vapour. Unknowns in the expression are the vapour pressure p_i in the zone, the vapour pressures p_j in the neighbouring zones, the vapour saturation pressure $p_{\text{sat},s,l}$ and relative humidity $\phi_{s,l}$ at the sorption-active surfaces, the vapour saturation pressure $p_{\text{sat},s,k}$ at the surfaces that suffer from condensation or drying, all airflows and the surfaces A_k and A_l. The suffixes l and k refer to surfaces l and k. Parameters that require preset or measured values in the equation are the vapour release or removal in the space, the vapour pressure outdoors, and the surface film coefficients for diffusion β_k and β_l.

3.3.1.2 Air

The inter-zone airflows follow from the air balance per zone (see Section 2.2.6). Their magnitude depends on the temperature outdoors, the wind velocity and direction, the building form, the building location, the envelope leakage, the type of ventilation system installed, the temperature in the separate zones, and so on. A really detailed study needs all intra-zone air movements. Simulating these requires solution of the Navier-Stokes equation in combination with the turbulence equations, conservation of mass and conservation of energy (CFD).

3.3.1.3 Heat

Fixing the vapour saturation pressures $p_{\text{sat},s,k}$ and $p_{\text{sat},s,l}$ requires knowledge of the surface temperatures. For that, the heat balances at the zone and assembly level have to be solved, the last defined by the thermal properties of the composing layers, their sequence and the ambient temperatures at both sides. A detailed approach is preferred, considering the building as a three-dimensional object, composed of spaces and fabric, and exposed to varying outdoor and user-defined indoor conditions. To be truly detailed, the air temperature distribution in each zone as well as the inside surface temperature anywhere on the envelope and the partitions should be known. For that, the building fabric and all zones must be meshed fine enough to allow a combination of CVM at fabric level with CFD at zone level.

Such full models are demanding, which is the reason why buildings are simplified to a sum of zones, separated by flat assemblies, where heat flows develop one-dimensionally and normal to all wall end faces, while the air in the zones is considered ideally mixed. In this way, one air temperature per zone suffices. Then the zone results are used to evaluate potential thermal bridge effects. Of course, simplifications always mean a loss of information, a fact often forgotten.

In the zone balances, heat flows are split into radiation and convection, with the convective injected in the zone nodes, where their sum equals the change in heat stored in the zone air, the furniture and all furnishings present:

$$\underbrace{\sum_{m=1}^{m}\left[h_{ci}A_m\left(\theta_{si,m}-\theta_j\right)\right]}_{(1)} + \underbrace{c_a\left[\sum_{l=1}^{l}\left(G_{a,\inf,x}\theta_x\right)-\theta_j\sum_{l=1}^{l}G_{a,\inf,x}\right]}_{(2)} + \underbrace{\sum\left(f_{c,i}\Phi_{i,j}\right)}_{(3)}$$

$$+\underbrace{f_{c,syst}\Phi_{heat/cool,net,j}}_{(4)} = \rho_a c V \frac{d\theta_j}{dt}$$

(3.24)

Term (1) represents the convective heat exchanged between all surfaces and the zone air. Term (2) stands for the air-coupled enthalpy flows exchanged with all neighbouring zones included the outdoors, and the enthalpy flow generated or withdrawn by any ventilation system in the zone ($x = e$ for outdoors, $x = l$ for a neighbouring zone in the building). Term (3) applies to the convective fraction ($f_{c,i}$) of the internal gains. Term (4) stands for the convective flow that the heating or cooling system delivers, with $f_{c,syst}$ being 1 for an air-based system and less than 1 for any other system.

Each opaque assembly has a face looking to the zone. For enclosure parts, the faces seeing the outdoors have as heat balance:

$$q_{c,se} + q_{LR,se} + a_{S,se}q_{S,se} + q_{T,se} = 0$$

with $q_{c,se}$ the convective heat flux with the outdoor air, $q_{LR,e}$ the radiant heat flux by long wave exchange with the terrestrial environment and the sky, $a_{S,se}$ the short wave absorptivity of the outside surface, $q_{S,se}$ the solar irradiation and $q_{T,se}$ the conduction flux near the outside surface.

Convection, long wave radiation and conduction are written as:

Convection: $q_{c,se} = h_{c,e}(\theta_e - \theta_{se})$
Long wave: $q_{se,LR} = 5.67 e_{Le}(F_{se}F_{Tse} + F_{ssk}F_{Tssk})(\theta_e - \theta_{se}) - 120 e_{Le}F_{ssk}F_{Tssk}(1-f_c)$
Conduction: $q_{T,si} = \frac{\lambda_n}{\Delta x_n}(\theta_{n-1} - \theta_{se})$

For the symbol notation, see Chapter 1. In the conduction equation, λ_n is the thermal conductivity of the material at the outside, Δx_n the mesh width used, θ_{n-1} the mesh temperature and θ_{se} the outside surface temperature.

For the partitions and enclosure part faces that see the zone, the heat balance is:

$$q_{c,si} + q_{R,si} + a_{S,si}q_{S,si} + q_{T,si} = 0$$

with $q_{c,si}$ the convective heat flux inside, $q_{R,si}$ the radiant heat flux by long wave exchange with all other surfaces in the zone, $a_{S,si}$ their short wave absorptivity, $q_{S,si}$ the solar radiation transmitted by the transparent parts and warming the surfaces,

3.3 Whole building level

and $q_{T,si}$ the conduction flux at the inside faces. Convection, long wave radiation, solar irradiation and conduction are written as:

Convection: $q_{c,si} = h_{c,i}(\theta_j - \theta_{si})$

Long wave: $q_{R,si} = \frac{e_L}{\rho_L}\left[\left(\sum_{k=1}^{m} a_{R,k} H_{b,k}\right) - H_b\right] + \frac{\sum[(1-f_{c,I})\Phi_I]}{\sum A} + \left(1 - f_{c,syst}\right)\frac{a_L \Phi_{heat/cool,net}}{\sum A}$

Solar: $q_{S,si} = \dfrac{0.9 \times 0.95 \sum (\tau_{S,w} f_w r_w A_w q_S)}{\sum\limits_{opaque} A - 0.9 \times 0.95 \sum (\tau_{S,w} f_w r_w A_w)}$

Conduction: $q_{T,si} = \frac{\lambda_1}{\Delta x_1}(\theta_1 - \theta_{si})$

In the long wave radiation equation, m is the number and ΣA the total inside surface of the opaque and transparent assemblies enclosing the zone, $a_{R,k}$ is the fraction of black body radiation emitted by all other surfaces absorbed by the surface considered, Φ_I the internal gains and $\Phi_{heat/cool,net}$ the heating or cooling load. Solar irradiation assumes that the short wave radiation entering the zone is reflected so many times between all surfaces that a diffuse field is left, whereby each opaque face absorbs radiation in proportion to its short wave absorptivity. All transparent assemblies instead are supposed to transmit part of that reflected solar radiation back to the outside. For the symbol notation, see Section 1.5. In the conduction equation, λ_1 is the thermal conductivity of the material at the inside, Δx_1 is the mesh width, θ_1 the mesh temperature and θ_{si} the inside surface temperature. As envelopes and partitions are assumed to be composed of flat assemblies, stepping to flows is simple: multiply all fluxes with the respective surfaces.

Transparent envelope parts are assumed to behave as steady state even over very short time intervals. One has to account for under-cooling, while the solar irradiation must be split in direct transmission and absorbed by the glass. To simplify things, the temperature across each pane is set as constant. This allows the description of each glass and solar shading combination by a system of as many equations as panes (see Section 1.6).

Moving to the meshed opaque assemblies, if air-tight, the balance per mesh is:

$$\frac{\lambda_i(\theta_{i-1} - \theta_i)}{\Delta x_i} + \frac{\lambda_{i+1}(\theta_{i+1} - \theta_i)}{\Delta x_{i+1}} = \left[(\rho c)_i \frac{\Delta x_i}{2} + (\rho c)_{i+1} \frac{\Delta x_{i+1}}{2}\right] \frac{\Delta \theta_i}{\Delta t}$$

The system of zone and mesh balances is solved per time step Δt using a Cranck-Nicholson calculation scheme, whereby each of the heat fluxes equals the average per time step.

3.3.1.4 Closing the loop

Once the air and heat balances per zone are solved, transposing the temperatures at the inside faces into saturation pressures is easily done. So, aside from all surfaces, all saturation pressures also become parameters in the zone vapour balance. The unknowns left are: the vapour pressure p_i in the zone, the vapour pressures p_j in the neighbouring zones, and the relative humidity $\phi_{s,l}$ on the sorption-active surfaces.

One inaccuracy, however, remains. Sorption, surface condensation and surface drying invoke latent heat release and uptake, which changes the inside surface temperatures. Not accounting for these allows for solution of the vapour balance without feedback to the heat balances. Otherwise, all sorption-active inside surfaces and those suffering from surface condensation or drying face an extra heat flux, requiring additional looping between the heat, air and vapour balances.

For a building of n zones, the related vapour balances generate a system of n equations:

$$|H_v\|p| + |1| \left| \sum_{l=1}^{n} \left(\beta_l A_l p_{sat,s,l} \phi_{s,l} \right) \right| = - \left| \sum_{k=1}^{n} \left(\beta_k A_k p_{sat,s,k} \right) + G_{vP} \right| \quad (3.25)$$

The terms on the diagonals in the array $|H_v|$ are:

$$H_{v,ii} = - \left[\left(6.21 \times 10^{-6} \sum G_{a,ij} + \frac{V_i}{RT} \mathbf{D} \right) + \sum_{k=1}^{n} \left(\beta_k A_k \right) + \sum_{l=1}^{n} \left(\beta_l A_l \right) \right] \quad (3.26)$$

with $\mathbf{D}$ the differential operator. These outside the diagonal are:

$$H_{v,ij} = H_{v,ji} = \left(6.21 \times 10^{-6} \sum G_{a,ij} \right) \quad (3.27)$$

This system of as many equations as zones in the building can be solved, on condition that the relationships between the vapour pressure in each zone and the relative humidity on all sorption-active surfaces are known, and all air- and heat-related quantities and values are given.

3.3.2 Sorption-active surfaces and hygric inertia

3.3.2.1 Generalities

Consider zone i. The zone's actual air and heat situation, the vapour pressure outdoors and the vapour pressures in all neighbouring zones are assumed to be known. Zone i has l sorption-active surfaces, which means that in the vapour balance the vapour pressure in the air and the relative humidity at these l surfaces are still unknown. Besides the zone vapour balance, a solution requires as many additional vapour balances as sorption-active surfaces:

$$\beta_l \left(p_i - p_{sat,s,i} \phi_s \right) + \partial_l \, \mathbf{grad}(p)_s = 0 \quad (3.28)$$

In other words, the sum of the vapour flux from the zone to a surface and the vapour flux from inside the assembly to that surface must be zero. The equation is of no use without knowing $\mathbf{grad}(p)$. This demands quantification of the inner vapour transfer close to each inside surface. In the absence of air transport, the related vapour balance is:

$$\mathbf{div} \left[\delta \, \mathbf{grad}(p) \right]_s = \frac{\partial w_{H,s}}{\partial t} \quad (3.29)$$

with δ the vapour permeability of the layer closest to the inside surface and w_H its sorption moisture content. Solving the system formed by the zone and all surface

3.3 Whole building level

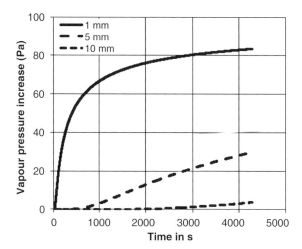

Fig. 3.2 Transient hygric response in gypsum board when facing a sudden 100 Pa vapour pressure increase at its surface

vapour balances gives the $1+l$ unknowns for zone i. With n zones, the system has $n + \sum_{n=1}^{n}(nl_j)$ equations.

3.3.2.2 Sorption-active thickness

In case of rather rapid changes in vapour pressure indoors, the correct solution can be approximated by attributing a sorption-active thickness (d_H) to each hygroscopic surface. Figure 3.2 underpins that possibility by showing how vapour pressure in a gypsum board evolves when its surface experiences a sudden 100 Pa increase in vapour pressure. After 1 hour, hardly any change happens at 1 cm depth, proving that activation of hygric inertia advances slowly.

What is called the sorption-active thickness is now defined as the distance between the inside surface and the interface where the vapour pressure amplitude due to a 1 Pa periodic oscillation at the inside surface dampens to 0.368 Pa. For a constant specific moisture content (ξ_ϕ), a constant vapour resistance factor (μ) and the same temperature as at the inside surface, that thickness equals:

$$d_H = \sqrt{\frac{\delta p_{sat,si} T}{\pi \xi_\phi}} \qquad (3.30)$$

with p_{sat} the saturation pressure at the inside surface and T the period considered. As long as the relative humidity stays between 30% and 80%, assuming constant specific moisture content is not too bad (see Figure 3.3). For a 1-hour and a 1-day period, that thickness is:

Hour: $d_H = 33.85\sqrt{\delta p_{sat,si}/\xi_\phi}$ Day: $d_H = 165.8\sqrt{\delta p_{sat,si}/\xi_\phi}$

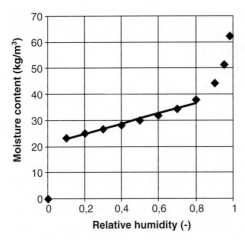

Fig. 3.3 Measured sorption curve, linear for relative humidity between 0.2 and 0.8

Related hygric capacitance equals:

Hour: $C_{\phi,H} = 33.85\sqrt{\delta\xi_\phi/p_{\text{sat,si}}}$ Day: $C_{\phi,H} = 165.8\sqrt{\delta\xi_\phi/p_{\text{sat,si}}}$

with the centre $0.422d_H$ metres away from the inside surface. When the sorption-active thickness includes different layers and each has a different specific moisture content and vapour resistance factor, the hygric capacitance per layer then is:

$$C_{\phi,H,i} = \xi_{\phi,i}d_i/p_{\text{sat,si}}$$

with d_i the layer thickness and $p_{\text{sat,si}}$ the vapour saturation pressure at the inside surface (Figure 3.4). The summed capacity, with its point of action where the resultant of the hygric capacitance of the different layers is situated for the separate capacitances are seen as vectors, is:

$$C_{\phi,H} = \frac{1}{p_{\text{sat,s,i}}}\sum(\xi_\phi d)$$

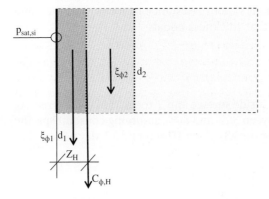

Fig. 3.4 Sorption-active thickness including different layers

3.3 Whole building level

Past that point, the steady-state monthly mean vapour flux passing the inside face replaces the correctly calculated flux across the rest of the assembly. This again is an approximation, although acceptable given the thinness of sorption-active layers.

Still unknown is the vapour pressure p_H in the point of action. A mass balance with the sorption active layer compressed into this one point gives:

$$\frac{p_i - p_H}{Z_H} + \frac{\bar{p}_x - \bar{p}_i}{Z_T} = \left(\frac{C_{\phi,H}}{p_{sat,si}}\right)\frac{dp_H}{dt} \tag{3.31}$$

with Z_H the diffusion resistance between indoors and the point of action, p_H the vapour pressure there, p_i the vapour pressure indoors, $\bar{p}_i$ the monthly mean vapour pressure indoors, $\bar{p}_x$ the monthly mean vapour pressure in the related neighbouring zone, the outdoors or the interface where condensate deposits, and Z_T the total diffusion resistance of the assembly or the diffusion resistance between indoors and the condensation interface. The diffusion resistance between indoors and the point of action equals:

Homogeneous: $Z_H = 1/\beta_l + 0.422\mu N d_H$ Composite: $Z_H = 1/\beta_l + \sum_{si}^{H}(\mu N d)$

The second term left in the mass balance equation just given is important because, indeed, for a given vapour release indoors and given ventilation flow, a vapour-permeable enclosure may help to lower the vapour pressure in the zone considered.

Once the vapour pressure at the point of action is known, the vapour pressure and relative humidity at the inside surface follow from:

$$p_{s,i} = p_i - \frac{(p_i - p_w)}{\beta_i Z_H} \qquad \phi_{s,l} = \frac{p_{s,i}}{p_{sat,s,i}} \tag{3.32}$$

Each hygroscopic surface so generates two equations. Combination of their vapour balance with the zone vapour balance allows calculation of the $l+1$ unknowns. In theory, for one zone the system can be solved analytically.

3.3.2.3 Zone with one sorption-active surface

The zone considered is ventilated with outside air (G_a), and the sorption-active surface area is A m². Neither surface condensation nor drying takes place. The vapour pressure outdoors is p_e, the vapour release indoors G_{vP}, the monthly mean vapour pressure at the other side of the sorption-active surface $\bar{p}_x$, and the monthly mean vapour pressure indoors $\bar{p}_i$. The balance equations become:

Zone $\quad -p_i\left[6.21 \times 10^{-6} G_a + \frac{A}{Z_H}\right] + \frac{p_H A}{Z_H} = \frac{V}{RT}\frac{dp_i}{dt} - G_{vP} - 6.21 \times 10^{-6} G_a p_e$

Surface $\quad \dfrac{p_i - p_H}{Z_H} = C_H \dfrac{dp_H}{dt} + \left(\dfrac{\bar{p}_i - \bar{p}_x}{Z_T}\right)\dfrac{A_e}{A}$

$$\tag{3.33}$$

The unknowns are the vapour pressure indoors (p_i) and the vapour pressure in the point of action of the surface's sorption-active thickness (p_H). Rearranging and writing 6.21×10^{-6} as 'α' gives:

$$\begin{vmatrix} \left[D + \dfrac{G_a}{\rho_a V} + \dfrac{A}{\alpha \rho_a V Z_H} \right] & -\left(\dfrac{A}{\alpha \rho_a V Z_H} \right) \\ -\left(\dfrac{1}{Z_H C_H} \right) & \left(D + \dfrac{1}{Z_H C_H} \right) \end{vmatrix} \begin{vmatrix} p_i \\ p_H \end{vmatrix} = \begin{vmatrix} \dfrac{G_a}{\rho_a V} p_e + \dfrac{G_{vP}}{\alpha \rho_a V} \\ \left(\dfrac{\bar{p}_x - \bar{p}_i}{Z_T C_H} \right) \dfrac{A_e}{A} \end{vmatrix}$$

or:

$$\begin{vmatrix} D + A_1 + A_2 & -A_2 \\ -B_1 & D + B_1 \end{vmatrix} \begin{vmatrix} p_i \\ p_H \end{vmatrix} = \begin{vmatrix} A_1 p_e + A_3 \\ B_2 (\bar{p}_x - \bar{p}_i) \end{vmatrix} \qquad (3.34)$$

with **D** the differential operator, ρ_a the density of air and A_1, A_2, A_3, B_1 and B_2 ratios equal to

$$A_1 = \frac{G_a}{\rho_a V} \quad A_2 = \frac{A}{\alpha \rho_a V Z_H} \quad A_3 = \frac{G_{vP}}{\alpha \rho_a V} \quad B_1 = \frac{1}{C_H Z_H} \quad B_2 = \left(\frac{1}{C_H Z_T} \right) \frac{A_e}{A}$$

For a sudden increase in vapour release indoors or vapour pressure outdoors, the solution is:

$$p_i = p_{i,\infty} + C_1 \exp(r_1 t) + C_2 \exp(r_2 t) \qquad p_H = p_{H,\infty} + C_3 \exp(r_1 t) + C_4 \exp(r_2 t)$$

with r_1 and r_2 the roots of the characteristic equation:

$$r_1 = \frac{1}{2} \left[-(A_1 + A_2 + B_1) + \sqrt{(A_1 + A_2 + B_1)^2 - 4 A_1 B_1} \right]$$

$$r_2 = \frac{1}{2} \left[-(A_1 + A_2 + B_1) - \sqrt{(A_1 + A_2 + B_1)^2 - 4 A_1 B_1} \right]$$

$p_{i,\infty}$ and $p_{H,\infty}$ are the asymptotes of the vapour pressure indoors and the vapour pressure in the point of action of the sorption-active surface. The integration constants C_1, C_2, C_3 and C_4 follow from the initial conditions, namely the vapour pressure values $p_{i,o}$ and $p_{H,o}$ at the moment of sudden increase, and from the relationship between the two as defined by the system of differential equations at time zero:

(1) $C_1 + C_2 = p_{i,o} - p_{i,\infty}$ (3) $C_1 r_1 + C_2 r_2 = A_1 p_{i,\infty} - (A_1 + A_2) p_{i,o} + A_2 p_{H,o}$

(2) $C_3 + C_4 = p_{H,o} - p_{H,\infty}$ (4) $C_3 r_1 + C_4 r_2 = B_2 (\bar{p}_x - \bar{p}_i) + B_1 p_{i,o} - B_1 p_{H,o}$

In case the vapour pressure indoors and its value in the point of action equalize, the initial conditions (3) and (4) simplify to:

$$(3) \ C_1(r_1 + A_1) + C_2(r_2 + A_1) = 0 \quad (4) \ C_3 r_1 + C_4 r_2 = B_2(p_x - \bar{p}_i)$$

with as the solution:

$$C_1 = \frac{p_{i,0} - p_{i,\infty}}{1 - \frac{r_1 + A_1}{r_2 + A_1}} \qquad C_2 = \frac{p_{i,0} - p_{i,\infty}}{1 - \frac{r_2 + A_1}{r_1 + A_1}}$$

Also a sudden change in ventilation rate at time zero occurs as a step excitation with linked asymptotes.

3.3.2.4 Zone with several sorption active surfaces

Each additional sorption-active surface adds an extra mass balance, linked to its point of action. Two surfaces thus give a system of three first-order differential equations, three give four, four give five, and so on. Per surface, the solution at the zone level gets an additional exponential and time constant. Calculating the roots of the characteristic equation quickly becomes impossible, unless the vapour pressure at all points of action is assumed equal (p_H). The presence of several sorption-active surfaces then reduces to solving a system of two first-order differential equations with coefficients A_1, A_2, A_3, B_1 and B_2:

$$A_1 = \frac{G_a}{\rho_a V} \qquad A_2 = \frac{1}{a\rho_a V} \sum_{l=1}^{n} \left(\frac{A_l}{Z_{H,l}}\right) \qquad A_3 = \frac{G_{vP}}{a\rho_a V}$$

$$B_1 = \frac{\sum_{l=1}^{n} \left(\frac{A_l}{Z_{H,l}}\right)}{\sum_{l=1}^{n} A_l C_{H,l}} \qquad B_2 = \frac{\sum_{k=1}^{m} \left(\frac{A_k}{Z_{T,k}}\right)}{\sum_{k=1}^{m} A_l C_{H,l}} \quad \text{if } |\bar{p}_i - p_x| > 0$$

3.3.2.5 Harmonic analysis

Over a one-year period the vapour pressure outdoors changes more or less harmonically. With the annually averaged ventilation rate and indoor vapour release known, a harmonic analysis seems an obvious choice. Assume all material properties are known. While the average vapour pressure indoors still follows from a steady-state balance, the complex vapour flow ($G_{v,s,l}$) to and in an envelope assembly then equals:

$$G_v = A_e \left(\frac{\mathbf{p_e}}{D_{g,e}^n} - \mathrm{Ad}_{v,e}^n \mathbf{p_i}\right) \tag{3.35}$$

with $\mathbf{p_e}$ and $\mathbf{p_i}$ the complex vapour pressures outdoors and indoors, $D_{g,e}^n$ the dynamic diffusion resistance of the assembly and $Ad_{v,e}^n$ its hygric admittance. For the vapour pressures in all zones equal, the equation for the partitions is:

$$G_v = A_i \left(\frac{\mathbf{p_i}}{D_{g,i}^n} - Ad_{v,i}^n \mathbf{p_i} \right) = A_i \, \mathbf{p_i} \left(\frac{1}{D_{g,i}^n} - Ad_{v,i}^n \right) \tag{3.36}$$

On average and per harmonic, the vapour pressure indoors thus becomes:

$$\bar{p}_i = \bar{p}_e + \frac{G_{pv}}{6.21 \times 10^{-6} G_a}$$

$$\hat{p}_i = \hat{p}_e \left(\frac{6.21 \times 10^{-6} G_a + \sum_e \left(\frac{A_e}{D_{g,e}^n} \right)}{6.21 \times 10^{-6} G_a + \sum_e \left(A_e Ad_{v,e}^n \right) + \sum_i A_i \left(Ad_{v,i}^n - \frac{1}{D_{g,i}^n} \right) + i \left(6.21 \times 10^{-6} \rho_a \frac{2\pi}{T} V \right)} \right)$$

3.3.3 Consequences

In steady state, hygric inertia, also called moisture buffering, has no impact, but in the transient state, it does. Consider the annual change in vapour pressure outdoors. A bedroom has a net floor area of 16 m² for an air volume of 40 m³. The window has an area of 2.25 m² and the door bay is 2 m². The cavity wall, the inside partitions and the ceiling are all finished with an unpainted gypsum plaster ($d = 1.5$ cm, $\mu = 5.8$, $\xi_H = 56$ kg/m³) while the floor has a vapour-tight cover. Thus the room has 51.8 m² of sorption-active surface, without counting the bed and all the furniture. Of these, 7.75 m² belong to the envelope, 28.05 m² are partition-related and 16 m² ceiling related. Vapour release totals 800 g per day for a ventilation flow of 22 m³/h or 54 m³/h. Indoors, the annual mean temperature is 20 °C, while the climate outdoors is temperate. Figure 3.5 illustrates the impact of moisture buffering on the monthly mean indoor to outdoor vapour pressure difference.

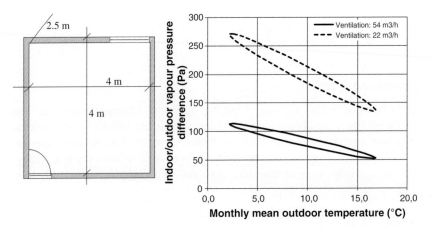

Fig. 3.5 Moisture buffering impacting the annual indoor/outdoor vapour pressure difference in a bedroom

3.3 Whole building level

As the slope of the oval representing it underlines, even on an annual basis, moisture buffering dampens the indoor vapour pressure amplitude, although with increased ventilation this occurs less (see the lowest oval). That such an oval is formed stems from the time shift between the vapour pressure indoors and outdoors. In fact, despite a constant vapour release and ventilation rate, buffering gives a greater difference in the autumn than in the spring. The fact that the oval flattens with more ventilation means that better airing decreases the time shift.

What about a daily returning change in vapour released indoors? Back to our example bedroom. A winter day has a daily mean temperature of 2.7 °C for a daily mean vapour pressure touching 663 Pa. If the 800 g of vapour released was equally spread over the day, a ventilation flow of 22 m^3/h would give a steady-state vapour pressure of 867 Pa. With 54 m^3/h ventilation flow the value drops to 746 Pa. But, in reality, the 800 g consists of 100 g per hour during 8 hours at night. Using the 9 mm sorption-active thickness of the plaster for a harmonic vapour release swing over one day, the transient for ventilation of 54 m^3/h over a few days is given in Figure 3.6.

When the release starts, air buffering gives a rapid rise in vapour pressure, followed by a strongly retarded additional rise due to sorption by the gypsum plaster. After 8 hours, the indoor vapour pressure is still far away from the steady-state 912 Pa, if the 100 g per hour were released day-long. Without surface buffering, vapour pressure reaches steady-state rapidly. The same effect occurs in the morning, when the release turns to zero. Yet, the average vapour pressure over several days remains 746 Pa.

Figure 3.7 shows the results for a ventilation flow of 22 m^3/h. The conclusions remain, although the average vapour pressure is higher.

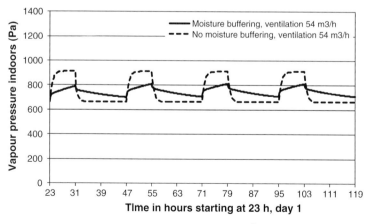

Fig. 3.6 Vapour pressure for a ventilation flow of 54 m^3/h and a vapour release of 100 g/h over 8 hours a night in a bedroom

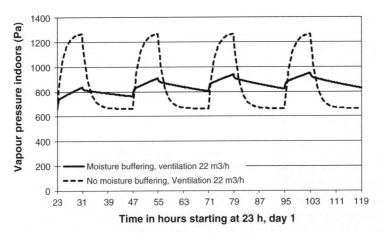

Fig. 3.7 Vapour pressure in a bedroom with a ventilation flow of 22 m³/h and a vapour release of 100 g/h during 8 hours a night

Figure 3.8 illustrates the effect of the window set ajar for one hour in the morning, giving a 10 ach ventilation peak, and closed the rest of the day, reducing the ventilation rate to 0.2 ach. A steep decrease in vapour pressure is seen, followed by a very slow climb to equilibrium with the hygric memory of the room. In reality, the increase after closing the window will be faster, because the sorption-active thickness approach underestimates the early vapour release by diffusion. So short peak ventilation beyond the hours of usage clearly has less effect in combating high indoor-to-outdoor vapour pressure differences.

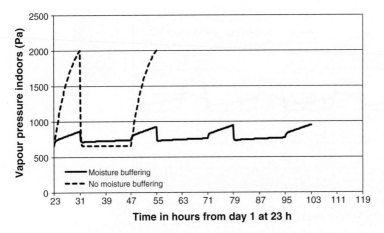

Fig. 3.8 Vapour pressure in a bedroom with vapour release of 100 g/h from 23:00 to 7:00. The window is ajar between 7 and 8 a.m., and closed the rest of the day

3.4 Problems and solutions

Problem 3.1

A social housing estate consisted of 48 two-storey houses with pitched roofs, coupled two by two:

Dwellings

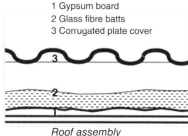
1 Gypsum board
2 Glass fibre batts
3 Corrugated plate cover

Roof assembly

The 2.5 m high ground floor measured 7.2×7.2 m^2, as did the first floor, which was equally high at the front and the rear sides. The long pitch of the asymmetric cathedralized roof had 17° slope, and the short pitch 10° slope. An open staircase linked both floors. None of the houses had a purpose-designed ventilation system. Instead, air leakage and window opening had to guarantee indoor air quality. The only difference between the 48 was the orientation: 9 had a front facade looking northwest, 4 north-northwest, 16 northeast, 6 east, 5 southeast and 8 southwest. A few years after occupation, 41 of the 48 houses showed moisture spots on the ceilings in the bedroom, while a number of tenants complained about moisture dripping onto their bed after a cold night.

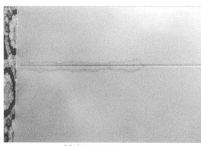
Moisture spots

Moisture dripping in the bed

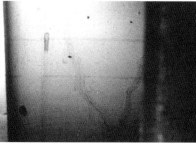
Run-off at the topside of the gypsum boards

PE air and vapour retarder mounted

Inspection of the roofs showed that the gypsum board ceilings were mounted with open joints. The glass fibre batts with bituminous paper backing, used as insulation, had open joints in between, while their flanges did not overlap at the rafter's underside. The backside of the corrugated fibre cement plate covers showed abundant traces of water runoff, as did the rafters and topsides of the gypsum boards, of which the inside surface suffered from discoloration along the open joints (see pictures).

A correlation was sought between the severity of the complaints and the average number of tenants in each house, whether or not a cooker hood was present, the annual end energy used for heating and the orientation of the front facade. Only heating showed relevance. On average, dwellings with severe complaints consumed 128 GJ/a, and those with moderate complaints 164 GJ/a. Their tenants apparently heated better or ventilated more while heating.

During winter, the inside temperature and relative humidity was monitored in two dwellings with severe problems and one with moderate problems (1 = moderate; 2 and 3 = severe). Before, dwelling 2 got a polyethylene (PE) air and vapour retarder mounted underneath the gypsum board lining, with all joints and overlaps carefully sealed (see picture). The table below condenses the logging results.

	Parents' bedroom		Children's bedroom		Bathroom	
	Temp., °C	Δp_{ie}, Pa	Temp., °C	Δp_{ie}, Pa	Temp., °C	Δp_{ie}, Pa
1	$13.6 + 0.42\theta_e$	$196 - 1.2\theta_e$	$14.1 + 0.42\theta_e$	$159 - 0.9\theta_e$		
2	$13.1 + 0.32\theta_e$	$373 - 14.7\theta_e$	$13.9 + 0.26\theta_e$	$237 - 2.5\theta_e$	$14.3 + 0.21\theta_e$	$457 - 17.7\theta_e$
3	$11.7 + 0.48\theta_e$	$324 - 10.8\theta_e$	$15.6 + 0.06\theta_e$	$411 - 34\theta_e$	$17.7 + 0.25\theta_e$	$395 - 19.4\theta_e$

The two with severe complaints clearly showed the highest indoor-to-outdoor vapour pressure difference, although the PE foil in dwelling 2 stopped dripping completely.

Additional data

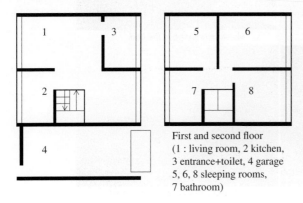

First and second floor
(1 : living room, 2 kitchen,
3 entrance+toilet, 4 garage
5, 6, 8 sleeping rooms,
7 bathroom)

3.4 Problems and solutions

The volume out-to-out is 344.9 m³, of which 149 m³ is the ground floor. Enclosed is an air volume of 248.3 m³, while the surface areas out-to-out of all envelope assemblies equal:

Assembly	Area, m²
Floor on grade	51.8
Cavity wall	
Sidewall 1	45.8
Sidewall 2	45.8
Front	26.7
Rear	24.3
Roof	
Large pitch	29.7
Small pitch	25.0
Windows and outer doors	
Front door	2.0
Toilet	0.2
Living room, front	5.4
Living room, back	5.4
Kitchen	5.3
Bedroom 1	4.5
Bedroom 2	4.5
Bedroom 3	4.5
Bathroom	1.8

The thermal transmittances and air permeances are:

Part	U value, W/(m².K)	K_a, kg/(m².Pa.s)
Façade: non-insulated cavity wall, plastered inside	1.66	0
Roof	0.49	$3.3 \times 10^{-4} \Delta P_a^{-0.33}$
Floor on grade	0.70	0
Window between both roof pitches	3.34	0
Glazing		
Double at the ground floor	2.70	
Single at the first floor	5.70	
Aluminium frames (20% of the window area)	5.90	

The overall air leakage obeys the following formula:

$$\dot{V}_a = a \Delta P_a^{0.67} \quad (m^3/s)$$

The roof in turn is composed of:

Layer (all air-permeable)	d, m	λ, W/(m.K)	R, m².K/W	μd, m
Corrugated plates	0.006	0.95		0.34
Air space	0.18		0.17	0.00
Thermal insulation	0.06	0.04		0.07
Vapour retarder	–	–		2.30
Air space	0.04		0.17	0.00
Gypsum board	0.0095	0.21		0.12

The inside surface film coefficients are 7.7 W/(m².K) for the outer walls, 6 W/(m².K) for the floor on grade and 10 W/(m².K) for the roof. Outside, the values are 25 W/(m².K) for the outer walls and the floor on grade, and 17 W/(m².K) for the roof. For the inside, a surface film coefficient for diffusion of 2.6×10^{-8} s/m can be used.

The inside temperature is 18 °C and the air change rate at 50 Pa (n_{50}) is 10.4 h^{-1}. The front and rear facade of the ground and first floor are considered equally air-permeable. The vapour release indoors reaches some 13.5 kg per day, while the outdoor climate during a cold week is:

Temperature, °C	Temperature roof, °C	Relative humidity at −2.5 °C, %	Mean wind velocity, m/s	Wind direction
−2.5	−3.9	98	3.8	NE

The mean wind velocity is the open field value at a height of 10 m. The estate has a closed landscape, with effective terrain roughness 1 m and friction velocity 0.47 m/s, changing the wind velocity into:

$$v = 1.12 \ln(h + 1) \quad \text{(m/s)}$$

with h being height above grade. Wind pressure in Pa follows from $0.6Cv^2$ with the pressure factor C equal to:

C	Wind angle normal to the front façade							
	0°	45°	90°	135°	180°	225°	270°	315°
Front	0.2	0.05	−0.25	−0.3	−0.25	−0.3	−0.25	0.05
Back	−0.25	−0.3	−0.25	0.05	0.2	0.05	−0.25	−0.3
Side left	−0.25	0.05	0.2	0.05	−0.25	−0.3	−0.25	−0.3
Side right	−0.25	−0.3	−0.25	−0.3	−0.25	0.05	0.2	0.05
Roof, <10°								
Front	−0.5	−0.5	−0.4	−0.5	−0.5	−0.5	−0.4	−0.5

Rear	−0.5	−0.5	−0.4	−0.5	−0.5	−0.5	−0.4	−0.5
Mean	−0.5	−0.5	−0.4	−0.5	−0.5	−0.5	−0.4	−0.5
Roof, 11–30°								
Front	−0.3	−0.4	−0.5	−0.4	−0.3	−0.4	−0.5	−0.4
Rear	−0.3	−0.4	−0.5	−0.4	−0.3	−0.4	−0.5	−0.4
Mean	−0.3	−0.4	−0.5	−0.4	−0.3	−0.4	−0.5	−0.4

The question is: what causes the complaints?

Solution

To come to an answer, the steady-state heat, air and moisture response during the cold week is analyzed. In order to do this, the dwelling is replaced by a three-node airflow system: one node 1 m high representing the ground floor, a second node 3.75 m high representing the second floor, and a third node 6 m high for the roof zone. For the thermal and vapour flow, a single node approach is maintained. Doors between rooms are assumed open all day, while the floor on grade, the outer walls, and the deck between both storeys are considered vapour-tight. All calculations are based on the dimensions out-to-out.

Considering air, in each of the three nodes, the sum of the airflows from neighbouring nodes and the outdoors must be zero, or:

$$\sum G_a = 0$$

Thermal stack for node 1 is null. At node 2, 2.75 m above node 1, for $\theta_i = 18\,°C$ and $\theta_e = -2.5\,°C$, thermal stack equals:

$$p_{T,2} = 2.75 \frac{gP_a}{R_a}\left(\frac{1}{T_i} - \frac{1}{T_e}\right) = -2.48\,\text{Pa}$$

At node 3, 5 m above node 1, for $\theta_i = 18\,°C$ and $\theta_{roof} = -3.9\,°C$, the value is

$$p_{T,3} = 5 \frac{gP_a}{R_a}\left(\frac{1}{T_i} - \frac{1}{T_e^*}\right) = -4.84\,\text{Pa}$$

As a reminder, g in these formulae is the acceleration due to gravity (9.81 m/s^2), P_a is atmospheric pressure (some 100,000 Pa) and R_a is the gas constant for air (287 Pa.m^3/(kg.K)). The ratio gP_a/R_a equals 3462 Pa.K/m.

Stack and wind give as vertical and horizontal airflows:

$$G_a = a\left(P_{a,x} + p_{T,x} - P_{a,y} - p_{T,y}\right)^b \qquad G_a = a\left(P_{a,x} - P_{a,y}\right)^b$$

The balances per node form a system of three equations with the node air pressures $P_{a,x}$ as unknowns:

Node 1:

$$a_{e1,1}A_{e1}\left[P_{a,e1} - P_{a,x1}\right]^{0.67} + a_{e2,1}A_{e2}\left[P_{a,e2} - P_{a,x1}\right]^{0.67}$$
$$+ a_{2,1}A_{2,1}\left[P_{a,x2} + p_{T,2} - P_{a,x1} - p_{T,1}\right]^{0.5} = 0$$

Node 2:

$$a_{e3,2}A_{e3}\left[P_{a,e3}-P_{a,x2}\right]^{0.67}+a_{e4,2}A_{e4}\left[P_{a,e4}-P_{a,x2}\right]^{0.67}$$
$$+a_{1,2}A_{1,2}\left[P_{a,x1}+p_{T,1}-P_{a,x2}-p_{T,2}\right]^{0.5}$$
$$+a_{3,2}A_{3,2}\left[P_{a,x3}+p_{T,3}-P_{a,x2}-p_{T,2}\right]^{0.5}=0$$

Node 3:

$$a_{e5,3}A_{e5}\left[P_{a,e5}-P_{a,x3}\right]^{0.67}+a_{e6,3}A_{e6}\left[P_{a,e6}-P_{a,x3}\right]^{0.67}$$
$$+a_{2,3}A_{2,3}\left[P_{a,x2}+p_{T,2}-P_{a,x3}-p_{T,3}\right]^{0.5}=0$$

Solving the system demands linearization, followed by a split between known and unknown terms. The resulting coefficients, called C_{ij}, and the known terms, called F_i, are:

Node 1:

$$C_{11}=-\left(\frac{a_{e1,1}A_{e1}}{\text{abs}(P_{a,e1}-P_{x1})^{0.33}}+\frac{a_{e2,1}A_{e2}}{\text{abs}(P_{a,e2}-P_{x1})^{0.33}}+\frac{a_{2,1}A_{2,1}}{\text{abs}(P_{a,x2}+p_{T,2}-P_{a,x1}-p_{T,1})^{0.5}}\right)$$

$$C_{12}=\frac{a_{2,1}A_{2,1}}{\text{abs}(P_{a,x2}+p_{T,2}-P_{a,x1}-p_{T,1})^{0.5}}$$

$$C_{13}=0$$

$$F_1=-\left(\frac{a_{e1,1}A_{e1}P_{a,e1}}{\text{abs}(P_{a,e1}-P_{x1})^{0.33}}+\frac{a_{e2,1}A_{e2}P_{a,e2}}{\text{abs}(P_{a,e2}-P_{x1})^{0.33}}+\frac{a_{2,1}A_{2,1}(p_{T,2}-p_{T,1})}{\text{abs}(P_{a,x2}+p_{T,2}-P_{a,x1}-p_{T,1})^{0.5}}\right)$$

Node 2:

$$C_{21}=\frac{a_{1,2}A_{1,2}}{\text{abs}(P_{a,x1}+p_{T,1}-P_{a,x2}-p_{T,2})^{0.5}}$$

$$C_{22}=-\left(\frac{a_{e3,2}A_{e3}}{\text{abs}(P_{a,e3}-P_{x2})^{0.33}}+\frac{a_{e4,2}A_{e4}}{\text{abs}(P_{a,e4}-P_{x2})^{0.33}}\right.$$
$$\left.+\frac{a_{1,2}A_{1,2}}{\text{abs}(P_{a,x1}+p_{T,1}-P_{a,x2}-p_{T,2})^{0.5}}+\frac{a_{3,2}A_{3,2}}{\text{abs}(P_{a,x3}+p_{T,3}-P_{a,x2}-p_{T,2})^{0.5}}\right)$$

3.4 Problems and solutions

$$C_{23} = \frac{a_{3,2}A_{3,2}}{\text{abs}\left(P_{a,x3}+p_{T,3}-P_{a,x2}-p_{T,2}\right)^{0.5}}$$

$$F_2 = -\left(\begin{array}{c}\dfrac{a_{e3,2}A_{e2}P_{a,e3}}{\text{abs}(P_{a,e3}-P_{x2})^{0.33}}+\dfrac{a_{e2,1}A_{e2}P_{a,e4}}{\text{abs}(P_{a,e4}-P_{x2})^{0.33}}+\dfrac{a_{1,2}A_{1,2}\left(p_{T,1}-p_{T,2}\right)}{\text{abs}\left(P_{a,x1}+p_{T,1}-P_{a,x2}-p_{T,2}\right)^{0.5}} \\ +\dfrac{a_{3,2}A_{3,2}\left(p_{T,3}-p_{T,2}\right)}{\text{abs}\left(P_{a,x3}+p_{T,2}-P_{a,x2}-p_{T,3}\right)^{0.5}}\end{array}\right)$$

Node 3:

$$C_{31} = 0$$

$$C_{32} = -\left(\frac{a_{e5,3}A_{e5}}{\text{abs}(P_{a,e5}-P_{x3})^{0.33}}+\frac{a_{e6,3}A_{e6}}{\text{abs}(P_{a,e6}-P_{x3})^{0.33}}+\frac{a_{2,3}A_{2,3}}{\text{abs}\left(P_{a,x2}+p_{T,2}-P_{a,x3}-p_{T,3}\right)^{0.5}}\right)$$

$$C_{33} = \left(\frac{a_{2,3}A_{2,3}}{\text{abs}\left(P_{a,x2}+p_{T,2}-P_{a,x3}-p_{T,3}\right)^{0.5}}\right)$$

$$F_3 = -\left(\frac{a_{e5,3}A_{e5}P_{a,e5}}{\text{abs}(P_{a,e5}-P_{x3})^{0.33}}+\frac{a_{e6,3}A_{e6}P_{a,e6}}{\text{abs}(P_{a,e6}-P_{x3})^{0.33}}+\frac{a_{2,3}A_{2,3}\left(p_{T,2}-p_{T,3}\right)}{\text{abs}\left(P_{a,x2}+p_{T,2}-P_{a,x3}-p_{T,3}\right)^{0.5}}\right)$$

The system thus becomes

$$\begin{vmatrix}C_{11} & C_{12} & C_{13} \\ C_{21} & C_{22} & C_{23} \\ C_{31} & C_{32} & C_{33}\end{vmatrix} \times \begin{vmatrix}P_{a,x1} \\ P_{a,x3} \\ P_{a,x3}\end{vmatrix} = \begin{vmatrix}F_1 \\ F_2 \\ F_3\end{vmatrix}$$

To solve this matrix equation, the known pressures and temperatures are inserted together with estimated values for the three unknown air pressures $P_{a,x1}$, $P_{a,x2}$ and $P_{a,x3}$. That allows calculation of the coefficients C_{ij} and known terms F_i, and a solution for the system. With the three new node pressures, the coefficients C_{ij} and known terms F_i are recalculated and the system solved again. Iteration goes on until the root of the summed quadratic deviations between the actual and previous solution is smaller than a preset value, or:

$$a_{preset} < \sqrt{\sum \varepsilon^2}$$

Calculating the air permeance of the front and rear façades, knowing that $n_{50} = 10$ ach, starts from the roof's mean air permeance coefficient:

$$K_{a,roof} = 3.3 \times 10^{-4}/1.2 = 2.75 \times 10^{-4} \, m^3/(m^2 \, sPa^b)$$

A ventilation rate of 10 ach at 50 Pa for a net air volume of 248.3 m³ now means a flow of $(10.4 \times 248.3/3600)/50^{0.67} = 0.0502 \, m^3/s$ at 1 Pa, of which $(3.3 \times 10^{-4}/1.2) \times 54.7 = 0.0154 \, m^3/s$ passes the roof, leaving $0.037 \, m^3/s$ for the front and rear façades. Assuming each quarter façade shows an equal leakage, the result is $0.037/4 = 9.25 \times 10^{-3} \, m^3/(s.Pa^b)$.

The flow equations for the open staircase and the link between nodes 2 and 3 follow from the conservation of energy equation, stating that the difference in pressure and stack should equal the change in kinetic energy of the airflow. This gives as permeance coefficients:

Staircase : $4.0 \, m^3/(s.Pa^{0.5})$ First floor to roof zone : $56.2 \, m^3/(s.Pa^{0.5})$

The stack pressures were calculated above. When the wind blows from the NE and the front façade looks NE, the wind-induced pressures become:

Height	P_w, Pa	
	Front façade	Rear façade
1 m	0.072	−0.09
3.75 m	0.37	−0.46
6 m (roof)	−0.85	−0.85

These are much smaller than stack, which means that the airflows are mainly buoyancy-related. The node system thus becomes:

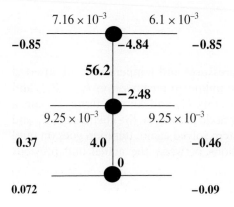

3.4 Problems and solutions

Solving the equations for $n_{50} = 10$ ach gives as air pressures and air flows (+ is inflow, − is outflow):

	Ground floor		First floor		Roof	
	Rear	Front	Rear	Front	Rear	Front
Air pressure (Pa)	−2.47		0.0086		2.37	
Flow, m³/h	59.5	62.2	−20.0	16.8	−54.2	−64.3

Turning to vapour now, in the single node dwelling, outside air enters and inside air leaves, while of the vapour released, some will condense. Without surface condensation, the vapour pressure indoors should be:

$$p_i = p_e + RT_i \frac{G_{v,P}}{\dot{V}_a}$$

Surface condensation, however, may deposit either on the window frames alone (1), the frames plus the single glass (2), or the frames, the single and the double glass (3):

(1) $$p_i = \frac{p_e + \frac{RT_i}{\dot{V}_a}\left(G_{v,P} + \beta A_{fr} p_{sat,fr}\right)}{1 + \frac{RT_i}{\dot{V}_a}\beta A_{fr}}$$

(2) $$p_i = \frac{p_e + \frac{RT_i}{\dot{V}_a}\left(G_{v,P} + \beta A_{fr} p_{sat,fr} + \beta A_{sg} p_{sat,sg}\right)}{1 + \frac{RT_i}{\dot{V}_a}\beta\left(A_{fr} + A_{sg}\right)}$$

(3) $$p_i = \frac{p_e + \frac{RT_i}{\dot{V}_a}\left(G_{v,P} + \beta A_{fr} p_{sat,fr} + \beta A_{sg} p_{sat,sg} + \beta A_{dg} p_{sat,dg}\right)}{1 + \frac{RT_i}{\dot{V}_a}\beta\left(A_{fr} + A_{sg} + A_{dg}\right)}$$

Saturation pressures at the inside surface of the frame, the single and the double glass equal 721, 749 and 1298 Pa. The average vapour pressure indoors (p_i) during the given cold week, whether or not there is surface condensation on the aluminium frames and the glass, the weekly amount of condensate deposited, and the percentage of the vapour released that condenses, are all listed in the table below.

	p_i, Pa	Surface condensation	Amount condensing, kg/week	% of vapour released
	852			
Aluminium		Yes	14.6	15.4
Single glass		Yes	22.3	23.6
Double glass		No		

Apparently large amounts deposit on the aluminium window frames on both floors and on the single glass on the ground floor.

Now to consider the roof. Combined diffusion and convection may cause condensation on the backside of the corrugated fibre-cement plates. First, the temperature there is calculated:

$$\theta_{cover} = 18 + (\theta_e^* - 18) \frac{1 - \exp(-1008\, g_a R_i^{cover})}{1 - \exp(-1008\, g_a R_i^e)} = -2.6\,°C$$

In this formula, g_a is the air flux in kg/(m².s) across the roof, equal to $\rho_a \dot{V}_a/(3600\, A_{roof})$. Related saturation pressure (p_{sat} cover) is:

θ_{cover}	p_{sat} cover	<p_i? (= 852 Pa)
−2.6 °C	493 Pa	Yes

The value found clearly underscores the inside vapour pressure, indicating that interstitial condensation against the corrugated fibre-cement plates is a fact. The amounts condensing total:

$$g_c = \frac{-6.21 \times 10^{-6}\, g_a}{1 - \exp(-6.21 \times 10^{-6}\, g_a Z_i^x)} \left(p_{sat,x} - p_i \exp(-6.21 \times 10^{-6}\, g_a Z_i^x)\right)$$

$$- \frac{-6.21 \times 10^{-6}\, g_a}{1 - \exp(-6.21 \times 10^{-6}\, g_a Z_x^{se})} \left(p_{sat,se} - p_{sat,x} \exp(-6.21 \times 10^{-6}\, g_a Z_x^{se})\right)$$

giving:

Deposit (g_c) kg/(m².week)	Total for the roof kg/week	% of the vapour released
0.98	53.4	56.6

3.4 Problems and solutions

Things are even worse, as the exterior surface of the plates will also suffer from condensation due to under-cooling.

What happens with this deposit? To have no dripping requires less than 10 kg of deposit a week for the whole roof. Here, dripping and wetting of the beds in the bedrooms will start from the second day onwards. Abundant surface condensation on the windows in the sleeping rooms will also annoy, although have less of an impact.

What causes the complaints? Clearly, the air-leaky roof and easy air coupling along the open staircase between ground and first floor are to blame.

What is the net energy demand during this one cold week? The calculations according to EN standards, using the dimensions out-to-out, but without counting the solar gains, give:

Net energy demand without exfiltration across the roof, MJ/week	Net energy demand with exfiltration across the roof, MJ/week	% less
4430	4224	4.6

Due to the high conductive losses and despite the high air leakage at 50 Pa, infiltration only accounts for some 15% of the total. Outflow through the roof even diminishes the conduction losses and related total demand, as if its thermal transmittance had dropped from 0.49 W/(m².K) to 0.21 W/(m².K). The impact remains marginal.

Problem 3.2

Keep the same dwelling and data as for Problem 3.1, and redo the analysis for an n_{50} value of 6 ach, with all other data and the methodology identical.

Solution

Thermal stack still represents:

Height above grade	P_T, Pa
1 m	0
3.75 m	−2.48
6 m (roof)	−4.84

Air pressures indoors and airflows are now:

	Ground floor		First floor		Roof	
	Rear	Front	Rear	Front	Rear	Front
Air pressure, Pa	1.29		−1.07		−3.55	
Flow, m³/h	31.1	32.1	9.7	17.3	−41.2	−49.0

The vapour pressure indoors and the amounts of surface condensate on the aluminium window frames on both floors and on the single glass in the bedrooms equal:

	p_i, Pa	Total, kg/week	% of vapour released
Aluminium	897	19.7	20.8
Single glass		32.2	34.0

In the roof, the temperature (θ_x) and saturation pressure at the backside of the corrugated fibre cement plates ($p_{sat,x}$) become:

θ_x, °C	$p_{sat,x}$, Pa	$<p_i$?
−2.8	486	Yes

The amount condensing there equals:

g_c, kg/(m².week)	Total roof, kg/week	% of vapour released
0.837	45.8	48.4

The net energy demand for heating in that one cold week reaches:

Without exfiltration, MJ/week	With exfiltration, MJ/week	% difference
4226	4059	4.0

The air change rate by infiltration does not exceed 0.36 ach.

Problem 3.3

Keep the same dwelling and data as for Problem 3.1, and redo the analysis for an n_{50} value of 14 ach, with all other data and the methodology identical.

Solution

Thermal stack represents:

Height above grade	P_T, Pa
1 m	0
3.75 m	−2.48
6 m (roof)	−4.84

3.4 Problems and solutions

Air pressures indoors and airflows are now:

	Ground floor		First floor		Roof	
	Rear	Front	Rear	Front	Rear	Front
Air pressure, Pa	−2.14		0.34		2.70	
Flow, m³/h	80.2	84.4	−42.8	4.7	−57.8	−68.7

The vapour pressure indoors and the amounts of surface condensate on the aluminium window frames on both floors and on the single glass in the bedrooms equal:

	p_i, Pa	Total, kg/week	% of vapour released
Aluminium	862	15.7	16.6
Single glass		24.5	25.9

In the roof, the temperature (θ_x) and saturation pressure at the backside of the corrugated fibre cement plates ($p_{sat,x}$) become:

θ_x, °C	$p_{sat,x}$, Pa	$<p_i$?
−2.5	495	Yes

The amount condensing there equals:

g_c, kg/(m².week)	Total roof, kg/week	% of vapour released
1.07	58.7	62.1

The net energy demand for heating in that one cold week reaches:

Without exfiltration, MJ/week	With exfiltration, MJ/week	% difference
4560	4345	4.7

The air change rate by infiltration reaches 0.68 ach.

Problem 3.4

Keep the same dwelling and data as for Problem 3.1. Assume the first floor is unheated, whereas the average inside temperature of the ground floor is 18 °C.

Vapour release reaches 7.88 kg a day on the ground floor and 5.62 kg a day on the first floor. Aside of the three air nodes the dwelling now has two heat and vapour nodes: a node 1 coinciding with air node 1, and a node 2 uniting the air nodes 2 and 3. The average internal gains on the first floor (node 2) equal 260 W, solar gains are overlooked, and the finished flooring between the ground and first floor has a thermal transmittance of 2.16 W/(m^2.K) for an area of 48 m^2.

Additional data:

First floor	Area, m^2	U, W/(m^2.K)
Cavity wall		
Front	12.65	1.66
Rear	13.35	1.66
Side wall	22.9	1.66
Roof		
Large pitch	29.7	0.49
Small pitch	25	0.49
Floor first floor	48	2.16

Solution

n_{50} equals 10.4 ach and the first floor remains unheated now. The temperature there drops to 6.8 °C and thermal stack left represents:

Height above grade	P_T, Pa
1 m	0
3.75 m	−1.82
6 m (roof)	−3.11

Air pressures indoors and airflows are now:

	Ground floor		First floor		[0,6-7]Roof	
	Rear	Front	Rear	Front	Rear	Front
Air pressure, Pa	−1.77		0.051		1.34	
Flow, m^3/h	47.2	50.2	−21.2	15.5	−41.9	−49.7

The vapour pressure indoors and the amounts of surface condensate on the aluminium window frames on both floors and on the single glass in the bedrooms equal:

	p_i, Pa	Total, kg/week	% of vapour released
Ground floor (aluminium)	879	7.7	8.1
First floor (aluminium and glass)	777	46.5	49.2

In the roof, the temperature (θ_x) and saturation pressure at the backside of the corrugated fibre cement plates ($p_{sat,x}$) become:

θ_x, °C	$p_{sat,x}$, Pa	$<p_i$?
−3.3	463	Yes

The amount condensing there equals:

g_c, kg/(m².week)	Total roof, kg/week	% of vapour released
0.602	32.9	34.8

The net energy demand for heating in that one cold week reaches:

Without exfiltration, MJ/week	With exfiltration, MJ/week	% difference
3822	3515	8.0

Partially heating looks an effective way to economize on net energy demand: −24%! Also, the air change rate by infiltration drops: 0.45 instead of 0.56 ach.

Further reading

Abuku, M. (2009) Moisture stress of wind-driven rain on building enclosures. PhD thesis, KU-Leuven.

Arfvidsson, J. (1998) Moisture transport in porous media, modelling based on Kirchhoffs potentials. Doctoral thesis, Lund University.

Blocken, B. (2004) Wind-driven rain on buildings, PhD thesis, KU-Leuven.

Blocken, B., Hens, H. and Carmeliet, J. (2002) Methods for the quantification of driving rain on buildings. *ASHRAE Transactions*, **108**, 338–350.

Brocken, H. (1998) Moisture transport in brick masonry: the grey area between bricks. Doctoral thesis, TU/e, Eindhoven.

Carmeliet, J., Hens, H. and Vermeir, G. (eds.) (2003) *Research in Building Physics*, Balkema, Lisse.

Chaddock, J.B. and Todorovic, B. (1991) *Heat and Mass Transfer in Building Materials and Structures*, Hemisphere Publishing Corporation, New York.

Desta, T. and Roels, S. (2010) *The influence of air on the heat and moisture transport through a lightweight building wall*. Proceedings of CESBP 2010 (eds D. Gawin and T. Kisilewicz).

Desta, T. and Roels, S. (2010) *Experimental and numerical analysis of heat, air, and moisture transport in a lightweight building wall*. Proceedings of Buildings XI (CD-ROM).

Duforestel, T. (1992) Bases métrologiques et modèles pour la simulation du comportement hygrothermique des composants et ouvrages du bâtiment. Doctoral thesis, Ecole Nationale des Ponts et des Chausses, Paris [in French].

Garrecht, H. (1992) Porenstrukturmodelle für den Feuchtehaushalt von Baustoffen mit und ohne Salzbefrachtung und rechnerische Anwendung auf Mauerwerk. Doctoral thesis, Universität Karlsruhe [in German].

Häupl, P. and Roloff, J. (Eds.) (2002) Proceedings of the 11th Bauklimatisches Symposium, Band 1 & 2. T.U. Dresden, Institut für Bauklimatik.

Hens, H. (1978, 1981) *Bouwfysica, Warmte en Vocht, Theoretische grondslagen, 1^e en 2^e uitgave*. ACCO, Leuven [in Dutch].

Hens, H. (1996) *Modelling*. Vol. 1 of the Final Report Task 1, IEA-Annex 24. ACCO, Leuven.

Holm, A. (2001) Ermittlung der Genauigkeit von instationären hygrothermischen Bauteilberechnungen mittels eines stochastischen Konzeptes. Doctoral thesis, Universität Stuttgart [in German].

Janssen, H. (2002) The influence of soil moisture transfer on building heat loss via the ground. Doctoral thesis, KU-Leuven.

Janssens, A. (1998) Reliable control of interstitial condensation in lightweight roof systems. Doctoral thesis, KU-Leuven.

Koci, V., Madera, J., Keppert, M. and Cerny, R. (2010) *Mathematical models and computer codes for modelling heat and moisture transport in building materials: a comparison*. Proceedings of the CESBP 2010 (eds D. Gawin and T. Kisilewicz).

Kohonen, R. and Ojanen, S. (1985) Coupled convection and conduction in two dimensional building structures. Proceedings of the 4th Conference on Numerical Methods in Thermal Problems, Swansea.

Kohonen, R. and Ojanen, T. (1987) Coupled diffusion and convection heat and mass transfer in building structures. Building Physics Symposium, Lund.

Krus, M. (1995) Feuchtetransport- und Speicherkoeffizienten poröser mineralischer Baustoffe. Theoretische Grundlagen und NeueMesstechniken. Doctoral thesis, Universität Stuttgart [in German].

Künzel, H.M. (1994) Verfahren zur ein- und zweidimensionalen Berechnung des gekoppelten Wärme- und Feuchtetransports in Bauteilen mit einfachen Kennwerten. Doctoral thesis, Universität Stuttgart [in German].

Pedersen, C.R. (1990) Combined heat and moisture transfer in building constructions. PhD thesis, Technical University of Denmark.

Roels, S. (2000) Modelling unsaturated moisture transport in heterogeneous limestone. Doctoral thesis, KU-Leuven.

Taveirne, W. (1990) *Eenhedenstelsels en groothedenvergelijkingen: overgang naar het SI*. Pudoc, Wageningen [in Dutch].

Time, B. (1998) Hygroscopic moisture transport in wood. Doctoral thesis, NUST Trondheim.

Trechsel, H.R. (ed.) (1994) *Moisture Control in Buildings*. ASTM Manual Series, MNL 18.

Van Mook, J.R. (2003) Driving rain on building envelopes. *TU/e Bouwstenen*, **69**, 198.

Welty, J., Wicks, C. and Wilson, R. (1969) *Fundamentals of Momentum, Heat and Mass Transfer*, John Wiley & Sons, New York.

Woloszyn, M. and Rode, C. (2008) *Modelling Principles and Common Exercises*. Final Report IEA ECBCS, Annex 41. ACCO, Leuven.

Xiaochuan, Q. (2003) Moisture transport across interfaces between building materials. PhD thesis, Concordia University, Montreal.

Zillig, W. (2009) Moisture transport in wood using a multiscale approach. PhD thesis, KU-Leuven.

"Solar Heating and Cooling" is a research programme initiated by the International Energy Agency. The programme's work is accomplished through the international collaborative effort of experts from Member countries and the European Union. The results are published in a series with Ernst & Sohn and Wiley.

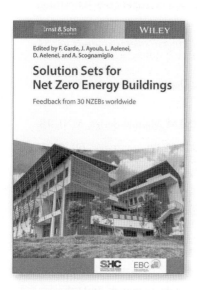

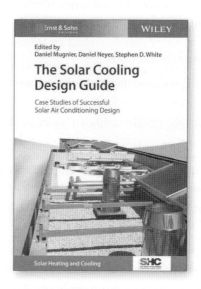

Ed.: François Garde, Josef Ayoub, Daniel Aelenei, Laura Aelenei, Alessandra Scognamiglio
Solution Sets for Net-Zero Energy Buildings
Feedback from 30 Buildings worldwide
2017. 252 pages
€ 79,–*
ISBN 978-3-433-03072-1
Also available as ebook

Ed.: Daniel Mugnier, Daniel Neyer, Stephen D. White
The Solar Cooling Design Guide
Case Studies of Successful Solar Air Conditioning Design
2017. 158 pages
approx. € 69,–*
ISBN 978-3-433-03125-4
Also available as ebook

Recommendations:

- Modeling, Design, and Optimization of Net-Zero Energy Buildings
- Solar and Heat Pump Systems for Residential Buildings

Order online:
www.ernst-und-sohn.de

Ernst & Sohn
Verlag für Architektur und technische Wissenschaften GmbH & Co. KG

Customer Service: Wiley-VCH
Boschstraße 12
D-69469 Weinheim

Tel. +49 (0)6201 606-400
Fax +49 (0)6201 606-184
service@wiley-vch.de

* € Prices are valid in Germany, exclusively, and subject to alterations. Prices incl. VAT. excl. shipping. 1112136_dp

Postscript

Any application-related thermal and moisture performance of building assemblies and whole building evaluations, with consequences for design and construction, should be based on a sound understanding of all heat, air and moisture facts and figures. Yet, this knowledge encompasses only a small part of what performance-based design, construction and retrofit of buildings requires. In fact, from a building physics point of view, a building and its parts must additionally guarantee good acoustics, should ensure correct but energy-efficient lighting, have low primary energy use, and care for excellent indoor environmental quality. Aside from this, architects, building engineers, builders and developers have to consider aesthetics, functionality, structural performance, durability, overall safety, fire safety, maintenance costs and sustainability.

Studying the heat, air and moisture tolerance certainly sharpens our understanding of some important realities. Heat transfer not only includes conduction but also convection and radiation. Convection combines conduction with gas and liquid movement. Because of its electromagnetic nature, radiation and related laws differ strongly from those shaping convection and conduction. Both define to a large extent the heat exchange between buildings and the outdoors, the heat exchange indoors, the heat exchange in air and gas layers and the heat flow through porous materials. The surface film coefficients, introduced to handle the heat exchange by convection and radiation at surfaces, are a mathematical expediency to facilitate calculations by transposing both heat flow modes into equivalent conduction across a fictitious surface-linked air layer.

Thermal insulation is very effective in lowering the heat losses and gains, although the added effectiveness drops with ever-increasing thickness. A precondition for efficiency, however, is air-tightness. When an air-permeable thermal insulation layer sits in an assembly that lacks air-tightness, it will not only lose thermal efficiency but may also induce severe condensation problems in layers and interfaces between layers at the cold side. Not preventing air inflow, air outflow, air looping and indoor air washing may have really negative consequences for the thermal performance and moisture tolerance of envelope assemblies. Therefore, when evaluating vapour-related moisture damage, lack of air-tightness and related humid airflow, rather than equivalent diffusion, needs most of the attention.

Moisture transport is not restricted to vapour only. At higher moisture contents, unsaturated water flow becomes much more important in open-porous materials. Without good knowledge of the laws governing its movement, one can neither understand nor cure phenomena such as rain penetration, rising damp, building moisture and drying.

Application of that knowledge is the subject of a discipline called performance-based building design. In fact, using performance metrics is the only way to ensure a correct overall engineered approach, as opposed to the 'construction as an art' approach.

The material discussed in the book has given birth to several software packages, which may help designers and constructers to make the right decisions, on condition that the limitations embedded are known. A profound understanding of heat, air and moisture transport in its full complexity is still based on the triumvirate of simulation/testing/practice. Most marketed software tools do not consider air inflow, air outflow, air looping and air washing, with all their nasty consequences. They cannot cope with gravity and pressure flow through cracks, leaks and voids, for example, due to rain runoff. Typically, the real geometry with its roughness, unknown cracks, invisible voids, and so on, is of much greater complexity than the virtual one-, two and three-dimensional geometries given as input. Moreover, a well-established, calculable linkage between heat, air and moisture transport and the many moisture-related durability issues encountered in practice is, for the most part, still lacking.